Carbonaceous Composite Materials

Editors

Dr. Gaurav Sharma
Dr. Amit Kumar

School of Chemistry, Shoolini University, India

Published by **Materials Research Forum LLC**
Millersville, PA 17551, USA

Published as part of the book series
Materials Research Foundations
Volume 42 (2018)
ISSN 2471-8890 (Print)
ISSN 2471-8904 (Online)

Print ISBN 978-1-945291-96-8
ePDF ISBN 978-1-945291-97-5

Distributed worldwide by

Materials Research Forum LLC
105 Springdale Lane
Millersville, PA 17551
USA
http://www.mrforum.com

Manufactured in the United States of America
10 9 8 7 6 5 4 3 2 1

Table of Contents

Preface

New materials revolutionize the human race and science as well. The modern age has found its present form by advancing through the Stone Age, Bronze Age, and Iron Age. The continuous research for technologies to elevate the standards of life and quest for getting into the unknown has led to "discovery" and "engineering" of novel materials.

During the last few decades, carbon and carbon based materials commonly called as "Carbonaceous materials" have grabbed a lot of attention from industry and academia. Carbon is a special material that is stable and chemically resistant to common corrosives. In addition carbon has low density which paves the way for its suitability in compact devices. The diverse morphologies in which carbon exists make it a suitable material for multi-pronged energy and environmental applications.

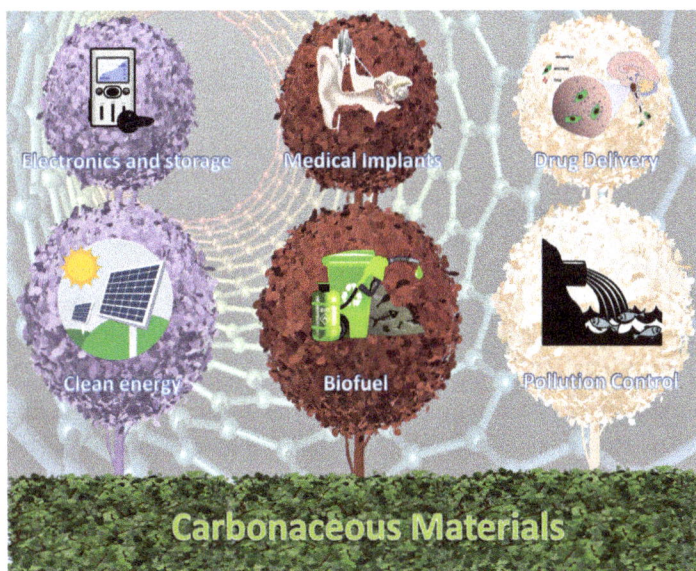

Representative image showing the fields where carbonaceous materials are in demand.

Advances in carbonaceous materials have led to shifts from carbons, graphite, biochars to "engineered carbons" which include fullerenes, graphene, carbon foam, nanotubes, graphene oxide, carbon aerogels, carbon matrix composites, reinforced polymers and many others. These materials have high economic value and environmental benign nature. In fact they can be functionalized variably for preparation of organic-inorganic hybrid materials.

Carbon based fibrous materials find their uses in construction and building materials in form of ceramic composites and polymer composites with a carbon matrix. In addition they are also used for moisture resistant walls and paints. Carbon nanotubes, conducting polymers and graphene based materials have found their role in energy storage in form of electrodes for batteries and supercapacitors which have better specific energy.

Due to increasing energy demands and dependence on non-renewable energy sources, demand for novel technologies and materials has increased. Hence owing to its extreme structural versatility in form of crystalline, amorphous, nano, bulk, functionalized, structured, reinforced, etc. a variety of materials are available for energy applications as in fuel cells. Carbon based materials are both used as sources as well as catalysts for bio-fuel production.

Activated carbon also referred to as charcoal is the most popular form of carbonaceous materials used for adsorption of organic pollutants, heavy metals and gaseous in addition to use as catalysts and catalytic supports. Another amorphous form biochar has been utilized for soil enrichment and greenhouse gaseous sequestration for climate control applications. Graphene a 2D carbon form possesses a large surface, electron mobility, strength and conductivity and is comprehensively used in electronics and bio-medical applications. Graphene oxide, reduced graphene oxide and graphitic carbon nitride are used as photocatalysts for hydrogen production and contaminants degradation. Carbon nanotubes have been used in various fields of engineering due to their high strength and low density.

The challenge lies in the development of new technologies to convert biomass and carbonaceous materials into more useful materials to prepare future smart hybrids for sustainable applications. This book focusses on current progress in development, designing and utilization of carbonaceous materials in various spheres. This will surely help in upgradation and innovations to be performed in the future for creating of an elite class of technologically advanced "carbonaceous materials"

We thank all the contributors. It's our fortune to grateful acknowledge Prof. P. K. Khosla, Vice chancellor Shoolini University, Dr. Inamuddin and Dr. Mu. Naushad for their support.

This work is dedicated to God, to our wonderful parents who formed part of our vision and taught us the good things that really matter in life. I am grateful for my siblings, for rendering me the sense and the value of togetherness. I am glad to be part of their life

Editors

Dr. Gaurav Sharma

Dr. Amit Kumar

School of Chemistry, Shoolini University, India

Carbonaceous Composite Materials
Materials Research Foundations **42** (2018) 1-33

Materials Research Forum LLC
doi: http://dx.doi.org/10.21741/9781945291975-1

Chapter 1

Graphene and Graphene/TiO₂ Nanocomposites for Renewable Dye Sensitized Solar Cells

Foo Wah Low, Chin Wei Lai*, Christelle Pau Ping Wong

Nanotechnology & Catalysis Research Centre (NANOCAT), Level 3, Block A, Institute of Postgraduate Studies (IPS), University of Malaya, 50603, Kuala Lumpur, Malaysia

*cwlai@um.edu.my

Abstract

Renewable solar cell energy is a key target for sustainable energies development, which are inexhaustible and non-polluting for our energy system. To bring more solar related technologies to the point of commercial readiness and viability in terms of performance and cost, substantial research on the development of high efficient renewable dye-sensitized solar cell (DSSCs) device is necessary. Recent studies have indicated that graphene is a relatively novel material with unique properties that could apply in photoanode/counter electrode component as efficient electrode. In fact, the atom-thick 2D structure of graphene (rGO) provides an extraordinarily high conductivity, repeatability, productivity, and prolong lifetime to the related solar cell applications. Continuous efforts have been exerted to further improve the graphene textural and electronic properties by loading an optimum content of titanium dioxide (TiO₂) as an efficient photocatalyst in DSSCs. In this chapter, different synthesis strategies and characterization analyses for TiO₂-rGO nanocomposites (NC) as well as its prospects in DSSCs will be reviewed in detail.

Keywords

Graphene, TiO₂, Graphene/TiO₂ Nanocomposites, Photoanode/Counter Electrode, Renewable Energy

Contents

1. Introduction

In recent years, lots of efforts have been paid to enhance the photocatalytic activity of TiO$_2$-rGO NC as the photoanode in DSSCs using the conventional synthesis techniques such as molecular grafting, spray and dispersion. Numerous researchers are giving their best to prepare an efficient approach on photoanode element to obtain a desired photocurrent density in order to obtain higher power conversion efficiency (PCE) for DSSCs application. However, there are still drawbacks on photocurrent density resulting in poor PCE performance. Accordingly, the one-step hydrothermal method is one of the common approaches utilized to enhance the photocatalytic activity and reduce the lattice matter of the TiO$_2$-rGO NC hybrid materials [1-3]. Perhaps, this technique is a way to shorten fabrication time. Previously, TiO$_2$-rGO NC as photoanode have attracted numerous researchers to explore and widely utilize them in DSSCs. The PCE performance based TiO$_2$-rGO NC is shown in Table 1.

Table 1: Summaries of PCE based TiO₂-rGONC photoanode in DSSCs

Photoanode materials	Preparation method	Reference Cell	Optimized content of rGO	Reference Cell of Photoanode				Improved of TiO₂-rGO NC Photoanode				Ref.
				J_{sc}, (mA/cm²)	V_{oc} (V)	ff	η (%)	J_{sc}, (mA/cm²)	V_{oc} (V)	ff	η (%)	
TiO₂ with rGO (from GO)	Molecular grafting	Ti(OBu)₄	With/without rGO incorporation	1.95	0.52	0.31	0.32	6.67	0.56	0.45	1.68	[5]
P25-rGO	n/a	P25	0.05 wt.%	5.04	n/a	n/a	2.70	8.38	n/a	n/a	4.28	[6]
TiO₂ with rGO (from GO)	Spray on TiO₂ film	TiO₂	Device comparison	16.40	0.60	0.52	5.09	18.20	0.58	0.58	6.06	[7]
P25-rGO	n/a	P25	1 wt.%	18.83	0.69	0.46	5.98	19.92	0.70	0.49	6.86	[8]
TiO₂ with rGO (from GO)	Solvothermal	P25	ultra-small 2 nm TiO2–rGO nanosheets	6.20	0.67	0.69	2.85	13.50	0.77	0.70	7.25	[9]
TiO₂ with rGO (from GO)	Dispersion	TiO₂ Nanosheets	0.75 wt.%	13.70	0.59	0.57	4.61	16.80	0.61	0.57	5.77	[10]
TiO₂-rGO	Hydrothermal	P25	1 wt.%	11.90	0.68	0.61	4.96	18.30	0.74	0.56	7.54	[11]
CdS-TiO₂-rGO	Hydrothermal	Ti(OBu)₄	8 mg	4.74	0.55	0.42	1.08	7.19	0.58	0.41	1.70	[12]
TiO₂-rGO	Grätzel	P25	0.75 wt.%	10.75	0.69	0.57	4.20	12.16	0.67	0.68	5.50	[13]

P25-rGO	Dispersion	P25	1 wt.%	11.90	0.69	0.61	4.96	15.40	0.67	0.64	6.58	[14]
TiO$_2$-rGO	CVD	TiO$_2$	Multilayer oxygenated rGO-TiO$_2$	12.70	0.70	n/a	5.60	16.0	0.73	n/a	6.70	[15]
TiO$_2$-rGO inverse opal	Infiltrated	TiO$_2$ inverse opal	3 wt.%	12.39	0.66	0.60	4.86	17.10	0.72	0.61	7.52	[16]
TiO$_2$-rGO	Hydrothermal	Ti(OBu)$_4$	n/a	7.85	0.66	0.60	3.11	10.07	0.75	0.57	4.28	[17]
TiO$_2$-rGO	n/a	TiO$_2$	0.005 wt.%	12.22	0.65	n/a	3.50	13.55	0.68	n/a	4.03	[18]
TiO$_2$-rGO	Hydrothermal	450 °C TiO$_2$-rGO	550 °C	13.65	0.69	0.51	4.85	14.17	0.74	0.51	5.34	[19]
Ti with rGO (from GO)	Hydrothermal	TTIP	Method comparison	15.26	0.68	0.59	5.82	18.46	0.73	0.69	8.62	[20]
Ti-rGO	Sonication	TiO$_2$	0.01 wt.%	6.27	0.71	0.60	2.68	8.42	0.68	0.64	3.69	[21]
TiO$_2$-rGO	Hydrothermal	TiO$_2$	0.5 mg/mL	6.93	0.62	0.65	2.82	12.13	0.65	0.64	5.08	[22]
rGO (from GO) with TiO$_2$	Thermal Reduction	TiO$_2$	n/a	10.14	0.68	0.67	4.60	11.06	0.67	0.74	5.50	[23]

To enhance the PCE performance, the continuous study regarding ion implantation technique are explored as novelty technique for DSSCs application. Notable, there are no researches reported on Ti ion implanted on rGO as efficient photoanode for DSSCs application. This approach is trying to build and endorsed most effective routes to improve the DSSCs performance is the fabrication of TiO_2-rGO NC (photoanode). Moreover, it is an enhanced approach which takes into consideration the photoinduced charge carrier, photogenerated electron-hole pairs, charge recombination and energy band gap of the TiO_2-rGO NC has been proposed to bring the PCE to the satisfactory level. On top of that, this technique allows the incorporation of accelerated high-energy Ti^{3+} ion species into the raw surface under high applied power in a short period of time with less structural defect/low interfacial defect possibility and good optical properties as compared to the literature study [4].

2. Historical overview of DSSCs

In the 1960s the idea developed that organic dye could function most efficiently to generate electricity at oxide electrode in electrochemical cells [24]. The preliminary concept was imitated as plants photosynthesis process and explored by the University of California at Berkeley with chlorophyll extracted from spinach (bio-mimetic or bionic approach) [25]. In 1972, the electric power generation experiments was started and demonstrated the principle of DSSCs [26]. For the following two decades, the instability in terms of PCE still remains a major challenge although porosity of the photo-electrode was optimized by fine oxide powders [27]. A modern DSSCs utilizes a porous layer of TiO_2 nanoparticles as photo-electrode and is covered with organic dye for sunlight absorption purpose. Eventually, it obeyes the chorophyl in green leaves concept. Besides, the electrolyte solution is placed between the anode (TiO_2) and cathode (Pt) in DSSCs. In other words, the principle is also similar to the conventional alkaline battery.

In 1991, the innovation of third generation of photovoltaic technology named DSSCs was inspired with a breakthrough of ~7.1 % [28] under solar illumination. The evolution has continued progressively in aspect of structural, substrate morphology, dye modification, and also electrolyte solution and thus obtained the latest improved performance with ~13 % [29]. Semiconductor materials of photoanodes played an essential component in DSSCs to perform the great conversion PCE with agreement to the thin nanostructured mesoporous film, maximum transparency of the thin layer, rapid electron transportation with low resistance, and the porous photoanode can be completely accessible for the electrolyte [30]. To date, DSSCs retained attention due to its large availability and low-cost material especially in term of low processing temperature. DSSCs also could integrated in portable devices as well as indoor facilities like chargers, solar key boards, and solar bags [31]. In 2006, Poortmans and co-researchers reported that the lifetime of DSSCs is around 20 years but the leakage of the liquid electrolyte still remain a challenge [32].

2.1   Material Selection for DSSCs

The DSSCs typically consists of photoelectrode, counter electrode, dye molecules, and organic electrolyte [33, 34]. Among these essential parts, the photoelectrode is the main factor in order to achieve the desired light PCE and maintain its stability. Taking into account of the processes involved in the photoactivity reaction under solar irradiation, the materials used as photoanode must satisfy several functional requirements with respect to high dye loading absorption, charge separation, and efficiency in charge carriers such as ZnO [35], La^{3+} & Mg [36], Cu & Ag [37], and rGO [38] etc. Among all of the available metal photoanode, rGO offers great promise for photovoltaic applications and it is perfectly suitable as a photoelectrode in DSSCs systems, which has a positive impact on the photo-induced charge transport and suppress the charge recombination in DSSCs. Moreover, rGO also will lead rapid charge carrier mobility (holes and electrons) in DSSCs by the ultrafast extraction of photo-generated carriers. However, one of the major drawbacks of rGO in DSSCs is related to its structural defects.

For further development, the metal oxide is required to enhance its photocatalytic activity. Some metal oxides such as TiO_2, zinc oxide (ZnO), and stannic oxide (SnO_2) have been used as photoanode in DSSCs. Among these metal oxide, TiO_2 nanocrystals gives the best PCE as compared to ZnO and SnO_2 due to the higher photocatalysis activity, abundance, and high quantum yield [39-46]. However, TiO_2 has limited solar light harvesting ability due to its wide band gap (3.0-3.2 eV) [47]. Henceforth, the incorporation of TiO_2-rGO NC has become a promising strategy for the enhancement in terms of PCE in DSSCs application.

3.   Reduced graphene oxide (rGO)

In 2004, a single layer rGO was first discovered by Geim and Novoselov using the "Scotch tape method" under ambient conditions. Accordingly, pristine graphene is an allotrope of carbon with honeycomb structure, in which the sp^2-hydridized carbon atoms are arranged in a basal plane of graphite lattice structure. However, pristine graphene remains a challenge where the aggregation or restacking occurred due to the van der Waals interactions. The interlayer spacing between graphite layers is around 0.34 nm [48]. The graphene, either in single-layer nanosheet or few stacking-layer nanosheets, has attracted extensive attention attributed to its rapid electron transport and good conductivity properties. On top of that, rGO acts as a conductive media, which is beneficial to the mechanism of the electrolyte liquid [49]. Additionally, rGO material is considered as a potential candidate to be introduced as photoanode in DSSCs due to its tunable bandgap and photon absorption, high visible light transparency, as well as ultra-fast charge carrier mobility. Figure 2.5 illustrates the carbon materials network, where a single-layer of 2D rGO is considered to be the dominant material for 0D fullerene (wrapped up), 1D nanotube (rolled up), and 3D graphite (stacked up). Among them, the function of fullerene is significantly different from the CNT and rGO. CNT and rGO are only slightly different in terms of the chemical and electronic properties, while CNT and fullerene are not exactly sp^2-hybridized. The outstanding properties of electronic, thermal stability, and optical of rGO will be described in the following sections.

3.1   Electronic properties of rGO based bilayer systems

Generally, the electronic properties of rGO provide high charge carrier concentrations and mobility, which make it a promising candidate for DSSCs. In particular, pristine graphene is considered as a zero band gap semiconductor due to its conduction and valance bands meeting at the Dirac points. However, it is still difficult to produce one atom thick carbon layer of pristine graphene. In rGO synthesis, the bilayer systems have 3 modifications called AA, AB or Bernal phase, and twisted bilayer. The AA bilayer where the second carbon atom layer is stacked on the first carbon atom layer [50]. The twisted bilayer structure is where the top carbon layer is rotated with the lower carbon layer by a specific angle, θ. The AB or Bernal phase is the most stable among the bilayer systems which half of the carbon atoms of the top layer are stacked on carbon atoms of the lower layer whereas the remaining are covered by different materials [51].

The electronic properties of rGO bilayer systems are built from the electronic properties of graphene single layer system. Indeed, the sp^2 hybridization (one s and two p obitals) formed σ bond between neighboring atoms in a single layer of pristine rGO. Carbon atoms have 6 electrons and its electron configuration is $1s^2\,2s^2\,2p^2$. In rGO, each carbon atom is bonded to three nearest neighbors of carbon atoms by strong covalent bonds. These σ bonds are formed from electrons in 2s, $2p_x$ and $2p_y$ valence orbitals and leaving one mobile electron in $2p_z$ orbital which aligned perpendicular to the rGO sheet [52]. The overlapping of $2p_z$ orbitals with neighbor atoms caused delocalization of valence (π) and conduction (π^*) bands. One electron from each carbon is donated to fill the valence band and leaving the conduction band empty.

For the AB bilayer configuration, the interlayer spacing, c_0 is estimated around 3.35 Å while AA and the twisted bilayer are slightly different [53]. Furthermore, the sp^2 hybridization of the bilayer is much heavier than a monolayer of pristine graphene. As an addition, Dirac-like effective Hamiltonian described that the charge carriers of pristine graphene are massless chiral quasiparticles with a linear dispersion [54]. Besides, the electronic spectrum of the bilayer of rGO maybe gapped or gapless and it could be controlled by a doping rate or other effective parameter. At last but not least, the electronic spectrum depends on the needs of the electronic application. Notably, bilayer of rGO is a more promising candidate than pristine graphene in DSSCs application due to the electronic properties being stacking-dependent and may change the electrodynamic features during hybridation processes with other semiconductor materials [55].

3.2   Thermal conductivity of rGO

Thermal conductivity is the ability of the material to conduct heat over a wide range of temperature. In rGO, the heat conductivity is dictated by phonons transport rather than electrons. This is due to the concentration of free carriers in rGO being relatively low as compared to metals. Phonon transport can be divided into two types, namely diffusive and ballistic. Diffusive conduction only occurs at high temperature while ballistic conduction occurs at low temperature [56]. The investigation of thermal conductivity of rGO through conventional method is extremely difficult as it requires measuring the temperature drop over the thickness of examined film and the fact that pristine graphene is only one carbon atom thick. The nature of the rGO in terms of thermal conductivity properties respective with DSSCs was reported by some researchers [57-59]. The rGO is a good heat conducting material with 5000 W/mK[52]. High thermal conductivity of rGO could be advantageous when high current density is loading that generate significant amount of heat within the DSSCs system.

3.3   Optical properties of rGO

Due to its unique optical properties, rGO is able to capture ~2.3 % of visible light per sheet [60, 61]. The amount of visible light absorption is proportional to the number of rGO layers [62]. Furthermore, a single layer of pristine graphene can only absorb 2.3% of light harvesting. This gives us a hint that the maximum light harvesting can go through ~5 sheets of pristine graphene or bilayer film provided the light transmittance is 90% [62]. For this phenomenon, rGO sheet/film is a dependent material and a supporter to reduce the charge mobility rate in which the photoexcited electron transfers from VB to CB passing through the outer circuit of DSSCs. However, the absorption per unit electron mobility is low. When Fermi level is located at the Dirac point, the number of charge carriers will decrease while the resistivity will increase untill its maximum (V_g = 0V). This is known as optical transition of rGO. The relaxation and recombination of photo-induced electron-hole pairs in rGO is highly dependent on the concentration of carrier in

Carbonaceous Composite Materials
Materials Research Foundations **42** (2018) 1-33

Materials Research Forum LLC
doi: http://dx.doi.org/10.21741/9781945291975-1

DSSCs [56]. Furthermore, rGO is also considered as a low-cost material and provides outstanding transparency properties to allow maximum light absorbance transfer along the FTO glass in DSSCs.

3.4   Electrochemical performance of rGO

In fact, rGO is a single layer sheet of sp^2-bonded carbon atoms arranged in honeycomb crystal lattice, has gained huge attention as electrode material due to its good electrochemical stability, high surface area (> 2600 m^2 g^{-1}), fast ions diffuse to its structure and good mechanical property [63]. Taking these facts into account, a recent literature review on the rGO electrode adopted in various forms of supercapacitor application is presented in Table 2. A wide variety of synthesis methods have been used in the preparation of rGO, including chemical, thermal, electrochemical, and microbial/bacterial as shown in Table 2 [64]. Prior reduction treatment, graphite powder, a precursor of rGO undergoes oxidation reaction to form graphite oxide using modified Hummers' method. This oxidation process helps to increase the interlayer spacing between graphene sheets in graphite powder by introducing oxygen functional groups such as hydroxyl, epoxides, carboxyl and carbonyl as illustration in Figure 1 [65].

Table 2: The electrochemical performance of rGO material.

Sample	Reduction method	Specific capacitance $(F\ g^{-1})$	Current density/ Scan rate $(A\ g^{-1})$	Remarks	Ref.
rGO	Thermal reduction	260.5	0.4	A lot of oxygen functional groups still present on the surface of rGO, which contribute the pseudocapacitance.	[66]
rGO	Electrochemical reduction	150.4	5	The enhanced specific capacitance of rGO due to more electrochemically active surface area.	[67]
rGO	Electrochemical reduction	128	212.16	rGO displayed high specific capacitance due to removal of functional group help to increase the surface contact for ions.	[68]
rGO	Alkaline hydrothermal reduction	145	0.5	High electrochemical capacitance may attributed to the large sp^2 domains of rGO which benefit the ion mobility and lower charge transfer resistance.	[69]
rGO	Chemical reduction (Hydrazine)	135	0.01	High electrical conductivity of rGO give rise to the stable electrochemical performance over a wide range of scan rates.	[70]
rGO	Chemical reduction (Hydrazine)	205	0.1	High specific capacitance owing to the high accessibility of electrolyte ions and high electrical conductivity (100 S m^{-1}).	[71]
rGO	Chemical reduction (Hydrazine)	154.1	1	High specific surface area that can be readily accessed by electrolyte ions.	[72]
rGO	Chemical reduction (HBr)	348	0.2	Remarkably specific capacitance due to rGO facilitates the penetration of aqueous electrolyte and the stable oxygen groups introduce pseudocapacitive effects.	[73]
rGO	Chemical reduction (NaBH$_4$)	135	0.75	rGO thin film electrode showed high specific capacitance due to high accessibility of surface area resulting in improved charge transfer kinetics.	[74]

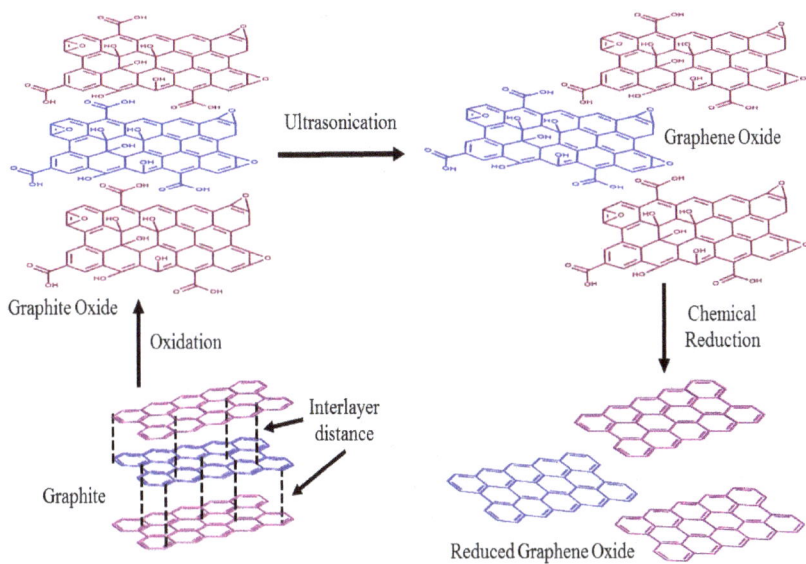

Figure 1: Synthesis of reduced graphene oxide via chemical reduction method [75].

Chemical reduction, a scalable technique, towards the production of reduced graphene oxide (rGO) from graphene oxide (GO), which normally sustaining the solution at low temperature in the range of 85-100 °C. GO is an exfoliated form of graphite oxide and prepared through ultrasonication methods in deionized water. The transformation of GO to rGO is normally indicated by a color change from brown to black. In this process, a vast number of reducing agents have been used to synthesis rGO, including thiourea, hydrazine, borohydrides, hydrohalic acid, metal-alkaline and others [75]. Such reducing agents are reported to produce high deoxygenation degrees of rGO. GO can be thermally treated to form rGO at high temperature of 1000 °C or more in a tube furnace in the presence of inert gas, which generated pressure to overcome the van der Waals force that occur between graphene sheets. However, large energy consumption is the major drawback of thermal reduction [76]. Moreover, electrochemical reduction technique, known as cyclic voltammetry (CV)-reduced GO, are reported to produce rGO by directly depositing it in GO suspension onto a substrate surface [77]. The resulted rGO showed a

Carbonaceous Composite Materials   Materials Research Forum LLC
Materials Research Foundations **42** (2018) 1-33   doi: http://dx.doi.org/10.21741/9781945291975-1

similar structure as compared to the pure graphene. Nevertheless, it is difficult to deposit a large amount of rGO onto a electrode surface. Based on the literature studies, it could be summarized that the experimentally measured capacitance of rGO are much lower than the expected value owing to the agglomeration of graphene sheets caused by its strong van der Waals force, which lower the contact surface area for the electrolyte ion. Unlike AC, the specific surface area of rGO is depends on the layers instead of the distribution of pores. Therefore, combining rGO and others material to form nanocomposite has gained great interest for application.

4.   TiO$_2$-rGO NC material

Based on literature review, rGO film is a potential candidate to improve the PCE of DSSCs, but it was normally studied and applied as a counter electrode [78]. Then, TiO$_2$ nanomaterials with superior photocatalytic activity have attracted great attention to be used in the DSSCs. However, photocatalysts suffer from drawbacks such as high electron-hole pair recombination resulting in a low PCE. Considering this fact, hybridization of TiO$_2$-rGO NC could enhance the photocatalysts activity by increasing the electron mobility and consequently reduce the charge recombination of the electron and hole. On top of that, agglomeration of TiO$_2$ can be overcome since the free electrons trapped at the active area are fully occupied by C-C bonding of rGO. This provides electron-hole separation and facilitates the interfacial electron transfer [79].

4.1   TiO$_2$-rGO NC material's properties

Over the past few years, hybrid of TiO$_2$-rGO NC has gained much attention and has been intensively studied because of the unique features of enhancement in photocatalyst activity and accelerated electron mobility to suppress the charge recombination. Among the vast number of different dopants, TiO$_2$ is one of the most capable candidates to be coupled with rGO for enhancement in numerous diverse applications, such as DSSCs photovoltaics. Several researchers have reported that the band gap of TiO$_2$ decreases with the tunable amount of rGO dopants in NC as shown in Table 3. This is due to the formation of Ti-O-C bond and the hybridization of C $2p^2$ orbitals and O $2p^4$ orbitals to form new valence band [80, 81].

According to electrical properties of TiO$_2$-rGO NC, Zhang and co-researchers clarified that the photocatalytic performance can be improved with enhancement of carrier concentration and mobility between the rGO and TiO$_2$ materials [80,86-89]. To enhance the photocatalytic activity of TiO$_2$-rGO NC, Khalid and co-researchers have shown that the function of TiO$_2$ can be easily enhanced in photocatalytic activity properties under visible light irradiation in terms of great absorptivity of dyes, extended light absorption range, and efficient charge separation with rGO [83,90]. Khalid and co-researchers demonstrated that the band gap energy is decreased from 3.20 eV of TiO_2 to 3.00 eV when incorporated with rGO, it is indicated that the influence of rGO on the optical properties where increasing rGO amount will result in the light absorption of TiO_2 [83]. Moreover, Khalid and co-researchers claimed that presence of rGO in TiO_2 composite could reduce the emission intensity in photoluminescence characterization and lead the enhancement of electron-hole pairs separation efficiency [83].

Table 3: Band gap energy values of TiO_2-rGO NC.

Methods	Results (eV)	Reference
Thermal	Pure TiO_2 = 3.10 TiO_2-rGO = 2.95	[82]
Hydrothermal	Pure TiO_2 = 3.20 1 wt% rGO-TiO_2 = 3.16 2 wt% rGO-TiO_2 = 3.13 5 wt% rGO-TiO_2 = 3.04 10 wt% rGO-TiO_2 = 3.00	[83]
Solvothermal	Pure TiO_2 = 3.28 TiO_2-rGO = 2.72	[84]
Hydrothermal	Pure TiO_2 = 3.03 TiO_2-rGO = 2.78	[85]
Sonication	0.01 wt% rGO-TiO_2 = 2.95	[21]

4.2 Formation mechanism of TiO_2-rGO NC material

Zhang H. and co-researchers formed TiO_2-rGONC under a simple liquid phase deposition method by utilizing titanium tetrafluoride (TiF_4) and electron beam (EB) irradiation-pretreated rGO [91]. He discovered that the preparation condition had an significantly effect on the structure and properties of TiO_2-rGONC. Through this method, it can be synthesized more uniformly, with smaller size TiO_2 nanoparticles and exhibited higher photocatalytic activities with EB irradiation-pretreated rGO. Figure 2 showns the mechanism of TiO_2-rGONC via the simple liquid phase deposition method.

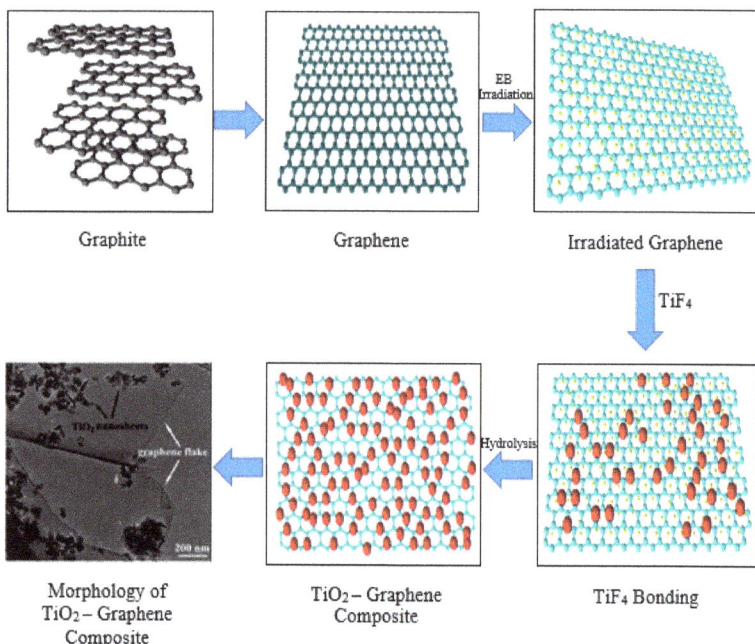

Type of Bond	Description
	Carbon in graphite
	Carbon in rGO
	Irradiated Carbon
	Titania Molecule
	TiO_2-rGO NC

Figure 2: Formation mechanism of TiO_2-rGONC material via liquid phase deposition method.

4.3   Mechanism of TiO_2-rGO NC applied in DSSCs

Figure 3 illustrates the electron flow when the rGO is loaded in between the TiO_2 molecules. The electron flow will be further enhanced if the rGO is well connected with

TiO_2. This phenomenon is caused by the suppression of back-transport electron from photoanode of FTO/ITO electrode to the I_3^- ions which subsequently increase the dye adsorption. Sung and co-researchers have mentioned that the presence of rGO oxide will reduce the back-transport in DSSCs and also assists in UV-reduction in TiO_2 [83,92].

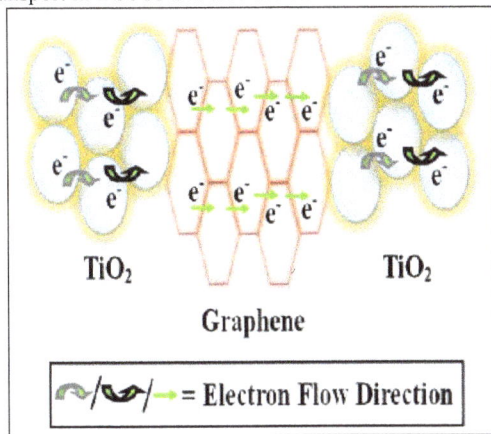

Figure 3: TiO_2-rGO NC bonding mechanism in DSSCs device.

4.4 Preparation of TiO_2-rGO NC

In this thin-film photovoltaic cell technologies, second generation solar cells are derived from the first generation solar cell by depositing one or more thin layers of semiconductors materials on the specified substrate such as metal, glass, or silicon wafer. According to Thien and co-researchers, a higher photocurrent density will attribute to a delayed recombination rate and longer electron lifetime [93]. The photocurrent response of a solar cell is defined as the photo-generated electron-hole pairs interaction between the photoanode and photocathode electrode [94-100]. The charge separation efficiency is increased due to the electronic interaction between rGO and photo-induced electrons of TiO_2 in the NC [101-107].

On top of low-cost and high reproducibility, TiO_2-rGO NC also show high interfacial contact and potential to enhance the photocatalytic activities of TiO_2. In these two decades, there are variety of techniques used to synthesize the TiO_2-rGO NC based materials to bright up photovoltaic technology especially in DSSCs application. For TiO_2-rGO NC, rGO could be easily synthesized from the graphite flakes through the intermediate product of GO [108]. This technique was beneficial to form the TiO_2 nanocrystals during the synthesis of TiO_2-rGO NC via the oxygenation of the functional groups from GO or rGO products [109]. Kim and co-researchers reported that GO could be reduced via the UV-assisted photocatalytic reduction process using a 450W xenon arc lamp forming the TiO_2-rGO NC with low surface roughness and good adhesion at the photoanode element [92]. Dubey and co-researchers also reported that the GO could be reduced by UV radiation in the presence of ethanol solvent and TiO_2 nanoparticles to form TiO_2-rGO NC [110]. Another efficient technique to prepare TiO_2-rGO NC is the direct growth process to enhance the photocatalytic activity. Recently, Xu and co-researchers reported that rGO quantum dots could directly grow on 3D micropillar/microwave arrays of rutile TiO_2 nanorods forming TiO_2-rGO NC [111]. Additionally, the pathway for large scale production of the TiO_2-rGO NC is the self-assembly approach of the in-situ grown nanocrystalline TiO_2 with stabilization of rGO in aqueous solutions by the anionic sulfate surfactants [112]. Furthermore, Liu and co-researchers reported an accessible synthetic route of solvothermal approach to form TiO_2-rGO NC with a better adsorption-photocatalytic activity than that of pure TiO_2 [113].

4.4.1 Sol-Gel synthesis

Sol-gel technique is widely used in the synthesis of rGO-based semiconductor composites. This method depends on the phase transformation of a sol obtained from metallic alkoxides or organometallic precursors. For instance, tetrabutyl titanate dispersed in rGO-containing absolute ethanol solution will gradually form a sol with continuous magnetic stirring, and eventually change into TiO_2-rGO NC after drying and post heat treatment [114,115]. The synthesis process is illustrated in Figure 4 (A) [115]. The resulting TiO_2 nanoparticles closely dispersed on the surface of 2D rGO NS (Figure 4 (B)) [115]. Wojtoniszak and co-researchers used a similar strategy to prepare the TiO_2-rGO NC via the hydrolysis of titanium (IV) butoxide in GO-containing ethanol solution [116]. The reduction of GO to rGO was performed in a post heat treatment process. Meanwhile, Farhangi and co-researchers prepared Fe-doped TiO_2 nanowire arrays on the surface of functionalized rGO sheets using a sol-gel method in the green solvent of supercritical carbon dioxide [117]. During the preparation the rGO NS act as template for nanowire growth through surface -COOH functionalities.

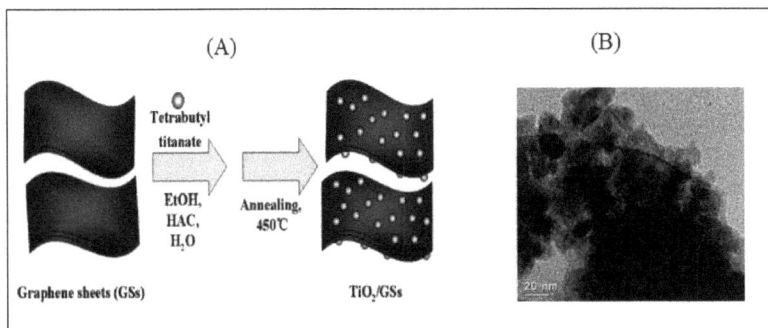

Figure 4: Schematic synthesis procedure (A) and typical TEM image of the TiO$_2$-rGO NC (B) [115].

4.4.2 Solution mixing synthesis

The solution mixing is a simple method used to fabricate rGO/semiconductor composite photocatalysts. The uniform distribution of photocatalysts is facilitated by the oxygenated functional groups on GO under vigorous stirring or ultrasonic agitation [118]. The rGO-based composites can be obtained after the reduction of GO in the composite. Bell and co-researchers fabricated TiO$_2$-rGO NC by ultrasonically mixing TiO$_2$ nanoparticles and GO colloids together, followed by ultraviolet (UV)-assisted photocatalytic reduction of GO to rGO [104]. Similarly, GO dispersion and N-doped Sr$_2$Ta$_2$O$_7$ have been mixed together, followed by reduction of GO to yield Sr$_2$Ta$_2$O$_{7-x}$N$_x$-rGO composites under xenon lamp irradiation [119]. Paek co-researchers have prepared the SnO$_2$ sol by hydrolysis of SnCl$_4$ with NaOH and then mixed with the prepared rGO dispersion in ethylene glycol to form the SnO$_2$-rGO composite [120]. On the other hand, Geng co-researchers have synthesized the CdSe-rGO quantum dots composites [121]. In their work, pyridine-modified CdSe nanoparticles were mixed with GO sheets, where pyridine ligands provide π-π interactions for the assembly of CdSe nanoparticles on GO sheets.

4.4.3 In-Situ growth synthesis

The in-situ growth strategy provides efficient electron transfer between rGO and semiconductor nanoparticles through their intimate contact. The functional GO and metal salts are commonly used as precursors. The presence of epoxy and hydroxyl functional groups on rGO can act as the heterogeneous nucleation sites and anchor semiconductor nanoparticles avoiding the agglomeration of the small particles [122]. Lambert co-researchers have reported the in situ synthesis of petal-like TiO$_2$-GO by the hydrolysis of

TiF_4 in the presence of aqueous dispersions of GO, followed by post thermal treatment to produce TiO_2-rGO NC [123]. With a high concentration of GO and stirring off, long-range ordered assemblies of TiO_2-GO sheets were self-assembled. Besides that, Guo co-researchers synthesized TiO_2-rGO NC sonochemically from $TiCl_4$ and GO in ethanol-water system, followed by a hydrazine treatment to reduce GO into rGO [124]. The average size of the TiO_2 nanoparticles were controlled at around 4-5 nm on the sheets, which is attributed to the pyrolysis and condensation of the dissolved $TiCl_4$ into TiO_2 by ultrasonic waves. Lastly, TiO_2-rGO were synthesized with various method not only apply in photovoltaic DSSCs application but it's other useful application and summarized in Table 4.

Table 4: Summary of TiO_2-rGO synthesis in various applications.

Synthesis Method	Materials	Application	Reference
Sol-Gel	Ce-TiO_2-rGO	Photoelectrocatalytic	[110, 125]
Sol-Gel	Anatase TiO_2-rGO	Photoelectrochemical Water Splitting	[126]
Solution Mixing	TiO_2-rGO	Photocatalytic selectivity	[127]
Solution Mixing	TiO_2-rGO	Hydrogen production	[128]
In-Situ Growth	TiO_2-rGO	Sodium/Lithium Ion Batteries	[129]
In-Situ Growth	TiO_2-rGO	Photocatalytic activity	[130]

5.   Conclusion

TiO_2 has been widely explored in decoration with rGO as photoanode in DSSCs device. With the low-cost and high reproducibility of TiO_2-rGO NC, it has high interfacial contact and potential to enhance the photocatalytic activity of TiO_2 and also charge carrier in DSSCs application. TiO_2-rGO NC based materials were synthesized via sol-gel, solution mixing, and in-situ growth method to bright up photovoltaic technology especially in DSSCs application. Innovative new approaches of synthesis of a high-purity hybrid of TiO_2 with rGO of photoanode are critical and crucial for determining the potential of the material as an efficient DSSCs system. Indeed, there are three main reasons that the promising of rGO coated with TiO_2 to form TiO_2-rGO NC in order to enhance the performance of DSSCs. First, the TiO_2 have the ability to improve the separation between the electron and hole since rGO has high electron mobility. Second, the corporation between TiO_2 and rGO can enlarge the absorption range of the visible region after the Ti-O-C was bonded. Third, the TiO_2-rGO NC can increase the interaction area and adsorption and result the N-719 dye to have the ability to create π-π interaction. In terms of the electron transfer, TiO_2 have good attraction with the epoxy and carboxylate groups of rGO. For the Fermi levels, rGO could be chosen as the electron shuttle in TiO_2 since the rGO has a higher value (-4.4 eV) as compared to the TiO_2 (-4.2 eV) at the conduction band (CB). Inaddition, 3D graphene-TiO_2 composite architecture warrants extended and in-depth study for next revolution of graphene science, comparable to the efforts that have been spent investigating the properties of 2D TiO_2-rGO NC. Certainly, the work has only just begun in exploring the science and engineering applications of this remarkable material platform.

6.   Acknowledgements

This research are supported by Fundamental Research Grant Scheme (FP008-2015A), Universiti Malaya Prototype Grant (RU005G-2016), and, Postgraduate Research Fund Grant, PPP (PG210-2014B), and Global Collaborative Programme-SATU joint research scheme (ST007-2017 & ST008-2017) from the University of Malaya.

References

[1]   D. Liang, C. Cui, H. Hu, Y. Wang, S. Xu, B. Ying, P. Li, B. Lu, H. Shen, One-step hydrothermal synthesis of anatase TiO_2/reduced graphene oxide nanocomposites with enhanced photocatalytic activity, Journal of Alloys and Compounds, 582 (2014) 236-240. https://doi.org/10.1016/j.jallcom.2013.08.062

[2]   J. Shen, B. Yan, M. Shi, H. Ma, N. Li, M. Ye, One step hydrothermal synthesis of TiO_2-reduced graphene oxide sheets, Journal of Materials Chemistry, 21 (2011) 3415-3421. https://doi.org/10.1039/c0jm03542d

[3]   B.Y.S. Chang, N.M. Huang, M.N. An'amt, A.R. Marlinda, Y. Norazriena, M.R. Muhamad, I. Harrison, H.N. Lim, C.H. Chia, Facile hydrothermal preparation of titanium dioxide decorated reduced graphene oxide nanocomposite, International Journal of Nanomedicine, 7 (2012) 3379-3387.

[4]   J. Chen, G. Zhang, B. Luo, D. Sun, X. Yan, Q. Xue, Surface amorphization and deoxygenation of graphene oxide paper by Ti ion implantation, Carbon, 49 (2011) 3141-3147. https://doi.org/10.1016/j.carbon.2011.03.045

[5]   Y. B. Tang, C. S. Lee, J. Xu, Z. T. Liu, Z. H. Chen, Z. He, Y. L. Cao, G. Yuan, H.
        Song, L. Chen, Incorporation of graphenes in nanostructured TiO_2 films via
        molecular grafting for dye-sensitized solar cell application, ACS Nano, 4 (2010)
        3482-3488. https://doi.org/10.1021/nn100449w

[6]   S. Sun, L. Gao, Y. Liu, Enhanced dye-sensitized solar cell using graphene-TiO_2
        photoanode prepared by heterogeneous coagulation, Applied Physics Letters, 96
        (2010) 083113. https://doi.org/10.1063/1.3318466

[7]   J. Song, Z. Yin, Z. Yang, P. Amaladass, S. Wu, J. Ye, Y. Zhao, W.Q. Deng, H.
        Zhang, X.W. Liu, Enhancement of photogenerated electron transport in dye-
        sensitized solar cells with introduction of a reduced graphene oxide–TiO_2 junction,
        Chemistry-A European Journal, 17 (2011) 10832-10837.
        https://doi.org/10.1002/chem.201101263

[8]   T. H. Tsai, S. C. Chiou, S. M. Chen, Enhancement of dye-sensitized solar cells by
        using graphene-TiO2 composites as photoelectrochemical working electrode,
        International Journal Electrochemical Science, 6 (2011) 3333-3343.

[9]   Z. He, G. Guai, J. Liu, C. Guo, J.S.C. Loo, C.M. Li, T.T.Y. Tan, Nanostructure
        control of graphene-composited TiO_2 by a one-step solvothermal approach for
        high performance dye-sensitized solar cells, Nanoscale, 3 (2011) 4613-4616.
        https://doi.org/10.1039/c1nr11300c

[10]   J. Fan, S. Liu, J. Yu, Enhanced photovoltaic performance of dye-sensitized solar
        cells based on TiO_2 nanosheets/graphene composite films, Journal of Materials
        Chemistry, 22 (2012) 17027-17036. https://doi.org/10.1039/c2jm33104g

[11]   B. Tang, G. Hu, Two kinds of graphene-based composites for photoanode
        applying in dye-sensitized solar cell, Journal of Power Sources, 220 (2012) 95-
        102. https://doi.org/10.1016/j.jpowsour.2012.07.093

[12]   J. Zhao, J. Wu, F. Yu, X. Zhang, Z. Lan, J. Lin, Improving the photovoltaic
        performance of cadmium sulfide quantum dots-sensitized solar cell by
        graphene/titania photoanode, Electrochimica Acta, 96 (2013) 110-116.
        https://doi.org/10.1016/j.electacta.2013.02.067

[13]   W. Shu, Y. Liu, Z. Peng, K. Chen, C. Zhang, W. Chen, Synthesis and photovoltaic
        performance of reduced graphene oxide–TiO_2 nanoparticles composites by
        solvothermal method, Journal of Alloys and Compounds, 563 (2013) 229-233.
        https://doi.org/10.1016/j.jallcom.2013.02.086

[14] B. Tang, G. Hu, H. Gao, Z. Shi, Three-dimensional graphene network assisted high performance dye sensitized solar cells, Journal of Power Sources, 234 (2013) 60-68. https://doi.org/10.1016/j.jpowsour.2013.01.130

[15] M. Shanmugam, C. Durcan, R.J. Gedrim, T. Bansal, B. Yu, Oxygenated-graphene-enabled recombination barrier layer for high performance dye-sensitized solar cell, Carbon, 60 (2013) 523-530. https://doi.org/10.1016/j.carbon.2013.04.083

[16] H.-N. Kim, H. Yoo, J.H. Moon, Graphene-embedded 3D TiO_2 inverse opal electrodes for highly efficient dye-sensitized solar cells: morphological characteristics and photocurrent enhancement, Nanoscale, 5 (2013) 4200-4204. https://doi.org/10.1039/c3nr33672g

[17] M. Zhu, X. Li, W. Liu, Y. Cui, An investigation on the photoelectrochemical properties of dye-sensitized solar cells based on graphene–TiO_2 composite photoanodes, Journal of Power Sources, 262 (2014) 349-355. https://doi.org/10.1016/j.jpowsour.2014.04.001

[18] N.T. Ha, P.D. Long, N.T. Trung, L.V. Hong, Graphene Effect on Efficiency of TiO_2-based Dye Sensitized Solar Cells (DSSC), Communications in Physics, 26 (2016) 43. https://doi.org/10.15625/0868-3166/26/1/7961

[19] J. C. Chou, C. H. Huang, Y. J. Lin, C. M. Chu, Y. H. Liao, L. H. Tai, Y. H. Nien, The Influence of Different Annealing Temperatures on Graphene-Modified TiO_2 for Dye-Sensitized Solar Cell, IEEE Transactions on Nanotechnology, 15 (2016) 164-170. https://doi.org/10.1109/TNANO.2015.2510081

[20] E. Nouri, M.R. Mohammadi, P. Lianos, Impact of preparation method of TiO_2-RGO nanocomposite photoanodes on the performance of dye-sensitized solar cells, Electrochimica Acta, 219 (2016) 38-48. https://doi.org/10.1016/j.electacta.2016.09.150

[21] U. A. Kanta, V. Thongpool, W. Sangkhun, N. Wongyao, J. Wootthikanokkhan, Preparations, Characterizations, and a Comparative Study on Photovoltaic Performance of Two Different Types of Graphene/TiO_2 Nanocomposites Photoelectrodes, Journal of Nanomaterials, 2017 (2017). https://doi.org/10.1155/2017/2758294

[22] R. Raja, M. Govindaraj, M.D. Antony, K. Krishnan, E. Velusamy, A. Sambandam, M. Subbaiah, V.W. Rayar, Effect of TiO_2/reduced graphene oxide composite thin film as a blocking layer on the efficiency of dye-sensitized solar cells, Journal of Solid State Electrochemistry, 21 (2017) 891-903. https://doi.org/10.1007/s10008-016-3437-7

[23] Y. Zhang, C. Wang, Z. Yuan, L. Zhang, L. Yin, Reduced Graphene Oxide Wrapped Mesoporous Hierarchical TiO_2-CdS as a Photoanode for High-Performance Dye-Sensitized Solar Cells, European Journal of Inorganic Chemistry, 2017 (2017) 2281-2288. https://doi.org/10.1002/ejic.201601535

[24] H. Gerischer, M. Michel-Beyerle, F. Rebentrost, H. Tributsch, Sensitization of charge injection into semiconductors with large band gap, Electrochimica Acta, 13 (1968) 1509-1515. https://doi.org/10.1016/0013-4686(68)80076-3

[25] H. Tributsch, M. Calvin, Electrochemistry of Excited Molecules: Photo-Electrochemical Reactions of Chlorophylls, Photochemistry and Photobiology, 14 (1971) 95-112. https://doi.org/10.1111/j.1751-1097.1971.tb06156.x

[26] H. TRIBUTSCH, Reaction of excited chlorophyll molecules at electrodes and in photosynthesis, Photochemistry and Photobiology, 16 (1972) 261-269.

[27] M. Matsumura, S. Matsudaira, H. Tsubomura, M. Takata, H. Yanagida, Dye sensitization and surface structures of semiconductor electrodes, Industrial & Engineering Chemistry Product Research and Development, 19 (1980) 415-421. https://doi.org/10.1021/i360075a025

[28] B. O'regan, M. Grfitzeli, A low-cost, high-efficiency solar cell based on dye-sensitized, nature, 353 (1991) 737-740.

[29] S. Mathew, A. Yella, P. Gao, R. Humphry-Baker, B.F. Curchod, N. Ashari-Astani, I. Tavernelli, U. Rothlisberger, M.K. Nazeeruddin, M. Grätzel, Dye-sensitized solar cells with 13% efficiency achieved through the molecular engineering of porphyrin sensitizers, Nature Chemistry, 6 (2014) 242-247. https://doi.org/10.1038/nchem.1861

[30] A. Hagfeldt, Brief overview of dye-sensitized solar cells, Ambio, 41 (2012) 151-155. https://doi.org/10.1007/s13280-012-0272-7

[31] C.C. Raj, R. Prasanth, A critical review of recent developments in nanomaterials for photoelectrodes in dye sensitized solar cells, Journal of Power Sources, 317 (2016) 120-132. https://doi.org/10.1016/j.jpowsour.2016.03.016

[32] J. Poortmans, V. Arkhipov, Thin film solar cells: fabrication, characterization and applications, John Wiley & Sons2006. https://doi.org/10.1002/0470091282

[33] L. Y. Lin, C. P. Lee, R. Vittal, K. C. Ho, Improving the durability of dye-sensitized solar cells through back illumination, Journal of Power Sources, 196 (2011) 1671-1676. https://doi.org/10.1016/j.jpowsour.2010.08.032

[34] Z. Wei, Y. Yao, T. Huang, A. Yu, Solvothermal growth of Well-Aligned TiO_2 Nanowire Arrays for Dye-Sensitized Solar Cell: Dependence of morphology and

vertical orientation upon substrate pretreatment, International Journal Electrochemical Science, 6 (2011) 1871-1879.

[35]   M. Giannouli, Nanostructured ZnO, TiO_2, and composite ZnO/TiO_2 films for application in dye-sensitized solar cells, International Journal of Photoenergy, 2013 (2013).

[36]   R.T. Ako, P. Ekanayake, A.L. Tan, D.J. Young, La modified TiO_2 photoanode and its effect on DSSC performance: A comparative study of doping and surface treatment on deep and surface charge trapping, Materials Chemistry and Physics, 172 (2016) 105-112. https://doi.org/10.1016/j.matchemphys.2015.12.066

[37]   S. Shakir, H.M. Abd-ur-Rehman, Enhancement in Photovoltaic Performance of Dye Sensitized Solar Cells Using Cu and Cu: Ag Co-Doped TiO_2 Photoanode, ASME 2016 Power Conference collocated with the ASME 2016 10th International Conference on Energy Sustainability and the ASME 2016 14th International Conference on Fuel Cell Science, Engineering and Technology, American Society of Mechanical Engineers, 2016, pp. V001T008A016-V001T008A016.

[38]   H. Ding, S. Zhang, J. T. Chen, X. P. Hu, Z. F. Du, Y. X. Qiu, D. L. Zhao, Reduction of graphene oxide at room temperature with vitamin C for RGO–TiO_2 photoanodes in dye-sensitized solar cell, Thin Solid Films, 584 (2015) 29-36. https://doi.org/10.1016/j.tsf.2015.02.038

[39]   M.R. Hoffmann, S.T. Martin, W. Choi, D.W. Bahnemann, Environmental applications of semiconductor photocatalysis, Chemical Reviews, 95 (1995) 69-96. https://doi.org/10.1021/cr00033a004

[40]   Y. Zhang, H. Feng, X. Wu, L. Wang, A. Zhang, T. Xia, H. Dong, X. Li, L. Zhang, Progress of electrochemical capacitor electrode materials: A review, International Journal of Hydrogen Energy, 34 (2009) 4889-4899. https://doi.org/10.1016/j.ijhydene.2009.04.005

[41]   Y. Fukai, Y. Kondo, S. Mori, E. Suzuki, Highly efficient dye-sensitized SnO_2 solar cells having sufficient electron diffusion length, Electrochemistry Communications, 9 (2007) 1439-1443. https://doi.org/10.1016/j.elecom.2007.01.054

[42]   C. Bauer, G. Boschloo, E. Mukhtar, A. Hagfeldt, Ultrafast studies of electron injection in Ru dye sensitized SnO_2 nanocrystalline thin film, International Journal of Photoenergy, 4 (2002) 17-20. https://doi.org/10.1155/S1110662X0200003X

[43] R. Bhattacharjee, I. M. Hung, A SnO_2 and ZnO Nanocomposite Photoanodes in Dye-Sensitized Solar Cells, ECS Solid State Letters, 2 (2013) Q101-Q104. https://doi.org/10.1149/2.013311ssl

[44] H. Seema, K.C. Kemp, V. Chandra, K.S. Kim, Graphene–SnO_2 composites for highly efficient photocatalytic degradation of methylene blue under sunlight, Nanotechnology, 23 (2012) 355705. https://doi.org/10.1088/0957-4484/23/35/355705

[45] S. Zhuang, X. Xu, B. Feng, J. Hu, Y. Pang, G. Zhou, L. Tong, Y. Zhou, Photogenerated Carriers Transfer in Dye–Graphene–SnO_2 Composites for Highly Efficient Visible-Light Photocatalysis, ACS Applied Materials&Interfaces, 6 (2013) 613-621. https://doi.org/10.1021/am4047014

[46] C. H. Hsu, C. C. Lai, L. C. Chen, P. S. Chan, Enhanced Performance of Dye-Sensitized Solar Cells with Graphene/ZnO Nanoparticles Bilayer Structure, Journal of Nanomaterials, 2014 (2014). https://doi.org/10.1155/2014/748319

[47] H. He, A. Chen, H. Lv, H. Dong, M. Chang, C. Li, Hydrothermal fabrication of Ni_3S_2/TiO_2 nanotube composite films on Ni anode and application in photoassisted water electrolysis, Journal of Alloys and Compounds, 574 (2013) 217-220. https://doi.org/10.1016/j.jallcom.2013.04.208

[48] D.D. Chung, Review graphite, Journal of Materials Science, 37 (2002) 1475-1489. https://doi.org/10.1023/A:1014915307738

[49] S. Yuan, Q. Tang, B. Hu, C. Ma, J. Duan, B. He, Efficient quasi-solid-state dye-sensitized solar cells from graphene incorporated conducting gel electrolytes, Journal of Materials Chemistry A, 2 (2014) 2814-2821. https://doi.org/10.1039/c3ta14385f

[50] Z. Liu, K. Suenaga, P.J. Harris, S. Iijima, Open and closed edges of graphene layers, Physical Review Letters, 102 (2009) 015501. https://doi.org/10.1103/PhysRevLett.102.015501

[51] J.L. Dos Santos, N. Peres, A.C. Neto, Graphene bilayer with a twist: electronic structure, Physical Review Letters, 99 (2007) 256802. https://doi.org/10.1103/PhysRevLett.99.256802

[52] J.H. Warner, F. Schaffel, M. Rummeli, A. Bachmatiuk, Graphene: fundamentals and emergent applications, Newnes2012.

[53] E. McCann, M. Koshino, The electronic properties of bilayer graphene, Reports on Progress in Physics, 76 (2013) 056503. https://doi.org/10.1088/0034-4885/76/5/056503

Carbonaceous Composite Materials   Materials Research Forum LLC
Materials Research Foundations **42** (2018) 1-33   doi: http://dx.doi.org/10.21741/9781945291975-1

[54] A.C. Neto, F. Guinea, N.M. Peres, K.S. Novoselov, A.K. Geim, The electronic properties of graphene, Reviews of Modern Physics, 81 (2009) 109. https://doi.org/10.1103/RevModPhys.81.109

[55] A. Rozhkov, A. Sboychakov, A. Rakhmanov, F. Nori, Electronic properties of graphene-based bilayer systems, Physics Reports, 648 (2016) 1-104. https://doi.org/10.1016/j.physrep.2016.07.003

[56] Y. Zhu, S. Murali, W. Cai, X. Li, J.W. Suk, J.R. Potts, R.S. Ruoff, Graphene and graphene oxide: synthesis, properties, and applications, Advanced Materials, 22 (2010) 3906-3924. https://doi.org/10.1002/adma.201001068

[57] X. Guo, G. Lu, J. Chen, Graphene-based materials for photoanodes in dye-sensitized solar cells, Frontiers in Energy Research, 3 (2015) 50. https://doi.org/10.3389/fenrg.2015.00050

[58] E. Pop, V. Varshney, A.K. Roy, Thermal properties of graphene: Fundamentals and applications, MRS Bulletin, 37 (2012) 1273-1281. https://doi.org/10.1557/mrs.2012.203

[59] G. Fugallo, A. Cepellotti, L. Paulatto, M. Lazzeri, N. Marzari, F. Mauri, Thermal conductivity of graphene and graphite: collective excitations and mean free paths, Nano Letters, 14 (2014) 6109-6114. https://doi.org/10.1021/nl502059f

[60] A. Aghigh, V. Alizadeh, H. Wong, M.S. Islam, N. Amin, M. Zaman, Recent advances in utilization of graphene for filtration and desalination of water: A review, Desalination, 365 (2015) 389-397. https://doi.org/10.1016/j.desal.2015.03.024

[61] J.D. Roy-Mayhew, I.A. Aksay, Graphene materials and their use in dye-sensitized solar cells, Chemical Reviews, 114 (2014) 6323-6348. https://doi.org/10.1021/cr400412a

[62] F. Bonaccorso, Z. Sun, T. Hasan, A. Ferrari, Graphene photonics and optoelectronics, Nature Photonics, 4 (2010) 611-622. https://doi.org/10.1038/nphoton.2010.186

[63] Z. Bo, Z. Wen, H. Kim, G. Lu, K. Yu, J. Chen, One-step fabrication and capacitive behavior of electrochemical double layer capacitor electrodes using vertically-oriented graphene directly grown on metal, Carbon, 50 (2012) 4379-4387. https://doi.org/10.1016/j.carbon.2012.05.014

[64] A. Bianco, H.-M. Cheng, T. Enoki, Y. Gogotsi, R.H. Hurt, N. Koratkar, T. Kyotani, M. Monthioux, C.R. Park, J.M. Tascon, All in the graphene family–a

recommended nomenclature for two-dimensional carbon materials, Carbon, 65 (2013) 1-6. https://doi.org/10.1016/j.carbon.2013.08.038

[65] T.F. Emiru, D.W. Ayele, Controlled synthesis, characterization and reduction of graphene oxide: A convenient method for large scale production, Egyptian Journal of Basic and Applied Sciences, 4 (2017) 74-79. https://doi.org/10.1016/j.ejbas.2016.11.002

[66] B. Zhao, P. Liu, Y. Jiang, D. Pan, H. Tao, J. Song, T. Fang, W. Xu, Supercapacitor performances of thermally reduced graphene oxide, Journal of Power Sources, 198 (2012) 423-427. https://doi.org/10.1016/j.jpowsour.2011.09.074

[67] Y. Shao, J. Wang, M. Engelhard, C. Wang, Y. Lin, Facile and controllable electrochemical reduction of graphene oxide and its applications, Journal of Materials Chemistry, 20 (2010) 743-748. https://doi.org/10.1039/B917975E

[68] X. Y. Peng, X. X. Liu, D. Diamond, K.T. Lau, Synthesis of electrochemically-reduced graphene oxide film with controllable size and thickness and its use in supercapacitor, Carbon, 49 (2011) 3488-3496. https://doi.org/10.1016/j.carbon.2011.04.047

[69] S.D. Perera, R.G. Mariano, N. Nijem, Y. Chabal, J.P. Ferraris, K.J. Balkus, Alkaline deoxygenated graphene oxide for supercapacitor applications: An effective green alternative for chemically reduced graphene, Journal of Power Sources, 215 (2012) 1-10. https://doi.org/10.1016/j.jpowsour.2012.04.059

[70] M.D. Stoller, S. Park, Y. Zhu, J. An, R.S. Ruoff, Graphene-based ultracapacitors, Nano Letters, 8 (2008) 3498-3502. https://doi.org/10.1021/nl802558y

[71] Y. Wang, Z. Shi, Y. Huang, Y. Ma, C. Wang, M. Chen, Y. Chen, Supercapacitor devices based on graphene materials, The Journal of Physical Chemistry. C, Nanomaterials and Interfaces, 113 (2009) 13103. https://doi.org/10.1021/jp902214f

[72] C. Liu, Z. Yu, D. Neff, A. Zhamu, B.Z. Jang, Graphene-based supercapacitor with an ultrahigh energy density, Nano Letters, 10 (2010) 4863-4868. https://doi.org/10.1021/nl102661q

[73] Y. Chen, X. Zhang, D. Zhang, P. Yu, Y. Ma, High performance supercapacitors based on reduced graphene oxide in aqueous and ionic liquid electrolytes, Carbon, 49 (2011) 573-580. https://doi.org/10.1016/j.carbon.2010.09.060

[74] A. Yu, I. Roes, A. Davies, Z. Chen, Ultrathin, transparent, and flexible graphene films for supercapacitor application, Applied Physics Letters, 96 (2010) 253105. https://doi.org/10.1063/1.3455879

Materials Research Forum LLC
doi: http://dx.doi.org/10.21741/9781945291975-1

[75]   C.K. Chua, M. Pumera, Chemical reduction of graphene oxide: a synthetic chemistry viewpoint, Chemical Society Reviews, 43 (2014) 291-312. https://doi.org/10.1039/C3CS60303B

[76]   H. B. Zhang, J. W. Wang, Q. Yan, W. G. Zheng, C. Chen, Z. Z. Yu, Vacuum-assisted synthesis of graphene from thermal exfoliation and reduction of graphite oxide, Journal of Materials Chemistry, 21 (2011) 5392-5397. https://doi.org/10.1039/c1jm10099h

[77]   Y. Zhang, H. Hao, L. Wang, Effect of morphology and defect density on electron transfer of electrochemically reduced graphene oxide, Applied Surface Science, 390 (2016) 385-392. https://doi.org/10.1016/j.apsusc.2016.08.127

[78]   S. Ghasemi, S.R. Hosseini, F. Mousavi, Electrophoretic deposition of graphene nanosheets: A suitable method for fabrication of silver-graphene counter electrode for dye-sensitized solar cell, Colloids and Surfaces A: Physicochemical and Engineering Aspects, 520 (2017) 477-487. https://doi.org/10.1016/j.colsurfa.2017.02.004

[79]   B. Bhanvase, T. Shende, S. Sonawane, A review on graphene–TiO_2 and doped graphene–TiO2 nanocomposite photocatalyst for water and wastewater treatment, Environmental Technology Reviews, 6 (2017) 1-14. https://doi.org/10.1080/21622515.2016.1264489

[80]   H. Zhang, X. Lv, Y. Li, Y. Wang, J. Li, P25-graphene composite as a high performance photocatalyst, ACS Nano, 4 (2009) 380-386. https://doi.org/10.1021/nn901221k

[81]   K. Li, J. Xiong, T. Chen, L. Yan, Y. Dai, D. Song, Y. Lv, Z. Zeng, Preparation of graphene/TiO_2 composites by nonionic surfactant strategy and their simulated sunlight and visible light photocatalytic activity towards representative aqueous POPs degradation, Journal of Hazardous Materials, 250 (2013) 19-28. https://doi.org/10.1016/j.jhazmat.2013.01.069

[82]   Y. Zhang, C. Pan, TiO_2/graphene composite from thermal reaction of graphene oxide and its photocatalytic activity in visible light, Journal of Materials Science, 46 (2011) 2622-2626. https://doi.org/10.1007/s10853-010-5116-x

[83]   N. Khalid, E. Ahmed, Z. Hong, L. Sana, M. Ahmed, Enhanced photocatalytic activity of graphene–TiO_2 composite under visible light irradiation, Current Applied Physics, 13 (2013) 659-663. https://doi.org/10.1016/j.cap.2012.11.003

[84]   Y. Wang, Z. Li, Y. He, F. Li, X. Liu, J. Yang, Low-temperature solvothermal synthesis of graphene–TiO_2 nanocomposite and its photocatalytic activity for dye

degradation, Materials Letters, 134 (2014) 115-118.
https://doi.org/10.1016/j.matlet.2014.07.076

[85]   R. Kumar, R.K. Singh, P.K. Dubey, D.P. Singh, R.M. Yadav, R.S. Tiwari,
       Hydrothermal synthesis of a uniformly dispersed hybrid graphene–TiO_2
       nanostructure for optical and enhanced electrochemical applications, RSC
       Advances, 5 (2015) 7112-7120. https://doi.org/10.1039/C4RA06852A

[86]   Y. Zhang, Z. R. Tang, X. Fu, Y. J. Xu, Engineering the unique 2D mat of graphene
       to achieve graphene-TiO_2 nanocomposite for photocatalytic selective
       transformation: what advantage does graphene have over its forebear carbon
       nanotube?, ACS Nano, 5 (2011) 7426-7435. https://doi.org/10.1021/nn202519j

[87]   Q. Xiang, J. Yu, M. Jaroniec, Graphene-based semiconductor photocatalysts,
       Chemical Society Reviews, 41 (2012) 782-796.
       https://doi.org/10.1039/C1CS15172J

[88]   B. Jiang, C. Tian, Q. Pan, Z. Jiang, J. Q. Wang, W. Yan, H. Fu, Enhanced
       photocatalytic activity and electron transfer mechanisms of graphene/TiO_2 with
       exposed {001} facets, The Journal of Physical Chemistry C, 115 (2011) 23718-
       23725. https://doi.org/10.1021/jp207624x

[89]   Y. Zhang, N. Zhang, Z. R. Tang, Y. J. Xu, Improving the photocatalytic
       performance of graphene–TiO_2 nanocomposites via a combined strategy of
       decreasing defects of graphene and increasing interfacial contact, Physical
       Chemistry Chemical Physics, 14 (2012) 9167-9175.
       https://doi.org/10.1039/c2cp41318c

[90]   D. Geng, H. Wang, G. Yu, Graphene Single Crystals: Size and Morphology
       Engineering, Advanced Materials, 27 (2015) 2821-2837.
       https://doi.org/10.1002/adma.201405887

[91]   H. Zhang, P. Xu, G. Du, Z. Chen, K. Oh, D. Pan, Z. Jiao, A facile one-step
       synthesis of TiO_2/graphene composites for photodegradation of methyl orange,
       Nano Research, 4 (2011) 274-283. https://doi.org/10.1007/s12274-010-0079-4

[92]   S.R. Kim, M.K. Parvez, M. Chhowalla, UV-reduction of graphene oxide and its
       application as an interfacial layer to reduce the back-transport reactions in dye-
       sensitized solar cells, Chemical Physics Letters, 483 (2009) 124-127.
       https://doi.org/10.1016/j.cplett.2009.10.066

[93]   G.S. Thien, F.S. Omar, N.I.S.A. Blya, W.S. Chiu, H.N. Lim, R. Yousefi, F. J.
       Sheini, N.M. Huang, Improved Synthesis of Reduced Graphene Oxide-Titanium

Dioxide Composite with Highly Exposed 001 Facets and Its Photoelectrochemical Response, International Journal of Photoenergy, 2014 (2014).

[94] K. Woan, G. Pyrgiotakis, W. Sigmund, Photocatalytic Carbon-Nanotube–TiO_2 Composites, Advanced Materials, 21 (2009) 2233-2239. https://doi.org/10.1002/adma.200802738

[95] S. Anandan, Recent improvements and arising challenges in dye-sensitized solar cells, Solar Energy Materials and Solar Cells, 91 (2007) 843-846. https://doi.org/10.1016/j.solmat.2006.11.017

[96] S.U. Khan, M. Al-Shahry, W.B. Ingler, Efficient photochemical water splitting by a chemically modified n-TiO_2, Science, 297 (2002) 2243-2245. https://doi.org/10.1126/science.1075035

[97] J.H. Park, S. Kim, A.J. Bard, Novel carbon-doped TiO_2 nanotube arrays with high aspect ratios for efficient solar water splitting, Nano Letters, 6 (2006) 24-28. https://doi.org/10.1021/nl051807y

[98] R. Sellappan, J. Sun, A. Galeckas, N. Lindvall, A. Yurgens, A.Y. Kuznetsov, D. Chakarov, Influence of graphene synthesizing techniques on the photocatalytic performance of graphene–TiO_2 nanocomposites, Physical Chemistry Chemical Physics, 15 (2013) 15528-15537. https://doi.org/10.1039/C3CP52457D

[99] B. Tryba, A. Morawski, M. Inagaki, Application of TiO_2-mounted activated carbon to the removal of phenol from water, Applied Catalysis B: Environmental, 41 (2003) 427-433. https://doi.org/10.1016/S0926-3373(02)00173-X

[100] H. Wang, X. Zhang, Y. Su, H. Yu, S. Chen, X. Quan, F. Yang, Photoelectrocatalytic oxidation of aqueous ammonia using TiO_2 nanotube arrays, Applied Surface Science, 311 (2014) 851-857. https://doi.org/10.1016/j.apsusc.2014.05.195

[101] P. Wang, Y. Ao, C. Wang, J. Hou, J. Qian, Enhanced photoelectrocatalytic activity for dye degradation by graphene–titania composite film electrodes, Journal of Hazardous Materials, 223 (2012) 79-83. https://doi.org/10.1016/j.jhazmat.2012.04.050

[102] Y. Min, K. Zhang, W. Zhao, F. Zheng, Y. Chen, Y. Zhang, Enhanced chemical interaction between TiO_2 and graphene oxide for photocatalytic decolorization of methylene blue, Chemical Engineering Journal, 193 (2012) 203-210. https://doi.org/10.1016/j.cej.2012.04.047

[103] J.S. Lee, K.H. You, C.B. Park, Highly photoactive, low bandgap TiO_2 nanoparticles wrapped by graphene, Advanced Materials, 24 (2012) 1084-1088. https://doi.org/10.1002/adma.201104110

[104] N.J. Bell, Y.H. Ng, A. Du, H. Coster, S.C. Smith, R. Amal, Understanding the enhancement in photoelectrochemical properties of photocatalytically prepared TiO_2-reduced graphene oxide composite, The Journal of Physical Chemistry C, 115 (2011) 6004-6009. https://doi.org/10.1021/jp1113575

[105] Y.H. Ng, A. Iwase, A. Kudo, R. Amal, Reducing graphene oxide on a visible-light $BiVO_4$ photocatalyst for an enhanced photoelectrochemical water splitting, The Journal of Physical Chemistry Letters, 1 (2010) 2607-2612. https://doi.org/10.1021/jz100978u

[106] Y.T. Liang, B.K. Vijayan, O. Lyandres, K.A. Gray, M.C. Hersam, Effect of dimensionality on the photocatalytic behavior of carbon–titania nanosheet composites: charge transfer at nanomaterial interfaces, The Journal of Physical Chemistry Letters, 3 (2012) 1760-1765. https://doi.org/10.1021/jz300491s

[107] W. Fan, Q. Lai, Q. Zhang, Y. Wang, Nanocomposites of TiO_2 and reduced graphene oxide as efficient photocatalysts for hydrogen evolution, The Journal of Physical Chemistry C, 115 (2011) 10694-10701. https://doi.org/10.1021/jp2008804

[108] D.C. Marcano, D.V. Kosynkin, J.M. Berlin, A. Sinitskii, Z. Sun, A. Slesarev, L.B. Alemany, W. Lu, J.M. Tour, Improved synthesis of graphene oxide, ACS Nano, 4 (2010) 4806-4814. https://doi.org/10.1021/nn1006368

[109] Y. Liang, H. Wang, H.S. Casalongue, Z. Chen, H. Dai, TiO_2 nanocrystals grown on graphene as advanced photocatalytic hybrid materials, Nano Research, 3 (2010) 701-705. https://doi.org/10.1007/s12274-010-0033-5

[110] P.K. Dubey, P. Tripathi, R. Tiwari, A. Sinha, O. Srivastava, Synthesis of reduced graphene oxide–TiO_2 nanoparticle composite systems and its application in hydrogen production, International Journal of Hydrogen Energy, 39 (2014) 16282-16292. https://doi.org/10.1016/j.ijhydene.2014.03.104

[111] Z. Xu, M. Yin, J. Sun, G. Ding, L. Lu, P. Chang, X. Chen, D. Li, 3D periodic multiscale TiO_2 architecture: a platform decorated with graphene quantum dots for enhanced photoelectrochemical water splitting, Nanotechnology, 27 (2016) 115401. https://doi.org/10.1088/0957-4484/27/11/115401

[112] D. Wang, D. Choi, J. Li, Z. Yang, Z. Nie, R. Kou, D. Hu, C. Wang, L.V. Saraf, J. Zhang, Self-assembled TiO_2–graphene hybrid nanostructures for enhanced Li-ion insertion, ACS Nano, 3 (2009) 907-914. https://doi.org/10.1021/nn900150y

[113] X. W. Liu, L. Y. Shen, Y. H. Hu, Preparation of TiO_2-Graphene Composite by a Two-Step Solvothermal Method and its Adsorption-Photocatalysis Property, Water, Air, & Soil Pollution, 227 (2016) 1-12. https://doi.org/10.1007/s11270-016-2841-z

[114] X. Zhang, X. Cui, Graphene/Semiconductor Nanocomposites: Preparation and Application for Photocatalytic Hydrogen Evolution, (2012).

[115] X. Y. Zhang, H. P. Li, X. L. Cui, Y. Lin, Graphene/TiO_2 nanocomposites: synthesis, characterization and application in hydrogen evolution from water photocatalytic splitting, Journal of Materials Chemistry, 20 (2010) 2801-2806. https://doi.org/10.1039/b917240h

[116] M. Wojtoniszak, B. Zielinska, X. Chen, R.J. Kalenczuk, E. Borowiak-Palen, Synthesis and photocatalytic performance of TiO_2 nanospheres–graphene nanocomposite under visible and UV light irradiation, Journal of Materials Science, 47 (2012) 3185-3190. https://doi.org/10.1007/s10853-011-6153-9

[117] N. Farhangi, R.R. Chowdhury, Y. Medina-Gonzalez, M.B. Ray, P.A. Charpentier, Visible light active Fe doped TiO_2 nanowires grown on graphene using supercritical CO_2, Applied Catalysis B: Environmental, 110 (2011) 25-32. https://doi.org/10.1016/j.apcatb.2011.08.012

[118] Q. Zhang, Y. He, X. Chen, D. Hu, L. Li, T. Yin, L. Ji, Structure and photocatalytic properties of TiO_2-graphene oxide intercalated composite, Chinese Science Bulletin, 56 (2011) 331-339. https://doi.org/10.1007/s11434-010-3111-x

[119] A. Mukherji, B. Seger, G.Q. Lu, L. Wang, Nitrogen doped $Sr_2Ta_2O_7$ coupled with graphene sheets as photocatalysts for increased photocatalytic hydrogen production, ACS Nano, 5 (2011) 3483-3492. https://doi.org/10.1021/nn102469e

[120] S. M. Paek, E. Yoo, I. Honma, Enhanced cyclic performance and lithium storage capacity of SnO_2/graphene nanoporous electrodes with three-dimensionally delaminated flexible structure, Nano Letters, 9 (2008) 72-75. https://doi.org/10.1021/nl802484w

[121] X. Geng, L. Niu, Z. Xing, R. Song, G. Liu, M. Sun, G. Cheng, H. Zhong, Z. Liu, Z. Zhang, Aqueous-Processable Noncovalent Chemically Converted Graphene–Quantum Dot Composites for Flexible and Transparent Optoelectronic Films, Advanced Materials, 22 (2010) 638-642. https://doi.org/10.1002/adma.200902871

[122] N. Li, G. Liu, C. Zhen, F. Li, L. Zhang, H.M. Cheng, Battery Performance and Photocatalytic Activity of Mesoporous Anatase TiO_2 Nanospheres/Graphene Composites by Template-Free Self-Assembly, Advanced Functional Materials, 21 (2011) 1717-1722. https://doi.org/10.1002/adfm.201002295

[123] T.N. Lambert, C.A. Chavez, B. Hernandez-Sanchez, P. Lu, N.S. Bell, A. Ambrosini, T. Friedman, T.J. Boyle, D.R. Wheeler, D.L. Huber, Synthesis and characterization of titania– graphene nanocomposites, The Journal of Physical Chemistry C, 113 (2009) 19812-19823. https://doi.org/10.1021/jp905456f

[124] J. Guo, S. Zhu, Z. Chen, Y. Li, Z. Yu, Q. Liu, J. Li, C. Feng, D. Zhang, Sonochemical synthesis of TiO_2 nanoparticles on graphene for use as photocatalyst, Ultrasonics Sonochemistry, 18 (2011) 1082-1090. https://doi.org/10.1016/j.ultsonch.2011.03.021

[125] M.R. Hasan, C.W. Lai, S. Bee Abd Hamid, W. Jeffrey Basirun, Effect of Ce doping on RGO-TiO_2 nanocomposite for high photoelectrocatalytic behavior, International Journal of Photoenergy, 2014 (2014).

[126] A. Morais, C. Longo, J.R. Araujo, M. Barroso, J.R. Durrant, A.F. Nogueira, Nanocrystalline anatase TiO_2/reduced graphene oxide composite films as photoanodes for photoelectrochemical water splitting studies: the role of reduced graphene oxide, Physical Chemistry Chemical Physics, 18 (2016) 2608-2616. https://doi.org/10.1039/C5CP06707C

[127] C. H. Yang, L. S. Wang, S. Y. Chen, M. C. Huang, Y. H. Li, Y. C. Lin, P. F. Chen, J. F. Shaw, K. S. Huang, Microfluidic assisted synthesis of silver nanoparticle–chitosan composite microparticles for antibacterial applications, International Journal of Pharmaceutics, 510 (2016) 493-500. https://doi.org/10.1016/j.ijpharm.2016.01.010

[128] D. Chen, L. Zou, S. Li, F. Zheng, Nanospherical like reduced graphene oxide decorated TiO_2 nanoparticles: an advanced catalyst for the hydrogen evolution reaction, Scientific Reports, 6 (2016). https://doi.org/10.1038/srep20335

[129] H. Liu, K. Cao, X. Xu, L. Jiao, Y. Wang, H. Yuan, Ultrasmall TiO_2 nanoparticles in situ growth on graphene hybrid as superior anode material for sodium/lithium ion batteries, ACS Applied Materials&Interfaces, 7 (2015) 11239-11245. https://doi.org/10.1021/acsami.5b02724

[130] H. Xing, W. Wen, J. M. Wu, One-pot low-temperature synthesis of TiO_2 nanowire/rGO composites with enhanced photocatalytic activity, RSC Advances, 6 (2016) 94092-94097. https://doi.org/10.1039/C6RA16484F

Carbonaceous Composite Materials   Materials Research Forum LLC
Materials Research Foundations **42** (2018) 33-56   doi: http://dx.doi.org/10.21741/9781945291975-2

Chapter 2

Carbon Based Nanomaterials for Energy Storage

P. Senthil Kumar*[1], K. Grace Pavithra[1] and S. Ramalingam[2]

[1]Department of Chemical Engineering, SSN College of Engineering, Chennai 603 110, India

[2]Department of Chemical Engineering, University of Louisiana at Lafayette, Lafayette, LA 70504, USA

senthilkumarp@ssn.edu.in*

Abstract

In order to store or transfer energy researchers have been focusing on solid state batteries, flow batteries, flywheels, compressed air energy storage, thermal and pumped hydropower. In recently nanomaterials have attracted attention in the field of energy storage due to its fast recharging capability, better durability, and high storage capacity. The smallest size and high surface area per unit volume or mass make nanomaterials unique in showing electric, magnetic, optical, structural, mechanical and chemical characteristics. This chapter discusses current status and future development trends of carbon nanomaterials in the field of energy storage systems.

Keywords

Non-Renewable, Nanomaterial, Recharging Capability, Physicochemical Properties, Energy Storage

Contents

1.   Introduction

Since 1990, the consumption of fossil fuels resulted in severe energy deficiency and the emission of unburnt carbon affects our environment in a larger manner. Many changes in engines were made but the pollution emitted was not nullified totally by the new techniques. The population increase and urbanization are other major sources for the utilization of fossil fuels in a drastic manner and due to increase in population, the energy demand has been increasing day by day. In the twenty-first century the need of alternative and renewable energy is an urging concept because in several parts of the world there is inequity in energy (i.e., electricity), for example in many suburban parts of India there is no electricity which is considered as basic need for every mankind and it was forecasted that in 2050 energy supply will be double the amount that of now. Energy is an unavoidable thing in developing society, in every step of life energy is the development of technologies like microbial fuel cells, hydrogen storage systems, polymer based energy storage system, photovoltaics, solar thermal storage, and membrane fuel cells are some of the examples of alternative technologies. Even though many technologies were introduced the pot light is on the development of eco-friendly storage system with low-cost techniques. Recently, nanomaterials are used for the purpose of storage of energy and it has been explored because of its unique properties that changes

Carbonaceous Composite Materials
Materials Research Foundations **42** (2018) 33-56

Materials Research Forum LLC
doi: http://dx.doi.org/10.21741/9781945291975-2

with its shape and size. Particularly carbon nanomaterial is found to be eminent technology in high-performance and in energy storage. The recent revolution in material science particularly in carbon nanomaterial has led to the research in energy technologies. The properties such as morphological, electrical, optical and mechanical which enhance the energy storage performance are found to be far superior in carbon nanomaterial than the conventional storage methods [1,2].

2. Carbonaceous nanomaterials

2.1 Origin

Carbonaceous nanomaterials are also produced from waste materials. Many techniques are followed in order to minimize the amount of waste from municipal and from industrial level and waste are recycled to value added products which in turn are used for the storage of energy. Many products such as gasoline, light oils, biofuels, alcohols, and ceramics are well known. The recent upcoming technique is the preparation of silicon carbide nanoparticles from automotive waste. The recycling of waste materials into high-value products paved way for synthesizing carbon nanomaterials from waste resources. Graphene and carbon nanotubes are found to be most important material in nanotechnology and their properties like electrical, physical, mechanical, chemical play a role in the field of supercapacitors, sensors, transistors, etc. The major challenge of using waste material is a eco-friendly production with low cost [3]. There are generally three methods of synthesizing CNTs and they are arc discharge, laser ablation, and chemical vapor deposition. Vegetable oils from coconut, neem, eucalyptus, palm, etc. possess carbon content for synthesizing pure CNTs. Among those palm oil provides good quality CNTs. Apart from vegetable oil, animal oil is used as raw material, mostly chicken fat is selected from the residue of chicken and it also has a low carbon to hydrogen ratio. Industrial waste products like red muds from Bayer process of alumina, printed circuit board, solid waste tires, and soot collected from bitumen waste are used for the synthesizing MWNTs. Chemical vapor deposition (CVD) method are used for the syncretization with ferrocene as a catalyst. Graphene is produced using by-products of plant materials, agricultural refuse. Carbon is considered as well-known solid state allotropes with diverse structures and properties and the fibrous carbon materials have attracted worldwide due to its practical applications. Carbon fibers are generally produced using pyrolyzing fibers spun from an organic precursor or by chemical vapor deposition (CVD). Recently ultra-thin carbon fibers are produced using electrospinning followed by thermal treatment. The advantages of using electro spinning include simplicity, efficiency, low cost, high yield, etc. [4,5].

2.2   Fullerenes

Due to the characteristics such as good acceptors of electrons, superconductivity, light absorption in the visible region, stability in carbon framework, low energies in electron transfer fullerenes are used in energy applications. The molecular structures of fullerenes and their derivatives are in the way of possessing photonic, electronic, superconducting and magnetic properties. Photochemical properties of fullerene make changes in a chemical reaction, luminescence, and absorption, thermal and electrical properties when compared with conventional materials. Superconductors based fullerene showed scientific interest towards energy application. Metallic behavior is seen in alkali doped films of C_{60} [6].

Types of fullerenes are as follows [7,8]:

C_{60}Fullerenes – It is a perfectly symmetric molecule with 60 carbon atoms and possesses sp^2 hybridization. These type of fullerenes are found to be highly reactive and stable and was called as Buckminsterfullerene.

Higher Fullerenes – More than 60 carbon atoms are available. In higher order fullerene C_{70} was the first member and the other members like C_{76}, C_{78}, C_{84}, C_{92} which are available in lesser quantity.

Fullerenes and Nano hybrids – C_{60} and higher fullerenes are paired with carbon or metal based nanomaterial. Two types of conjugation such as exohedral and endohedral are possible. When fullerene are conjugated with carbon nanotubes in endohedralnanohybrids are formed and this type of structures are referred as peapods. Fullerenes are conjugated with either graphene or carbon nanotubes in order to increase the performing efficiency. Some of the applications of fullerene are found in solar cells, photovoltaic materials, hydrogen storage and electronic components storage.

Fullerenes and Photovoltaic materials – The photoexcitation properties of C_{60} replaces the conventional photovoltaic material and paved way for the production of inexpensive solar cells and photovoltaic devices. The first plastic photovoltaic cells are made from the conjugation of polymers and C_{60}. The inorganic semiconductors such as silicon, gallium arsenide, selenide, amorphous silicon etc. are replaced by conjugated polymers and C_{60}fullerence. Many studies are reported on C_{60} conjugation with polymer combinations and the some are methyl-ethyl-hydroxyl-poly propyl vinyl/ C_{60} thin film, ITO/ polyalkylthiophene (PAT)/ C_{60}/ A, poly(3-alkylthiophenes)/C_{60} etc. Material stability was found to be lower when compared with hybrid systems.

2.3 Carbon nanotubes

Two forms of carbon nanotubes are generally seen. They are, single- walled nanotubes (SWNTs) and multi walled nanotubes (MWNTs). Rolled- up graphene sheet with sp^2 hybridization are seen in SWNTs. When additional graphene coaxial tubes are around the SWNT core they are considered as multi-walled carbon nanotubes (MWNT). The diameter of the nanotubes is from few angstroms to tens of nanometers and length are from micrometers to centimeters containing the structure of pentagons. The semi-conducting or metallic behavior depends upon the diameter and helicity arrangements of carbon atoms in the walls [9,10]. In CNTs the conductivity is directly proportional to the charge transport capability. Due to high length-to-diameter ratio, sp^2 C-C bonds CNTs have shown wider applications in the field of electronic, chemical and biomedical. The introduction of CNTs into the field of sensors, transistors, electrochemical cells has increased the performance of those devices. CNTs have good catalyst due to its high surface area [11,12]. CNTs have good electrical and mechanical properties due to its large surface area and a coating of CNTs in any material will increase its corrosion resistance and the basic properties of carbon nanoparticles are as follows:

Electrical properties – When the structure is altered CNTs exhibit semiconducting or metallic properties. Apart from superlative quantum and electronic properties CNTs have magnetic properties [13].

Mechanical properties – Stiffness, strength, and toughness are achieved through strong σ bonding and upto 100GPa tensile strength is achieved using Young's modulus.
Optical and Thermal properties –High heat capacity and thermal conductivity are offered by CNTs in ambient temperature. It is suggested that for the temperatures over 100˚C CNTs are considered to be the better choice over conventional materials and for the wavelength of 300 to 3000 nm CNTs provide good optical properties [14,15].

2.4 Graphene

Graphene is considered as a basic unit for fullerenes and CNTs. Honeycomb lattice of monolayer carbon atoms. Due to its unique structure and features like specific surface area, high carrier concentration, mobility, high thermal conductivity and optical transparency has made graphene attractive for energy storage. The physicochemical properties such as thermal conductivity, elasticity, stiffness, carrier mobility make graphene unique. Methods like chemical vapor deposition, bottom-up approach, top-down approach, mechanical exfoliation, arc discharge are used for synthesizing graphene. Graphene sheets are used as electrode materials like super capacitor and batteries.

Graphene is a one atom thick layer that is arranged in hexagonal patterns [16,17,18]. This arrangement makes graphene having amazing physical, thermal and electrical properties Mechanical properties – Due to its strong covalent bonds and compact structure Graphene was considered as a strongest material with Young's modulus of 1TPa and the intrinsic strength was considered to be 130GPa [19].

Optical properties – Single layer graphene are transparent which is due to its 2D structure and zero band-gap and the transparency is indirectly proportional to the number of layers and addition of one layer is equal to the transparency decrease of 2.3%. Due to its good optical features graphene has a chance in the field of optoelectronic devices [20,21,22]. Thermal properties - Because of robust covalent bonds graphene shows tremendous thermal properties under elevated temperature. Thermal conductivity is inversely proportional to the temperature [23,24].

2.5   Nitrogen doped carbon nanomaterial

The interaction of surface heteroatom into carbon nanomaterials provides many desirable electronic structures for potential applications. Two methods are available and they are in-situ doping and post-doping. The former incorporate nitrogen atoms into CNT structure homogeneously and the latter post-doping leads to surface fictionalization without altering bulk properties. The synthesization of N-doped CNTs by arc-discharge, laser ablation. N-doped carbon nanotubes were researched for many years but large scale synthesization were done on graphene sheets by CVD. Generally, heteroatoms are introduced into nitrogen-doped graphene sheets. The resultant N-graphene are used as a metal-free electrode with higher electro catalytic activity, operation stability, and tolerance towards crossover. Research being done in the production of large-area N-graphene films, this can be achieved using solution casting followed by subsequent heat treatment. N-doped carbon shows significant behavior in oxygen reduction reactions (ORR) in fuel cells. The catalytic effect and the adsorption of O_2 in Li-O_2 batteries are good for N-doping in carbon than conventional nanomaterials [25,26].

2.6   Carbon gels

There are considered as porous carbon material with interconnected network structure. The porous structure of carbon gels are nano sized so they are called as carbon nanogels. The unique features of nanogels are high surface area, tunable porosity, and low apparent density and can be used in different fields such as adsorbent, catalyst support, as gas storage media, electrode materials for super capacitors and batteries. There are three kinds of carbon nanogels such as carbon aerogels, carbon xerogels and carbon cryogels. Among them carbon aerogels are synthesized by carbonization of organic gels and from the extraction of pore solvent. Because of their low mass densities, chemical stabilities and high surface areas carbon aerogels are receiving increased attention for the past several years. In environmental remediation and in energy conservation and storage, carbon aerogels receive special interest due to 3D interconnected network with open pores. They are collectively called as open-cell foams which exhibits properties like low mass densities, porosities, high surface area, high electrical conductivity, mechanical strength, etc., The porous carbon material are obtained by the pyrolysis of organic aerogels at different temperatures.

Benefits of using carbon aerogels:

- When compared to conventional capacitors very low impedance electric double layer super capacitors can be produced.
- They can act as excellent insulators, this is due to a reduction in heat transfer by gaseous components.
- The properties like electrical, thermal and mechanical have better performance due to the addition of dopants.
- Carbon aerogels provides high specific surface area with three-dimensional structure
- Carbon aerogels are synthesized using sol-gel chemistry.
- Carbon aerogels are used for hydrogen and electrical storage in the future that is due to its unique structure

Carbon hydrogel- Hydrogels are basically cross-linked 3D hydrophilic solid networks by physical or chemical means. It is considered as soft materials and used in the field like artificial tissues, electrode material, drug carrier, adsorbents, sensors and actuators. The hydrogels are classified on various characteristics such as source, cross-linking nature and its structural features. Conventional hydrogels are associated with poor mechanical properties and stability [27,28]

3.   Energy storage system

In the twenty-first century, challenges are faced such as energy storage and there is a need for environmental friendly and low-cost energy storage systems and their performance mainly depend upon the properties of the material chosen. Nanostructured materials properties namely mechanical, electrical and optical properties make them attractive to the field of energy storage.

Carbonaceous Composite Materials   Materials Research Forum LLC
Materials Research Foundations **42** (2018) 33-56   doi: http://dx.doi.org/10.21741/9781945291975-2

3.1 Electrochemical storage system

The demand for portable electronic devices is increased due to the use of smart phones, laptops, tablets, etc. In order to meet such demand compact, light weighted and efficient energy storage systems need to be manufactured. Super capacitors, lithium-sulfur batteries, lithium-oxygen batteries, lithium metal batteries are considered to be the most commonly used electric energy storage because of its long lifespan, reversibility, low cost and high energy power density. The drawback such as side reactions between electrolytes and electrode exist, it is due to high contact area, low volumetric energy and density of the electrode. Advanced functional materials such as polypyrrole, polyaniline and metal oxides like RuO_2, MnO_2, NiO and Co_3O_4 were widely used as super capacitors. But the drawbacks such as poor solubility and mechanical brittleness are seen in this type of super capacitors. Using specific power and stability super capacitors are categorized as electric double-layer capacitors based on physical charge separation and redox pseudo capacitors based on reversible redox reactions at the electrode surface. When compared to lithium ion batteries, supercapacitors exhibit low specific energy. Significant process is seen in the field of electrochemical energy conversion and storage devices due to the use of nanostructured materials [29,30].

3.1.1 Binder free electrodes

Binder free electrodes are generally used where electroactive surface area is inaccessible and the electrolyte is blocked. Recently binder-free electrodes are used in the area of energy storage because of its electrical conductivity and accessible reaction surface. Electrodes are generally binder-enriched and they are prepared by slurry-coating technique and the performance of energy storage depends on the properties and structures of the electrodes. It's porous and interconnected structures for electron and ion transfer makes graphene a binder free electrode [31]. The advantages of using graphene as binder free electrodes are as follows

- It can be used as flexible electrodes.
- They are referred as free movable electrodes avoiding the use of metal collector.
- The conductivity and specific capacity of the electrodes increases without the intervention of non-conductive and inactive binders.
- The graphene interconnected sheet increases the conductivity by increasing the electron transfer inside the electrodes.

Generally, binder free electrodes are synthesized using chemical exfoliation, CVD, mechanical cleavage and graphitization. Among all the processes chemical exfoliation was found to be an effective technique in the production of graphene oxide in a larger manner. Graphene in the form of foams, networks, gels are used as binder materials. In a research, graphene foams are synthesized using CVD method on nickel foams as a template and after the removal of nickel foam template, the porous structure can be restored. Graphene gels are synthesized by freeze drying method in presence of graphene oxide. Graphene's are made into 1-dimensional, 2-dimensional and 3-dimensional structures such as features like wire-like, paper-like and monolithic-like structures by various preparation methods. In recent times, 3D graphene nanomaterials show extraordinary mechanical, electrical and electrochemical as well as surface properties [32-35].

3.1.2 Super capacitors

Super capacitors are generally referred as electrical-energy-storage device and it is based on charge accumulation from an electrolytic solution by means of electrostatic attraction from polarized electrodes which increases its attention due to its unique characteristics. Super capacitors are also named as electrochemical capacitors, ultra capacitors or electrochemical double layer capacitors and possess long shelf life as well as long life cycle and fast charging/discharging capabilities when compared to conventional capacitor, super capacitors have high power density so that they can be used in power backup systems, electronic appliances, industrial power and energy management, electrical vehicles and in other devices [36,37]. The ions accumulated on the electrodes are resulted as capacitance C and it is described by Helmholtz with the following formula

$$C = \frac{\varepsilon r \varepsilon 0 A}{d}$$

ε, ε - electrolyte and vacuum dielectric constants
d - the effective thickness of the electrical double layer
A - interface surface area

The essential components found in super capacitors are the electrodes, the electrolyte, and the separator and the overall performance lies on the physical properties of both electrode and the electrolyte materials. Charge storage and delivery are done by electrodes and it plays a major role in energy and power densities of super capacitors. In the last decade activated carbon are used as electrode material but due to its low mesoporosity limited capacitance are obtained which shows low electrolyte accessibility. Due to large surface area, high mesoporosity, electrical properties and electrolyte accessibility carbon nanotubes, mainly graphene and carbon nanotubes replaced activated carbon. CNTs with high aspect ratio, large specific surface area and good mechanical and

Carbonaceous Composite Materials   Materials Research Forum LLC
Materials Research Foundations **42** (2018) 33-56   doi: http://dx.doi.org/10.21741/9781945291975-2

electrical properties are used as active electrodes in super capacitors. Graphene, one-atom-thick sheets with 2Dplanar geometry was found to be superior and advantageous than CNTs in acting as electrodes. Multiwalled nano carbontubes and nitrogen enriched carbon porous nano structured carbon which possesses 3 Dimensional hierarchical morphology are known to have enhanced electrochemical performance. Recent research has proven that vertically aligned carbon nanotubes with open tips show high performance as electrodes. Graphene and CNT hybrid systems increase the efficiency of the electrodes. 1D CNTs are used to physically separate 2D graphene sheets in order to preserve the graphene surface area and to provide a well-controlled architecture for efficient charge transport. Theoretical studies indicates that 3D pillared architectures consisting of parallel graphene layers with vertically aligned CNTs shows good structural tunability and desirable transport for energy storage. It not only mechanically supports the graphene layer it also provides good conductive paths for electron and ion transport and provides high capacitance and capability. Ultra-thin in-plane micro-super capacitors are developed in order to utilize the properties of carbon nanotubes fully. Ultrahigh power densities of several orders of magnitude higher than conventional super capacitors are provided. The electrode materials including CNTs, carbide-derived carbon, polymers and metal oxides are investigated for the newly developed micro-super capacitors. Due to their in-plane electrical conductivity, large surface area and easy fabrication, graphene are used as electrode materials for micro-super capacitors. In recent time wearable fiber like super capacitors attracted attention in the electronics' field. Materials like Kevlar fiber, metal fiber, carbon fiber, CNT fiber and graphene fiber are used as electrode material. Due to their excellent electrical conductivity, mechanical properties and outstanding flexibility CNBT fibers are used as electrodes in wearable fiber like super capacitors. Stretchable electronics which includes transistors, light-emitting diodes, polymer solar cells and active matrix displays are developed to maintain the electronic performance with high levels of mechanical deformations. MnO_2 based pseudo capacitive super capacitors have attractive features, which include environmental beneficial and high theoretical capacitance [38,39]. By tuning the deposition voltage, current and electrolyte control the mass loading densities and structure of deposited MnO_2 which makes fabrication of flexible MnO_2 super capacitors easy. CNTs and graphene are found to be highly conductive and porous substrates which are usually deposited with higher densities for MNO_2 excellent behavior as super capacitors. The MnO_2-CNT-sponge super capacitors show 45 degradations after 1000 cycles at 5A/g of specific current, the specific power of 64 KW/Kg and Energy at 31 Wh/Kg [40,41].

Carbonaceous Composite Materials   Materials Research Forum LLC
Materials Research Foundations **42** (2018) 33-56   doi: http://dx.doi.org/10.21741/9781945291975-2

3.1.3 Lithium-ion batteries

Rechargeable lithium batteries have made a revolution in portable electronic devices and they have become the predominant power source for cell phones, digital camera, laptops, etc., because of their higher energy density. In battery system usage of carbon are identified in one or the other form. In order to eliminate lithium deposition, Li-ion cells are introduced. While charging lithium metal electrode deposits dendrite like lithium are deposited on the surface of the metal electrode which induces internal shorting. The use of the carbonaceous material as anode will completely prevent the circumstances and it is a key factor in increasing the capacitance of the batteries. The properties of carbon materials such as irreversible capacities, high cycle-ability and faster mobility with lithium make comfortable in using carbon material as anode. Carbon material such as highly oriented pyrolytic graphite (HOPG), graphite, artificial graphite, hard carbons, mesophase low-temperature fibers. The performance of HOPG and natural graphite are one and the same. When natural graphite powders are coated on the current collector, they are aligned parallel to the collector surface which directly induces the ohmic resistance which is perpendicular to the electrode surface and it is responsible for high rate charging and discharging. Mesophase carbon are produced by melting the mesophase pitch and it is produced by blowing molten pitch through tiny holes in a heater chamber and finally, well-graphitized bead or fiber form. The formed fibers have two advantages, they are 1) the radial crystal orientation of the fiber which accepts Li ions from all the sides and 2) the side walls are covered with porous thin graphene skin which saves the fiber from shape change during Li interaction/intercalation. Hard carbons are produced from resins like phenol, epoxy resins, cellulose, sugars or proteins heated under inert gas or vacuum condition which decomposes to release water and gas to form carbon and the carbon particles are joined together forming complicated structure other than graphite. The disadvantages of using hard carbon in batteries field are 1) Poor high rate performance; 2) The discharge curves are not flat but gradual slopes are seen: 3) It has a large initial irreversible capacity. Because of the slow rate in Li doping/ undoping and large irreversible capacity hard carbons doesn't fit for batteries. Better performance and reversible capacity are achieved by the use of nanostructured electrodes in lithium ion batteries and the use of nanostructures also increases the insertion/removal rate of lithium, increases the contact area with the electrolyte and enhances electron transport. Generally, $LiCoO_2$ replaces lithium metal and kept as a cathode, a non-aqueous liquid electrolyte, graphite electrode as anode [42,43]. On charging, lithium ions are deintercalated from the $LiCoO_2$and pass across the electrolyte to intercalate with graphite layers.

Advantages in using nanomaterials for lithium batteries [44,45],
- The rate of lithium insertion/removal increases with a decrease in dimensions and the characteristics time constant for diffusion are given by $t = L^2/D$ where L is the diffusion length and D is the diffusion constant.
- The nanometer sized particles enhance the electron transport within the particles.
- The high surface area of nanomaterial permits high contact area within the electrolyte.
- For very small particles, modifications can be done on lithium ions and electrons which result in a change of electrode potential.

Disadvantages of using nanomaterial for lithium batteries
- Synthesization and the dimensions of nanoparticles are very difficult.
- The high surface area may lead to more side reactions in maintaining difficulties in interparticle contact.

3.2 Nanomaterials as electrodes

Transparent conductive electrodes are considered as an essential feature in solar cells, liquid-crystal displays, alight-emitting diode and in touch screens. Tin-doped indium oxide (ITO) are generally used as a transparent electrode in polymer solar cell. Due to high cost, low availability and brittle nature of ITO alternative materials such as polymers, nanowires, CNTs, and graphene were used. Among them, CNTs and graphene were considered to be an efficient transparent electrode which shows high performance. High electrical conductivity, low optical absorption, surface smoothness, chemical stability, and flexibility are the qualifying properties which makes graphene a transparent electrode. The disadvantages like low conductivity are the drawbacks of graphene in using as transparent material. CNTs in flexible form were found to be a trustable material for the transparent electrode. The thickness of about 5-50 nm and optical transmittance of 60-90% were found in CNTs. To make CNTs flexible they are deposited on the top of the plastic substrate and can be used as a flexible material in electronics like flexible organic light-emitting diodes (OLEDs). A Korean research team from Pohang University has developed flexible OLEDs with high efficiency using graphene as a transparent electrode. The combination of graphene and CNT shows the synergistic effect and it improves the conductivity. At high current density OLED Graphene quantum dots, graphene and CNTs are recently used as acceptor/donormaterials. The use of carbon nanomaterials improves the charge transport as well as a uniform dispersion in donor or acceptor. Carbon paste electrode (CPE) is a kind of heterogeneous carbon electrode with a mixture of graphite, glassy carbon and other carbonaceous materials, water-immiscible and non-conducting binder and they are applicable for electrochemical as well as electro analysis due to their low cost, feasibility, miniaturization, easy preparation, etc. The advantageous of using CPE is that desired compositions with determined properties can be prepared. The method in which the CPE is prepared are important for its property determination and the characteristics such as structure, particle size, conductivity, porosity, sorption, purity, the surface composition also strongly influencing the properties of CPE. In some papers CPE are also used for detection of antioxidants and used in cyclic voltammetry experiments which are used for identification, characterization, and quantification. Screen printed electrodes (SPE) are modified using single or multi-walled carbon nanotubes, nanofibers, fullerenes, microspheres, etc. A number of works were done using SPE based sensors full characteristics and the comparison with different material was not fully done. In many applications like analytical chemistry, biology, medicine, pharmacy SPEs are used. Counter electrodes are used as a mediator in the regeneration of sensitizer after electron injection into the photo-anode. Generally, platinum is used as CEs but due to its high cost, platinum is replaced by carbonaceous material and another important factor to use the carbonaceous material as CEs is its corrosion resistivity towards iodine and high electrical conductivity. Compared to platinum, carbon compounds are found to be lower in catalytic activity but they are compensated by increasing the active surface area of the electrode by using a porous electrode structure. Graphite powder is generally used in the preparation of porous carbon electrodes. It was found that the catalytic activity of graphite counter electrodes may be enhanced by adding about 20% of carbon black [46,47].

3.3   Hydrogen storage system

Hydrogen meets the goal for renewable and cleaner energy option. It was considered as a safe and versatile source of fuel which can be converted to the desired form without any release of harmful emissions. It is considered as ideal fuel for the future as it controls the emission of greenhouse gases and it also reduces the dependence on fossil fuels. Hydrogen is mainly used in the form of fuel cells during usage, hydrogen is directly converted to water, electricity, and heat

$H_2 + \frac{1}{2} O_2 \rightarrow$ water molecule + electricity + heat

Hydrogen has been considered as an alternative source by major countries and hydrogen storage becomes a key factor for hydrogen policies and it is important a hydrogen needs a crucial method for storage [48,49,50]. Hydrogen are used as fuel for transportation and for stationary as well as portable applications such as power supply back-up and to power generators. Hydrogen is stored in various forms like

- Compressed form ranging from 20 MPa to 100 MPa.

- Liquefied cryogenic form when large amount of hydrogen are transported
- Chemisorption by intermetallides
- Carbon nanotubes, fullerenes, zeolites, metal organic frameworks by adsorption process
- Hydrogen bounded in aromatic substrates with cycles of hydrogenation and dehydrogenation.

The use of liquid hydrogen is unimaginable due to its cost, maintenance and storage at cryogenic storage temperature [51,52]. When comparing to liquefied hydrogen, compressed hydrogen is found to be cheaper. Nanotechnology was found to be an efficient method for storing hydrogen, particularly carbon nanotubes capture hydrogen by means of chemisorption and physisorption. Physisorption was found to be an easy technique as it supports reversible technique, adsorption-desorption kinetics and it is easy to handle. In materials with porous carbon structures, zeolites hydrogen are physisorbed on the pores and it mainly depends on the factors such as pore size, surface area, accessible surface area, surface chemical composition, pressure and temperature. Among them, surface area and the pore structure are the important parameters [52,54].

Fullerene and its related material show good storage capacities towards hydrogen. Fullerene application towards storage of hydrogen molecules was found to be a promising technology but rare amount of research was done regarding hydrogen storage and fullerene. Due to low volumetric energy density of hydrogen, hydrogen is stored in compressed form in alloys and CNTs. Coated fullerene plays an ideal role in hydrogen storage applications. Ni-coated fullerene can store three hydrogen molecules and hydrogenated silicon fullerence are also suitable for hydrogen storage [55,56]. Ca-coated fullerene, Ca-coated boron fullerene, Mg-coated boron fullerene are reported for hydrogen storage. Recently a electrochemical compressor with a membrane electrode has been found to be an efficient method than conventional compressing methods. But the drawbacks such as humidity affects the compressed hydrogen. It has been noted that coated fullerenes show ideal performances towards hydrogen storage applications. Calcium coated fullerenes show high capacity in storing hydrogen [57-59].

3.4 Thermal energy storage

Solar is a supreme source of energy and it is a vital and abundant resource. The biggest challenge behind the utilization of this kind of renewable energy is the geographical location, atmospheric conditions, cost, etc. The use of solar collectors was considered to be one of the efficient ways and it converts solar to thermal energy. Three types of solar collector technologies are in use which are,

1. photo-voltaic collector

2. Solar thermal collector

3. Photo-voltaic and thermal collector

Photovoltaic collector converts solar energy into electricity using semiconductor material like Si, SiO_2, GaAs, CuS, etc. and the produced electricity can be used for various applications. In solar thermal collector, solar energy is converted directly into heat energy and the photovoltaic thermal collector is a combination of both thermal and photo-voltaic which converts solar energy to electricity and thermal energy. Generally, water is used as a thermal storage element in household applications apart from this the energy collected using solar cells has to be stored efficiently. Abundant of solar energy is available during day time therefore efficient thermal storage material are to be identified in order to store the excess thermal energy during the night. Thermal energy storage materials should have following parameters for its better storage which are:

1. Eco-friendly nature

2. High thermal conductivity

3. It should have low thermal energy loss

4. Low cost

5. Thermal storage capacity should be high

Sensible heat storage, latent heat storage and chemical heat storage are three types of techniques used by thermal storage material. The sensible heat storage are divided into liquid storage media and solid storage media. Nanomaterials are used in liquid storage media by dispersing the nanoparticles into conventional base fluids like water, engine oil, etc. The thermal conductivity of nanomaterials depends on particle size, particle shape, material, type of fluid and temperature. Solar cells have the capability to convert solar energy into electricity. Silicon based solar cells are dominating the market for its stability and efficiency but the drawbacks such as high cost, inflexibility, lack of transparency and color tunability lead to the search of new combinations. The qualities of graphene such as low cost, flexibility, optical transparency and high conductivity are the ideal characteristics to choose as electrodes in organic solar cells. Spin coating, spray coating, hot pressing and CVD are the methods for the deposition of graphene into electrodes [60-62].

Carbonaceous Composite Materials
Materials Research Foundations **42** (2018) 33-56

Materials Research Forum LLC
doi: http://dx.doi.org/10.21741/9781945291975-2

3.5   Nanomaterials as Fuel cells

Fuel cells convert chemical energy directly into electricity without burning. Based on different electrolyte materials different varieties of fuel cells can be synthesized. All fuel cells have an anode, the electrolyte and a cathode. Research is done to improve the performance of proton exchange fuel cells (PEMFCs). Hydrogen and methanol are used as a source in the direct hydrogen and direct methanol PEMFCs. It has been noted that in order to make DMFCs commercially available anode catalyst doping needs to drop below 1.0mg/cm^2 and Pt-alloys used as anode material. In order to avoid precious metal as anodes carbon material are used as catalyst support as well as free catalyst [63,64]. Many types of research were done in this field in order to lower the load of metals and to enhance the catalyst activity carbon nanomaterials are used as catalyst support in organic solar cells. Carbon blacks are used as support materials for PEMFC, their characteristics like specific surface area, electrical conductivity, mechanical strength, corrosion resistance. Carbon support of large surface area is needed in order to provoke high-performance of PEMFCs. It is found that Vulcan XC-72 have high surface area among all carbon black materials and is widely used as catalyst support in PMFCs. The pore structure of the carbon catalyst also affects catalytic activity of the catalyst [65,66]. Three classifications are seen in VulcanXC-72 they are micropores (greater than 2 nm), mesopores (2-50 nm) and macropores (lesser than 50 nm). High electrical conductivity is required for high catalytic activity of supported catalyst and Vulcan XC-72 proved to be a good catalyst supporting material by having high electrical conductivity. CNTs are used as supporting material for its high surface area, mesoporosity, electrical conductivity, mechanical strength and corrosion resistivity. Comparing to carbon black CNTs enables dispersed and uniformly distributed deposition of catalyst nanoparticles and facilitates the formation of triple-phase-boundaries. For hydrogen PEMFCs, it has been researched that CNT supported Pt catalyst with 12 wt% Pt loading would give 10% fuel voltage and twice the power density than the previous non supported Pt catalyst. High mechanical strength and corrosion-resistance of CNTs provide excellent durability for nanotube electrode materials [67,68]. Nitrogen doped-CNTs have promising technology either as supporting material or as a metal-free catalyst. N-doping strengthens the binding between the catalyst and CNT surface to increase the durability of the catalyst. Aligned structures are also preferred for NCNTs for the application of fuel cells than conventional CNTs. N-doped Carbon nanotube fiber show 10-fold increase in catalytic activity for the decomposition of H$_2$O$_2$ in both neutral and alkaline conditions and still it shows less electroactive properties when compared with a platinum catalyst. According to quantum mechanics calculations, the carbon atoms adjacent to nitrogen dopants of vertically aligned nitrogen doped CNTs possess high positive charge density to counter balance the strong electron affinity of the nitrogen atom. Like CNTs graphene also show attractiveness in using graphene as a metal free catalyst. The graphene material and their N-doped derivatives are produced using methods like chemical vapor deposition, chemical reduction of graphite oxide, exfoliation of graphite, microwave plasma reaction and atmospheric pressure graphitization of silicon carbide [69-71].

3.6 Capture of carbondioxide and methane

Greenhouse gases, carbon dioxide, and methane, nitrous oxide, sulfur hexafluorocarbon, hydrofluorocarbon, etc. are the resultant for the increase in the concentration of global warming. Carbon dioxide and methane are abundantly emitted greenhouse gases. The largest source of CO_2 production is from fossil fuel plants and purification of hydrogen from biomass. The anthropogenic sources for CH_4 are from livestock, natural gas production and distribution, landfill and from coal mining. Cleaner and environmental friendly processes are to be introduced in order to reduce CO_2 emission. It was noted that conventional adsorbents are not as efficient in storing methane. The number of physicochemical properties of nanomaterials supports gas purification and gas capture. The size of the nanomaterial makes the surface reactive and nature provides stability and robustness. Advanced nanoporous materials, porous organic polymers received attention in adsorption storage applications [72,73]. Due to wide availability, low cost, electrical and heat conductivity, thermal and chemical stability, low sensitivity to moisture carbonaceous adsorbents are used for CO_2 capture or storage. The capacity of activated carbon depends upon the textural properties, surface groups of the carbon-based adsorbents. CNTs have been reported for the storage of natural gas for its large pore volume, high specific area, thermal stability, mechanical stability. It has been reported that the mixture of CNTs is used for the adsorption of CH_4, CO_2 and H_2S. Based on the conceptual models CNTs are reported for its adsorption towards natural gas storage. Graphene is also considered as lighter and cheaper material in the production of solid-state gas adsorption materials. For CO_2 sequestration, biogas upgrading, air dehumidification nanoporous graphene materials are applied. Fe_3O^4-graphene shows CO_2 sorption capacities greater than any other materials and continuous researches are done on this topic. In order to enhance the adsorption properties, graphene and CNTs are modified using different compounds. The sorption of gas molecules on the surface of carbonaceous material depends upon electrostatic attraction, dispersion materials, van der Waals interaction or charge transfer. Due to exceptional and unimaginable properties nanomaterials are found to be suitable materials in adsorbing natural gases and the modified form of carbon nanomaterials with metals give promising results toward gas adsorption [74].

4.   Conclusion and future development

Energy is considered as an important element in today's life style. Energy conservation and storage are the areas where alternative strategies related research has to be carried out. Nanotechnology, their synthesis and characterization tools excite us drastically. It is evident that there will be continuous improvement in the field of nanotechnology for energy storage systems. In nanomaterials particularly carbonaceous nanomaterials receive greater attention because of its unbelievable properties and its abundant availability. A great future is seen for biomass related carbon nanomaterials and in hybrid materials.

References

[1]   L. Dai, D.W. Chang, J. BeomBaek, W. Lu, Carbon Nanomaterials for Advanced Energy Conversion and Storage, Sci, 8 (2012) 1130-1166.

[2]   J.Gibbs, A.A. Pesaran, P.S. Sklad, L.D. Marlino, In Fundamentals of Materials for Energy and Environ. Sustain.in: D.S. Ginley, D. Cahen (Eds.), Cambridge Univ. Press., 2012, pp. 426-444.

[3]   R. aito,M.S. Dresselhaus, Optical Properties of Carbon Nanotubes, in Carbon Nanotubes and Graphene, K.T. Iijima (Eds.), Elsevier: Oxford., 2014, pp. 77-98.

[4]   Uriani, A.R. Dalila, A. Mohamed, M.H. Mamat, M.Salina, M.S. Rosmi, J.Rosly, R.MdNor, Vertically aligned carbonnanotubes synthesized from waste chicken fat, Mater. Lett. 101 (2013) 61-64. https://doi.org/10.1016/j.matlet.2013.03.075

[5]   Suriani, R. Mdnor, M. Rusop, Vertically aligned carbon nanotubes synthesized from waste cooking palm oil, J. Ceram. Soc. Jpn.118 (1382) (2010) 963-968. https://doi.org/10.2109/jcersj2.118.963

[6]   W. Kratschmer, The story of making fullerenes, Nanoscale, 3 (2011) 2485-9. https://doi.org/10.1039/c0nr00925c

[7]   N. Aich, J. Plaza-Tuttle, J.R. Lead, N.B. Saleh N B, A critical review of nanohybrids: synthesis, applications and environmental implications, Environ. Chem. 11 (2014) 609-23. https://doi.org/10.1071/EN14127

[8]   N.B. Sleh, N. Aich N, J. Plaza-Tuttle, T. Sabo-Attwood, Research strategy to determine when novel nanohybrides pose unique environmental risks, Environ. Sci. Nano. 2 (2015) 11-18. https://doi.org/10.1039/C4EN00104D

[9]   D. Vukicevic, M. Randic, On kekule structures of buckminsterfullerence, Chem. Phy. Letters. 401 (2005) 446-50. https://doi.org/10.1016/j.cplett.2004.11.098

[10] A.D. Franklin, Electronics: The road to carbon nanotube transistors, Nature 498(7455) (2013) 443-444. https://doi.org/10.1038/498443a

[11] C. Fisher, A.E. Rider, Z.J. Han, S. Kumar, I. Levchenko, K. Ostrikov, Applications and nanotoxicity of carbon nanotubes and graphene in biomedicine, J.Nano. mater. 2012 (2012) 3. https://doi.org/10.1155/2012/315185

[12] Y. Fan, Q. Zhang, C. Lu, Q. Xiao, X. Wang, B. kangTay, High performance carbon nanotube–Si core–shell wires with a rationally structured core for lithium ion battery anodes, Nanoscale 5(4) (2013) 1503-1506. https://doi.org/10.1039/c3nr33683b

[13] Q. Zhang, J.Q. Huang, W.Z. Qian, Y.Y. Zhang, F. Wei, The road for nanomaterials industry: A review of carbon nanotube production, post-treatment, and bulk applications for composites and energy storage, Small 9(8) (2013) 1237-1265. https://doi.org/10.1002/smll.201203252

[14] k. anagi, Differentiation of Carbon Nanotubes with Different Chirality, in Carbon Nanotubes and Graphene, K.T. Iijima (Eds.), Elsevier: Oxford., 2014, 19-38.

[15] Y.H. Lu, Z.X. Lu, Mechanical Properties of Carbon Nanotubes and Graphene, in Carbon Nanotubes and Graphene, K.T. Iijima(Eds.), Elsevier: Oxford., 2014,165-200.

[16] R. aito, M.S. Dresselhaus, Optical Properties of Carbon Nanotubes, in Carbon Nanotubes and Graphene, K.T. Iijima (Eds.), Elsevier: Oxford., 2014, 77-98.

[17] L.W. Fan, Z.Q. Zhu, Y. Zeng, Q. Lu, Z.T. Yu, Heat transfer during melting of graphene based composite phase change materials heated from below, Int. J. Heat Mass Transfer 79 (2014) 94–104. https://doi.org/10.1016/j.ijheatmasstransfer.2014.08.001

[18] P. Zhang, X. Xiao, Z.W. Ma, A review of the composite phase change materials: Fabrication, characterization, mathematical modeling and application to performance enhancement, Appl. Energy 165 (2016) 472–510. https://doi.org/10.1016/j.apenergy.2015.12.043

[19] A.A. Al-Abidi, S. Mat, K. Sopian, M.Y. Sulaiman, A.T. Mohammad, Internal and external fin heat transfer enhancement technique for latent heat thermal energy storage in triplex tube heat exchangers, Appl. Therm. Eng. 53 (1) (2013) 147–156. https://doi.org/10.1016/j.applthermaleng.2013.01.011

[20] Y. hu, S. Murali, W. Cai, X. Li, J.W. Suk, J.R.Potts, R.S. Ruoff, Graphene and graphene oxide: synthesis, properties, and applications, Adv. Mater. 22(35) (2010) 3906-3924. https://doi.org/10.1002/adma.201001068

[21]   P. vouris, C. Dimitrakopoulos, Graphene: synthesis and applications. Mater. Today. 15(3) (2012) 86-97. https://doi.org/10.1016/S1369-7021(12)70044-5

[22]   D.S.L. bergel, V. Apalkov, J. Berashevich, K.Ziegler, T. Chakraborty, Properties of graphene: atheoretical perspective, Adv. Phys. 59(4) (2010) 261-482.

[23]   T. Denaro, V. Baglio, M. Girolamo, V. Antonucci, A. S. Arico, F. Matteucci, R. Ornelas, Investigation of low cost carbonaceous materials for application as counter electrode in dye-sensitized solar cells, J. Appl. Electrochem.39 (2009)2173–2179. https://doi.org/10.1007/s10800-009-9841-2

[24]   P. Avouris, C. Dimitrakopoulos, Graphene: synthesis and applications. Mater. Today. 15(3) (2012)86-97. https://doi.org/10.1016/S1369-7021(12)70044-5

[25]   A.A. alandin, Thermal properties of grapheneand nanostructured carbon materials. Nat. Mater. 10(8) (2011) 569-581. https://doi.org/10.1038/nmat3064

[26]   K.E. Whitener, P.E. Sheehan, Graphene synthesis. Diamond Relat. Mater. 46 (2014) 25-34. https://doi.org/10.1016/j.diamond.2014.04.006

[27]   Z. Xiang, Q. Dai, J. Chen, L. Dai, Edge functionalization of graphene and two-dimensional covalent otganic polymers for energy conversion and storage, Adv. Mater. 28 (2016) 6253-6261. https://doi.org/10.1002/adma.201505788

[28]   Z. Xing, X. Luo, Y. Qi, W.F. Stickle, K. Amine,, J. Lu, X. Ji, Nitrogen-doped nanoporous graphenic carbon: An efficient conducting support for O2 cathode, Chem. Nano. Mat. 2 (2016) 692-697. https://doi.org/10.1002/cnma.201600112

[29]   M. Mirzaeian, P.J. Hall, Preparation of controlled porosity carbon aerogels for enrgy storage in recgargeable lithium oxygen batteries, Electrochimica. Acta. 54(28) (2009) 7444-7451. https://doi.org/10.1016/j.electacta.2009.07.079

[30]   C.H.J. Kim, D. Zhao, G. Lee, J.Liu, Strong, machinable carbon aerogels for high performance supercapacitors, Adv. Mater. 26(27) (2016) 4976-4983.

[31]   Y.F. Ma, Y.S. Chen, Three-dimensional graphene networks: synthesis, properties and applications. Natl. Sci. Rev. 2 (2015) 40-53. https://doi.org/10.1093/nsr/nwu072

[32]   X.H. Cao, Z.Y. Yin, H. Zhang, Three-dimensional graphene materials: preparation, structures and application in supercapacitors, Energy. Environ. Sci. 7 (2014) 1850-65 https://doi.org/10.1039/C4EE00050A

[33]   Z.P. Chen, W.C. Ren, L.B. Gao, B.L. Liu, S.F. Pei, H.M. Cheng, Three-dimensional flexible andconductive interconnected graphene networks grown by chemical vapour deposition. Nat. Mater. 10 (2011) 424-8. https://doi.org/10.1038/nmat3001

[34]   M. Zhou, T.Q. Lin, F.Q. Huang, Y.J. Zhong, Z. Wang, Y.F. Tang, Highly conductive porous graphene/ceramic composites for heat transfer and thermal energy storage, Adv. Funct. Mater. 23 (2013)2263-9. https://doi.org/10.1002/adfm.201202638

[35]   W. Wei, S.B. Yang, H.X. Zhou, I. Lieberwirth, X.L. Feng., K. Müllen, 3D graphene foams cross-linked with pre-encapsulated Fe3O4nanospheresfor enhanced lithium storage. Adv. Mater. 25 (2013) 2909-14. https://doi.org/10.1002/adma.201300445

[36]   W. Lu, L. Qu, K. Henry and L. Dai, Capacitive performance of ordered mesoporous carbons with tunable porous texture in ionic liquid electrolytes, J. Power Sources. 189 (2009) 1270–1277. https://doi.org/10.1016/j.jpowsour.2009.01.009

[37]   H. Zhang, G. Cao, Z. Wang, Y. Yang, Z. Shi and Z. Gu, Vertically-Aligned Carbon Nanotubes for Electrochemical Energy Conversion and Storage, Nano.

[38]   W. Lu, L. Qu, L. Dai and K. Henry, Electrochemistry of Novel Electrode Materials for Energy Conversion and Storage, ECS Trans. 6 (2008) 257– 261.

[39]   K. Geim, Graphene: Status and Prospects, Science, 324 (2009) 1530–1534. https://doi.org/10.1126/science.1158877

[40]   C.Fisher, A.E. Ridetr, Z.J.Han, S. Kumar, I. Levchenko, K. Ostrikov, Applications and nanotixicity of carbon nanotubes and graphene in biomedicine, J. Nanomater.(2012) 3.

[41]   J. N. Coleman, M. Lotya, A. O'Neill, S. D. Bergin, P. J. King, U. Khan, K. Young, A. Gaucher, S. De and R. J. Smith, Two-Dimensional Nanosheets Produced by Liquid Exfoliation of Layered Materials, Sci. 331 (2011) 568–571. https://doi.org/10.1126/science.1194975

[42]   J.P. Alper, M. Vincent, C. Carraro, R. Maboudian, Siliconcarbidecoatedsi- licon nanowiresasrobustelectrodematerialforaqueousmicro-supercapacitor, Appl. Phys. Lett. 100 (2012) 1639-01.

[43]   J.P. Alper, S. Wang, F. Rossi, G. Salviati, N. Yiu, C. Carraro,R. Maboudian, Selective ultra thin carbons heath on porous silicon nanowires:materials for extremely high energy density planar micro-supercapacitors.NanoLett.14 (2014) 1843–1847. https://doi.org/10.1021/nl404609a

[44]   N. Berton, M. Brachet, F. Thissandier, J. LeBideau, P. Gentile, G. Bidan, T. Brousse, S. Sadki,Wide voltage window silicon nano wire electrodes from microsupercapacitors via electrochemical surface oxidation in ionic liquid

electrolyte. Electro. chem. Commun.41 (2014) 31–34.
https://doi.org/10.1016/j.elecom.2014.01.010

[45] S. Boukhalfa, K. Evanoff, G. Yushin, Atomic layerd deposition of vanadium oxide
      on carbon nanotubes for high power super capacitor electrodes,
      EnergyEnviron.Sci.5 (2012) 6872–6879. https://doi.org/10.1039/c2ee21110f

[46] T. Takamura, R.J. Brodd, New Carbon Based Materials for Electrochemical
      Energy Storage Systems, I.V. Barsukov et al. (eds.), Springer., (2006) 157–169.

[47] P.G. Bruce, B. Scrosati, J. Tarascon, Angew,Nanomaterials for Rechargeable
      Lithium Batteries, Chem. Int. Ed. 47 (2008) 2930 – 2946.

[48] D. Sen, R. Thapa, K.K. Chattopadhyay, Small Pd cluster adsorbed double vacancy
      defect graphene sheet for hydrogen storage: A first-principles study. Int. J.
      HydrogenEnergy. 38 (2013) 3041–3049.
      https://doi.org/10.1016/j.ijhydene.2012.12.113

[49] Y. Wang, J.H. Liu, K. Wang, T. Chen, X. Tan, C.M. Li, Hydrogen storage in Ni-B
      nanoalloy-doped 2D graphene. Int. J. Hydrogen Energy.36(2011) 12950–12954.
      https://doi.org/10.1016/j.ijhydene.2011.07.034

[50] M. Zhou, Y.H. Lu, C. Zhang, Y.P. Feng, Strain effects on hydrogen storage
      capabilityof metal-decorated graphene: A first-principles study, Appl. Phys. Lett.
      97 (2010) 103109–103111. https://doi.org/10.1063/1.3486682

[51] Y.F. Zhao, Y.H. Kim, L.J. Simpson, A.C. Dillon, S.H. Wei, M.J. Heben, Opening
      space for H2 storage: Cointercalation of graphite with lithium and small organic
      molecules, Phys. Rev. B.78 (2008) 144102–144106.
      https://doi.org/10.1103/PhysRevB.78.144102

[52] C. Ataca, E. Akturk, S. Ciraci, H. Ustunel, High-capacity hydrogen storage by
      metalized Graphene, Appl. Phys. Lett. 93 (2008) 043123–043125.
      https://doi.org/10.1063/1.2963976

[53] T. Hussain, B. Pathak, T.A. Maark, C.M. Araujo, R.H. Scheicher, R. Ahuja, Initio
      study of lithium-doped graphane for hydrogen storage. EPL. 96 (2011) 27013–
      27016. https://doi.org/10.1209/0295-5075/96/27013

[54] J. ang, A. Sudik, C. Wolverton, D.J. Siegel, High capacity hydrogen storage
      materials: attributes for automotive applications and techniques for materials
      discovery, Chem. Soc. Rev. 39(2010) 656–675. https://doi.org/10.1039/B802882F

[55] Paster, M. D. et al. Hydrogen storage technology options for fuel cell vehicles:
      well-to-wheel costs, energy efficiencies, and greenhouse gas emissions. Int. J.

Hydrogen Energy, 36(2011)14534–14551.
https://doi.org/10.1016/j.ijhydene.2011.07.056

[56]   S. Barman, P. Sen, G. P. Das, Ti-decorated doped silicon fullerence: a possible hydrogen-storage material, J. Phy. Chem C. 112 (2008) 19963-8.

[57]   J. L. Li, Z. S. Hu, G. W. Yang, High-capacity hydrogen storage of magnesium decorated boron fullerence, Chem. Phy. 392 (2012) 16 – 20. https://doi.org/10.1016/j.chemphys.2011.08.017

[58]   M. Li, Y. F. Li, Z. Zhou, P.W. Shen, Z. F. Chen, Ca-coated boron fullerences and nanotubes as superior hydrogen storage materials, Nano Letters. 9 (2009) 1944 – 8. https://doi.org/10.1021/nl900116q

[59]   Y. Hou R. Vidu, P. Stroeve P. Solar Energy Storage Methods. Ind Eng Chem Res. 50 (2011) 8954–64. https://doi.org/10.1021/ie2003413

[60]   S. Kuravi, J. Trahan, D.Y. Goswami, M.M. Rahman, E.K. Stefanakos, Thermal energy storage technologies and systems for concentrating solar power plants. Prog Energy Combust Sci. 39 (2013) 285–319. https://doi.org/10.1016/j.pecs.2013.02.001

[61]   Z. Wang, W. Yang. F. Qiu, X. Zhang, X. Zhao, Solar water heating: From theory, application, marketing and research. Renew Sustain Energy Rev. 41 (2015) 68–84. https://doi.org/10.1016/j.rser.2014.08.026

[62]   H.J. Ahn, W. J. Moon, T.Y. Seong, D. Wang, In situ X-ray spectromicroscopy study of bipolar plate material stability for nano-fuel-cells with ionic-liquid electrolyte, Electrochem. Commun. 11 (2009) 635. https://doi.org/10.1016/j.elecom.2008.12.056

[63]   C. Santoro, K. Artyushkova, S. Babanova, P. Atanassov, I. Ieropoulos, M. Grattieri, P. Cristiani, S. Trasatti, B. Li, A.J. Schuler, Parameters characterization and optimization of activated carbon (AC) cathodes for microbial fuel cell application, Bioresour. Technol. 163 (2014) 54–63.] https://doi.org/10.1016/j.biortech.2014.03.091

[64]   A. Iwan, M. Malinowski, G. Pasciak, Polymer fuel cell components modified by graphene: electrodes, electrolytes and bipolar plates, Renew. Sustain. Energy Rev. 49 (2015) 954–967. https://doi.org/10.1016/j.rser.2015.04.093

[65]   L.T. Soo, K.S. Loh, A.B. Mohamad, W.R.W. Daud, W.Y. Wong, An overview of the electrochemical performance of modified graphene used as an electrocatalyst and as a catalyst support in fuel cells, Appl. Catal. A 497 (2015) 198–210. https://doi.org/10.1016/j.apcata.2015.03.008

Materials Research Forum LLC

doi: http://dx.doi.org/10.21741/9781945291975-2

[66]   Y.-C. Chiang, M.-K. Hsieh, H.-H. Hsu, The effect of carbon supports on the performance of platinum/carbon nanotubes for proton exchange membrane fuel cells, Thin Solid Films 570 (Part B) (2014) 221–229.

[67]   M. Borghei, G. Scotti, P. Kanninen, T. Weckman, I.V. Anoshkin, A.G. Nasibulin, S. Franssila, E.I. Kauppinen, T. Kallio, V. Ruiz, Enhanced performance of a silicon microfabricated direct methanol fuel cell with PtRu catalysts supported on few-walled carbon nanotubes, Energy 65 (2014) 612–620. https://doi.org/10.1016/j.energy.2013.11.067

[68]   P. Kanninen, M. Borghei, O. Sorsa, E. Pohjalainen, E.I. Kauppinen, V. Ruiz, T.Kallio, Highly efficient cathode catalyst layer based on nitrogen-doped carbon nanotubes for the alkaline direct methanol fuel cell, Appl. Catal. B 156–157 (2014) 341–349. https://doi.org/10.1016/j.apcatb.2014.03.041

[69]   Z. Xie, G. Chen, X. Yu, M. Hou, Z. Shao, S. Hong, C. Mu, Carbon nanotubes grown in situ on carbon paper as a microporous layer for proton exchange membrane fuel cells, Int. J. Hydrogen Energy 40 (29) (2015) 8958–8965. https://doi.org/10.1016/j.ijhydene.2015.04.129

[70]   J. Song, G. Li, J. Qiao, Ultrafine porous carbon fiber and its supported platinum catalyst for enhancing performance of proton exchange membrane fuel cells,Electrochim. Acta 177 (2015) 174–180. https://doi.org/10.1016/j.electacta.2015.03.142

[71]   A.Garcia-Gallastegui, D. Iruretagoyena, V. Gouvea, M. Mokhtar, A.M. Asiri, S.N. Basahel, S.A. Al-Thabaiti, A.O. Alyoubi, D. Chadwick, M.S.P. Shaffer, Graphene oxide as support for layered double hydroxides: enhancing the CO_2 adsorption capacity. Chem. Mater. 24 (2012) 4531 –4539. https://doi.org/10.1021/cm3018264

[72]   Z. Kang, M. Xue, D. Zhang, L. Fan, Y. Pan, S. Qiu, Hybrid metal-organic framework nanomaterials with enhanced carbon dioxide and methane adsorption enthalpy by incorporation of carbon nanotubes. Inorg. Chem. Commun. 58 (2015) 79 –83. https://doi.org/10.1016/j.inoche.2015.06.007

[73]   K.C. Kemp, H. Seema, M. Saleh, N.H. Le, K. Mahesh, V. Chandra, K.S. Kim, Environmental applications using graphene composites: water remediation and gas adsorption. Nano 5 (2013) 3149 –3171.

[74]   N.H. Khdary, M.A. Ghanem, M.A., 2012. Metal-organic-silica nanocomposites: copper, silver nanoparticles-ethylenediamine-silica gel and their CO_2 adsorption behaviour. J. Mater. Chem. 22 (2012) 12032 –12038 https://doi.org/10.1039/c2jm31104f

Carbonaceous Composite Materials　　　　　　　　　Materials Research Forum LLC
Materials Research Foundations **42** (2018) 57-92　　doi: http://dx.doi.org/10.21741/9781945291975-3

Chapter 3

Molecular Dynamics Simulation of Capped Single Walled Carbon Nanotubes and their Composites

Sumit Sharma[1]*, Mandeep Singh[2]

[1]Assistant Professor, Department of Mechanical Engineering, Dr. B R Ambedkar National Institute of Technology Jalandhar, Punjab, India

[2]Research Scholar, School of Mechanical Engineering, Lovely Professional University, Phagwara, India

sumit_sharma1772@yahoo.com, sharmas@nitj.ac.in*

Abstract

Molecular dynamics simulation has been used to study the effect of capping carbon nanotubes (CNT) with hemispherical caps on mechanical properties of CNTs. Simulation has also been performed for studying the effect of volume fraction (V_f) on the mechanical properties of CNT reinforced poly methyl metha-acrylate (PMMA) composites. Materials Studio 8.1 has been used as a tool for finding the longitudinal (E_{11}), transverse (E_{22}) and shear moduli of the composites. Results show that capped CNTs have lower moduli in comparison to the uncapped CNTs. Adding CNTs into PMMA increases E_{11} till V_f of 12%.

Keywords

Carbon Nanotube, Mechanical Properties, Elasticity, Molecular Dynamics, Polymer

Contents

1.   Introduction

The fullerenes, named after Buckminster Fuller were brought to light in 1985 by Richard Smalley, Robert Curl, James Heath, Sean O'Brien and Harold Kroto [1], all researchers from the Rice University. C60, the allotrope of carbon was the first discovered fullerene. The carbon atoms in C60 are sp^2 hybridized and their arrangement is in such a manner that it takes a shape similar to that of a soccer ball. The discovery of C60 was also an accident like many other scientific breakthroughs. Smalley and Curl found a way to examine the atom clusters formed by laser vaporization with mass spectroscopy. It caught the attention of Kroto [1] subsequently as he was researching on interstellar dust, the long chain polyenes formed by red giant stars. When they vaporized graphite using laser, they could create and evaluate extended chain polyenes. Two major peaks were observed in mass 720 and a little lower at mass 840 corresponding to 60 and 70 carbon atoms, respectively. The band structure of graphite was first discovered by Wallace [2] in 1947 but multi -walled nanotubes (MWCNTs) were not discovered up to that time. In this article, we have concentrated on single walled carbon nanotubes (SWCNTs).

In 1990, Richard Smalley [3] proposed the presence of a tube-shaped fullerene and a bucky tube that could be finished by extending a C60 molecule. In 1991, Dresselhaus [4] recommended for CNT capped at both end by fullerene hemispheres at a fullerene workshop in Philadelphia. But only after Iijma [5] imaged MWCNTs using transmission electron microscope (TEM), the presence of CNTs was experimentally recognized. After two years of his surveillance on MWCNTs, Iijima [5] along with his coworkers and Bethune with his coworkers simultaneously and self-sufficiently spotted SWCNTs.

Though Iijma [5] is recognized for the official discovery, CNTs were already observed 30 years ago by Bacon [6] at Union Carbide in Parma. He probably found CNTs in his samples during a study of melting point of graphite under high temperature and pressure. He stated the presence of carbon nano whiskers and proposed a scroll like structure in his paper published in 1960. Endo [7] also imaged nanotubes via high resolution TEM in 1970. Although CNTs were found four decades ago, but their significance was not known until the discovery of fullerenes. CNT research has established into an important area in nanotechnology growing at a tremendously fast rate.

Several mechanical properties of capped and uncapped SWCNTs have been studied by researchers. Zhou and Shi [8] used a bond order potential for molecular dynamics (MD) simulations to predict the mechanical properties of SWNTs under tensile loading with and without hydrogen storage. (10,10) armchair and (17,0) zigzag CNTs were studied. Up to the necking point of the armchair CNT, two twisting stages were recognized. In the principal stage, the elongation of the nanotube was mainly owing to the changing of angles among two neighboring carbon bonds. Young's Modulus obtained at this stage was similar with earlier research. In another stage, the lengths of carbon bonds were protracted up to the fracture point. The tensile strength at this stage was more advanced than that detected in the principal stage. Comparable outcomes were also established for the zigzag CNT with a minor tensile strength. Griebel and Hamaekers [9] performed MD simulation to evaluate the elastic moduli of polymer-CNT composite. SWCNT was embedded in polyethylene and MD was used to develop a stress strain curve by Parrinello-Rahman [10] approach. It was observed that CNT can tolerate extreme distortion without fractures. The results of elastic moduli were compared with the rule of mixture (ROM) predictions. Three periodic systems, i.e. a finite CNT embedded in polyethylene, polyethylene mixture itself and infinite CNT were studied. In case of short CNT the result was in agreement with ROM.

Mylvaganam and Zhang [11] discussed essential topics of MD simulation for examining CNT and their mechanical properties. On the basis of discussed parameters the structural changes in both armchair and zigzag CNTs and also their elastic modulus were examined. The elastic modulus of armchair was 3.96 TPa and Poisson's ratio was 0.15 and Young's modulus for zigzag was 4.88 TPa and Poisson's ratio was 0.19. The ultimate tensile strain of a CNT was about 40% before atomic bond damage. The armchair tube experienced a greater tensile stress as in comparison to the zigzag tube. Zhou et al. [12] (2004) examined the interfacial bonding of SWCNTs/epoxy composites. MD simulation was used to predict the interfacial bonding between the cured epoxy resin & CNT. On the basis of the simulation, the interfacial shear strength between nanotubes and cured epoxy was determined to be 75MPa which indicated a strong and efficient stress transfer from resin to nanotube. The scattering and strong interfacial bonding of the nanotubes with epoxy resin resulted in a 250-310 % increase in the modulus with an addition in 20-30 weight% nanotubes.

Bao et al. [13] used MD simulation to determine the mechanical properties of the CNT. The interatomic short-range interaction and long-range interaction of the CNT were modeled with the help of REBO potential and LJ potential. All three types of SWCNTs i.e. armchair, zigzag and chiral were examined. It was observed that elastic modulus of SWCNTs was 929.871 GPa. Gao and Li [14] studied shear lag model for CNT/polymer composite using a multiscale approach. Molecular structural mechanics was used to determine the Young's modulus of capped CNT. It was observed that capped CNT represents an effective fiber having the same diameter and length, but having different Young's modulus by putting nanotube under an iso-strain condition. It was also detected that aspect ratio of nanotube was the governing parameter for CNT/polymer composites. Young's modulus for capped and uncapped SWCNTs increased quickly when length was less than 10 Å and after that become almost constant.

Deng et al. [15] assembled MWCNT reinforced 2024 Al composite by mixing 2024 Al powder and CNTs. Fabrication was achieved with the help of both hot and cold extrusion processes. It was discovered that when the temperature range was 400 °C, the storage modulus achieved was 82.3 GPA and damping capacity of composite with frequency 0.54 Hz reaches up to 975×10^3. So, to achieve a high range of damping capacities at very high temperature without affecting mechanical properties and stiffness, CNTs should be used as a desirable reinforcement. The damping capacities were uniform with increasing temperature, but when the temperature was 200 °C the damping capacities increase with increase in temperature and frequency was inversely proportional to damping capacities which means it decreases with increase of damping capacity.

Adnan et al. [16] discussed the effect of filler size on the mechanical properties of CNT/polymer composites using MD simulation. It was observed that Young's modulus was inversely proportional to the bucky ball size. With decrease in the Bucky ball size the elastic modulus of CNT/polymer increased. Han and Elliott [17] evaluated the Young's modulus of two different amorphous polymer matrices of PMMA/SWCNT and PmPV/SWCNT with different volume fraction. It was also observed that when the

Carbonaceous Composite Materials   Materials Research Forum LLC
Materials Research Foundations **42** (2018) 57-92   doi: http://dx.doi.org/10.21741/9781945291975-3

interaction between CNT/polymer was strong the interfacial ordering effect can't be ignored. Simulation results were in agreement with ROM results that show there was strong interfacial interaction between CNT/polymer. Mokashi et al. [18] performed MD simulation on SWCNT reinforced polyethylene (PE) composites to evaluate the mechanical properties of the model. It was observed that Young's moduli of crystalline CNT/PE were 212-215 GPa, which was in agreement with the experimental data. It was concluded that elastic stiffness of amorphous PE was 3.19-3.69 GPa and tensile strength of amorphous PE was 0.21-0.25 GPa. Significant rise was detected in the tensile properties when amorphous PE was reinforced in longitudinal direction. Reduction in both the properties was witnessed when amorphous PE was reinforced by inserting CNTs.

Cho and Sun [19] examined Young's modulus of polymeric nanocomposite with MD simulation. Spherical nano particles were added to create a model. With the help of numerical performed tensile test it was observed that elastic modulus of nanocomposites were affected by the size of nano-particles and also by the interaction strength between nano-particles and polymer chains. It was observed that elastic modulus of the nanocomposite was inversely proportional to the size of the nanoparticles. As long as the strength was more or equal to the polymer-polymer interaction the elastic modulus increase with a decrease in the size of nano particles. Gan and Zhao [20] performed Hartree-Fockland density functional analysis on capped SWCNT with the number of atoms up to 400. The hypothetical analysis of capped SWCNT was done. The average bond length determined for SWCNTs with smaller diameter was larger as compared to the SWCNTs with larger diameter. It was also observed that the average bond length decrease with the number of atoms. Ganji et al. [21] studied the effect of curvature, elastic modulus of armchair SWCNTs and the average energy of atoms under axial strains using self-consistent charge density function. It was concluded that by increasing the amount of curvature the average density and elastic modulus of (7-7) SWCNTs was decreased and equilibrium carbon-carbon distance was increased. It was also discovered that (5-5) SWCNTs did not show any variation in average energy of atoms and Young's modulus with increase in curvature. The Young's modulus and the average energy of atoms of SWCNTs armchair was less for small diameter as compared to large diameters. The elastic modulus of straight (7,7) CNT was 1.22 TPa , for 1-curve was 1.20 TPa, for 2-curve was 1.12 TPa and for 3-curve was 0.98 TPa.

Fereidoon et al. [22] studied SWCNTs in both periodic and non-periodic system. Elastic properties of (6,6) SWCNTs were investigated with the help of density functional theory. Axial and torsional strains were applied on both periodic and C-capped CNTs. It was observed that elastic modulus of the CNT was directly proportional to the length of the nanotubes. As the length of the nanotube increases the modulus of periodic CNT also increases and after a certain value it become constant. It was concluded that C-capped CNT display opposite behavior to a periodic CNT under compression. The elastic modulus for capped CNTs was always high than other types of CNTs. Gao et al. [23] compared the electrical resistance of capped CNT/Cu interface and open end CNT/Cu interface. It was concluded that capped CNT/Cu interface has lack of dangling bonds of carbon atoms at the capped end of CNTs which leads to a decrease in the interfacial bonding. The electrical resistance of capped CNT/Cu interface was much higher. Haghighatpanah and Bolton [24] investigated a (6,6) CNT/PE system for minimum energy structure as well as binding energy between CNT/PE with molecular mechanics and with first principle method. It was concluded that force fields were in qualitative promise with the first principles results and PE chains may wrap everywhere on SWCNTs. It was calculated using the COMPASS force field for (5, 5) SWCNTs relating to a PE chain. This force field was utilized to calculate the mechanical properties of (5, 5) SWCNT/PE nano composites. It was discovered that interfacial shear stress of SWCNT/PE was 141.09 MPa and interfacial bonding energy of SWCNT/PE was 0.14 N/m. The simulations indicated that mechanical properties were not affected when small size SWCNTs were used in CNT/PE whereas using large SWCNTs enlarged the Young's modulus of CNT/PE axial direction.

Jeong and Kim [25] performed MD simulations of SWCNTs filled with fullerenes C60. The different behavior of SWCNTs completely filled with C60 fullerenes, under compressive, combined tensile-torsional, tensile, and torsional loads were observed. Multiple failure modes in combined tension-torsion were examined. In the cases of uniaxial loading, merging the CNTs with C60 fullerenes prominently increase their compressive and torsional buckling loads (not tensile failure loads), and the amount of rise was advanced in torsional loading. The explanations under mutual tensile-torsional loading depict that as the tensile failure load falls with combined torsion, the torsional buckling load rises with combined tension. Mahboob and Islam [26] performed MD simulation to study the effect of stone-wales (SW) flaws on the mechanical properties of composites reinforced with SWCNTs. It was discovered that the longitudinal modulus of composites was powerfully reliant on the number of SW defects and CNT volume fraction. It was concluded that as the number of SW increases the elastic modulus of SWCNT falls which means both were inversely proportional to each other. Elastic modulus of the PE achieved from the MD simulation was in agreement with PE properties available in the literature. Shao et al. [27] performed density functional theory to study the effect of hydrogen, oxygen and nitrogen atomic chemisorption on (5-5) capped armchair SWCNTs. It was concluded that oxygen and nitrogen chemisorption could break the C-C bond to form doping type structures and C bonds became weaker by H chemisorption which favor hydrogen storage. The total amount of charge transferred between N or O atom and the carbon atoms was expected to be higher with N adsorption around the cap portion. It was concluded that due to small local curvature radius atomic chemisorption was much stable than on the cap as compared to tube. The work function increased to 5.0 eV with the adsorption of N and O and decreased to 4.8 eV for H adsorption as compared to 4.89 eV for the clean tube.

Sharma et al. [28] concluded that CNTs containing no defect exhibited excellent mechanical properties. CNTs show many defects like vacancies, doping, SW defect, hybridization which come into existence during purification and chemical treatment. MD simulation was performed to evaluate the effect of these defects on the elastic moduli of CNTs. Effect of the above mentioned defects on the mechanical properties of the SWCNTs was studied using MD simulation. A step by step reduction in elastic moduli was noticed with increasing diameter and no of defects in SWCNT. The elastic moduli of CNTs with SW defects were found to decrease with increase in number of defects. Sharma et al. [29] studied polymer/CNT composites by embedding SWCNT in two different amorphous polymer matrices of poly methyl metha acrylate (PMMA) and poly m-phenylene vinylene (PmPV) with the help of MD simulation. The obtained results were compared with the macroscopic rule of mixtures (ROM) for composite systems which showed that there was a large deviation of MD results from the ROM for strong interfacial interaction.

Based on the literature review, following conclusions could be drawn:
(i) Most of the work was carried out on open ended SWCNTs. There was a negligible amount of work done on closed end or capped SWCNTs. There was no literature review available about the modeling of elastic properties of capped SWCNTs.
(ii) No data was available about the mechanical properties of capped SWCNTs reinforced poly methyl metha acrylate (PMMA) nanocomposites.
(iii)Most of the work has been done for continuous CNTs. In most of the cases, only longitudinal modulus and the corresponding loss factor have been evaluated and other moduli have not been given much importance.

In this study, in addition to longitudinal modulus, transverse and shear moduli have been calculated.

2.   Materials and method

MD is a computer model for learning the corporeal actions of atoms. The atoms are permitted to collaborate for a stable time, to generate a vision of the dynamical

Carbonaceous Composite Materials   Materials Research Forum LLC
Materials Research Foundations **42** (2018) 57-92   doi: http://dx.doi.org/10.21741/9781945291975-3

development of the structure. In many cases, the routes of atoms are obtained by mathematically explaining Newton's equations of motion for a structure of relating particles and molecular mechanics (MM) was utilized to describe the forces between atoms and their potential energies. The technique was first established in the area of assumed physics in the late 1950s, but was applied in the study of materials science. It was a challenge to examine the properties of such a complex molecular system comprising of a large number of particles. By using mathematical methods this problem can be resolved. However, extended MD simulations are precisely ill-conditioned, generating growing errors in numerical integration that can be diminished with proper choice of algorithms and parameters, but not removed entirely.

The development of a single MD simulation may be used for the systems which follow the ergodic hypothesis to describe thermodynamic possessions of the structure. The time arithmetic mean of an ergodic system resembles to Micro-canonical ensemble (NVE) joint averages. MD was also known as "statistical mechanics by figures" and "Laplace's vision of Newtonian mechanics" of forecasting the future by enlivening nature's forces and permitting idea into molecular motion on an atomic gauge. In the NVE ensemble, the system is inaccessible from variations in moles (N), volume (V) and energy (E). It resembles to an adiabatic process with no heat exchange. A micro canonical MD route may be understood as an argument of potential and kinetic energy, with overall energy being conserved. For a structure of N particles with coordinates X and velocities V, the subsequent duo of initial order differential equations may be inscribed with Newton's notation given by Eq. (1) and Eq. (2).

$$F(X) = -\nabla U(X) = M\dot{V}(t) \tag{1}$$

$$V(t) = \dot{X}(t) \tag{2}$$

For every single time step, all particles position (X) and velocity (V) may be joined with a simplistic method such as Verlet. The time evolution of X and V is called a route. Given the original positions and velocities we can compute all upcoming positions and velocities. In the canonical collective (NVT) ensemble, amount of substance (N), volume (V) and temperature (T) are preserved. It is also known as constant temperature molecular dynamics (CTMD). In NVT, the energy of endothermic to exothermic processes is swapped with a thermostat. A variation of thermostat processes is accessible to improve and eliminate energy from the borders of MD simulation in a supplementary way, reminiscent of the NVT ensemble. General methods to govern temperature include velocity rescaling Nose-Hoover chains, the Berendsen thermostat, the Andersen thermostat and Langevin dynamics. Note that the Berendsen thermostat influences the hovering ice cube effect, which leads to un-physical conversions and revolutions of the assembly. It was not easy to obtain a canonical supply of conformations and velocities by using above algorithms. In the isothermal-isobaric collaborative or NPT, amount of substance (N), pressure (P) and temperature (T) are preserved. A barostat was required along with the thermostat. It resembles most faithfully to laboratory circumstances by a flask open to ambient temperature and pressure. There are different ways by which position of a particle in successive time intervals can be calculated:

(i) Verlet method: When velocity relations are not necessary for calculating molecular position at the succeeding step, then this method was known as Verlet method.

(ii) Velocity Verlet method: When the system uses both velocities and positions side by side by ensuring that the system temperature should remain constant. Any system in which both velocity and position of the particle are required to calculate the position of the particle at the next step, then the method is known as a velocity Verlet method.

(iii)Leapfrog method: The name was derived from the estimation of position and forces and the velocities by using all of them in a leapfrog manner. This method is the best method among all and provides better stability and accuracy than velocity Verlet method.

The MD method can be used for both equilibrium and non-equilibrium physical occurrence which makes it a dominant device which can be further utilized to simulate numerous physical occurrences. The procedure of using the MD method by using the velocity Verlet method is given below:

(i) Provide the initial position and velocity of all particles.

(ii) Determine the force acting on particles.

(iii)Determine the position of all particles at the next step.

(iv)Determine the velocities of all particles at the next time step.

(v) Now repeat the procedure from step 2.

In the above method the velocities and positions are determined at every time interval in the MD simulation.

2.1   CNT

The single sheet of graphite (called graphene) when rolled into a cylinder forms a CNT. The diameter was generally considered in the nanometer and length in micrometers. Due to their huge aspect ratio, CNTs lead to unusual electrical support. Many of the CNTs behave as metals and some of them behave as semiconductors.

Carbonaceous Composite Materials   Materials Research Forum LLC
Materials Research Foundations **42** (2018) 57-92   doi: http://dx.doi.org/10.21741/9781945291975-3

CNT has cylindrical nanostructure with a length to diameter fraction up to 132,000,000:1 which was considerably greater for any further metal. These CNTs were allotropes of carbon and have unique properties. It has extraordinary thermal and mechanical properties. The various types of CNTs are shown in Figure 1. SWCNTs can be further classified into two categories, i.e. open end and closed end SWCNTs. Open end SWCNTs are those tubes which are open from both ends and also known as uncapped SWCNTs. The closed end SWCNT which is closed from both ends by hemi-spherical caps. Both these types are shown in Figure 2.

2.2 Polymer

In this study, MD simulation has been performed to study the polymer/CNT composite system. The polymer used in this study is poly (methyl metha-acrylate) or PMMA. The structure of PMMA is shown in Figure 3. PMMA is the polymer of MMA with the chemical formula $(C_5H_8O_2)_n$. PMMA is a linear thermoplastic polymer and has low elongation at break. It does not destroy on rupture and is the toughest thermoplastic. Due to its low water absorption volume, PMMA is very appropriate for an electrical engineering drive. It exhibits very good dielectric properties. It can survive temperatures as low as -70 °C. Its resistance to temperature variation is very decent. PMMA can transmit more light than glass. PMMA is an inexpensive multipurpose material. It is accessible in extruded form or cast material in sheet, rod and tubes, as well as custom outlines. Numerous forms of acrylics have been used in an extensive range of applications such as in vehicles, medicines, optics, office equipment, electrical engineering and many others.

2.3 Simulation strategy

The simulation approach used in this study has been discussed below.

After forming the polymer and CNT, the next step is to pack the polymer around the CNT. This is done using the "Amorphous cell" module. Under "Amorphous Cell", click on the "packing" option, keeping the"quality" as "fine". Give the essential density. Give the "force field" under "energy" option. Give the file name of the polymer which is to be packed outside the CNT by selecting from the pull-down menu under "option" in" Amorphous cell". To end click "Run" . At the end of this step, we will get a periodic cell in which the polymer will be packed round the CNT.

(i) The next step is performing the geometry optimization using the"For cite module". There is "more" option in the for cite module. On clicking the option, we can give the type of algorithm which is to be used for optimization and we also give the number of interations for which the simulation will run.

Materials Research Forum LLC
doi: http://dx.doi.org/10.21741/9781945291975-3

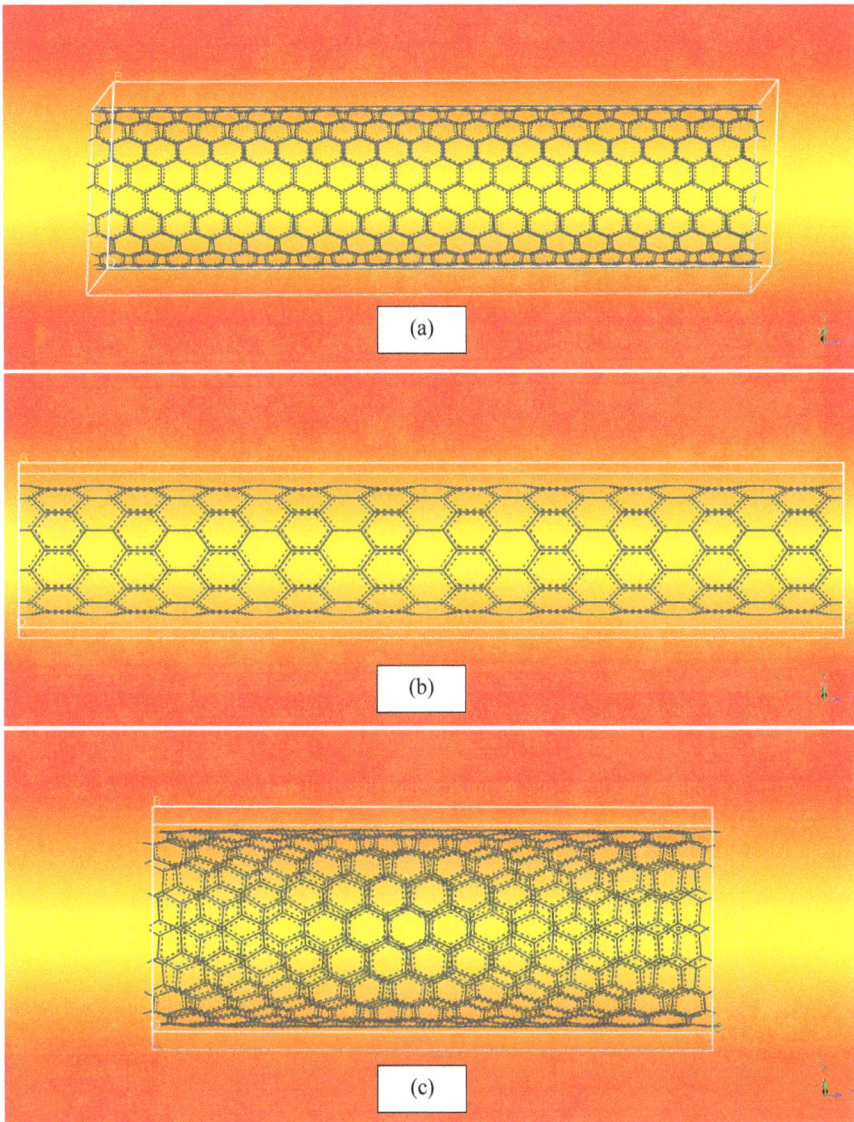

Figure 1. Structure of CNT (a) Armchair (10,10), (b) Zigzag (0,10) and (c) Chiral (7,10)

Materials Research Forum LLC

doi: http://dx.doi.org/10.21741/9781945291975-3

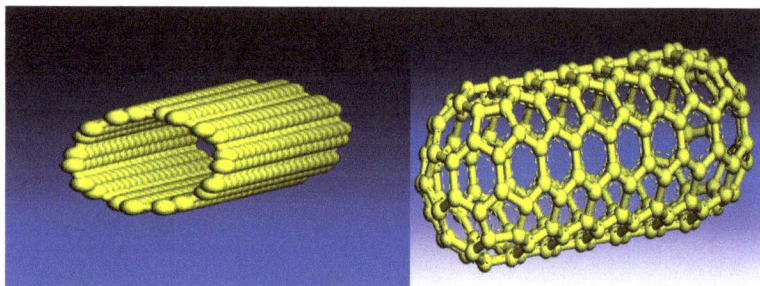

Figure 2. (Left) uncapped SWCNT (5,5) and (Right) capped SWCNT (5,5).

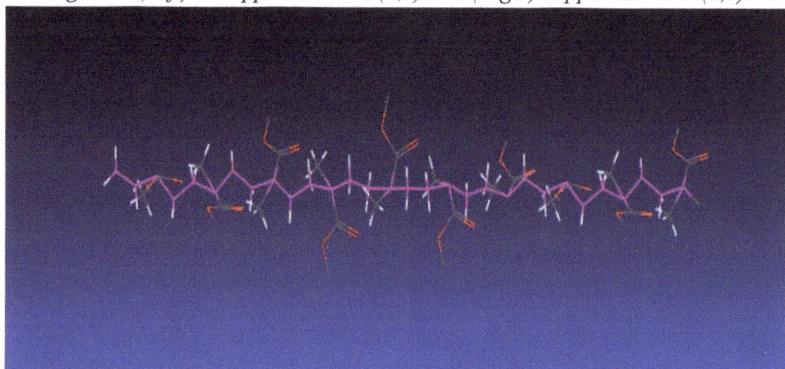

Figure 3. Structure of PMMA

Parameters for optimization of the geometry have been listed in every chapter. Run the optimization till the energy is minimized.

(ii) After geometry optimization, next step is dynamic run. Again, go to the "For cite" module and select "dynamics" run instead of geometry optimization. Clicking the "More" option will open a new window in which we specify the ensemble, velocities, temperature, time step and the total number of steps. Select the quality as "fine" and click "Run". After the completion of this step, we can view the energy plot and the temperature plot to see whether the system has been stabilized or not.

(iii)The next step is calculating mechanical properties. Under "For cite" module, select the "Mechanical Properties" task. Specify the number of steps and maximum strain amplitude. Here, we can again specify whether we want to optimize our structure or not. Click "Run" to begin the task of mechanical properties calculation. At the end of this step, we will get as tiffness matrix and the value of Young's modulus. The simulation strategy discussed above has been shown in the form of the flowchart in Figure 4.

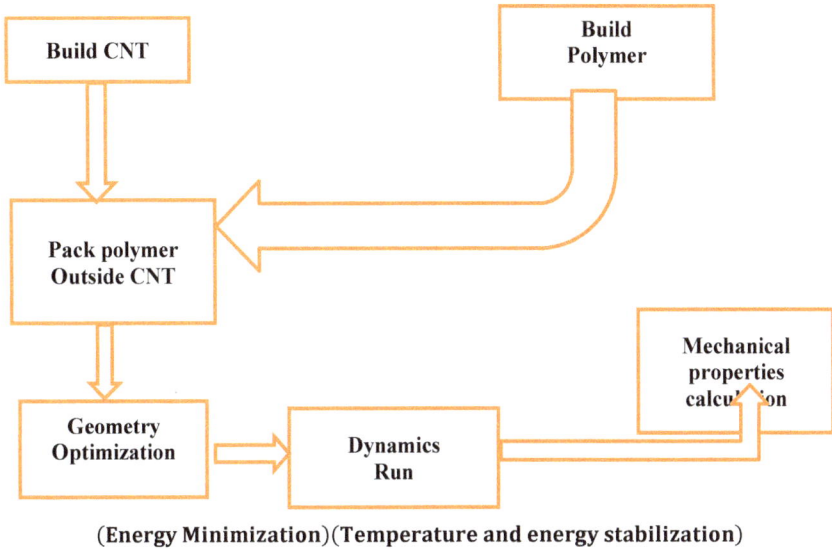

Figure 4. Flowchart of simulation strategy.

In MD simulation mutual atomic interaction as defined by force potentials related with bonding and non bonding occurrences. The inter-atomic potential energy was the addition of bonding energy and nonbonding energy:

$$U = U_{bonding} + U_{nonbonding} \quad (3)$$

For CNT, the non bonding word is generally the energy of Van-der-Walls force, which usually has a weak impact on the mechanical conduct among the atomic connection of the carbon microstructure. The governing part of the overall potential energy and the bonding energy is asummation of three dissimilar interactions among atoms: bond stretching, bond bending and bond torsion.

$$U_{bonded} = U_{bind-stretch} + U_{angle\,bend} + U_{torsion} \quad (4)$$

$$U_{bond-stretch} = \sum_b [k_2(b - b_0)^2 + k_3(b - b_0)^3 + k_4(b - b_0)^4] \quad (5)$$

$$U_{angle-bend} = \sum_\theta [k_2(\theta - \theta_0)^2 + k_3(\theta - \theta_0)^3 + k_4(\theta - \theta_0)^4] \tag{6}$$

$$U_{torsion} = \sum_\varphi [k_1(1 - cos\varphi) + k_2(1 - cos2\varphi) + k_3(1 - cos3\varphi)] \tag{7}$$

Here,

$k_1, k_2, k_3,$ and k_4 = Force constant evaluated experimentally

b, θ = bond angle and bond strength after stretching

b_0, θ_0 = Balance bond length and bond angle correspondingly

φ = Bond torsion angle

In all the simulations, periodic border condition (PBC) was used for reproducing the bulk phase of the nanocomposite. This method of discussing the CNTs as a bulk solid has been accepted by Griebel and Hamaekers [9]. The elastic moduli were calculated by adding the normal mechanical forces recognized between carbon atoms in the CNT. The actual elastic moduli, using the force method, can be calculated straight away from the viral theorem given by Swenson [30] in which the expression of the stress tensor in a macroscopic structure was given as the determination of atom coordinates and interatomic forces. The method delivers a continuous portion of the internal mechanical interaction among atoms in an atomistic calculation. The internal stress tensor can be achieved using the so-called virial expression given by Swenson [30], as shown below:

$$\sigma = -\frac{1}{V_0}\left[\left(\sum_{i=1}^n m_i(v_i v_i^T)\right) + \left(\sum_{i<j} r_{ij} f_{ij}^T\right)\right] \tag{8}$$

Where index i runs overall particles 1 through N; $m_i v_i$ and f_i means the mass, velocity and force applied to particle i and v_0 and signifies the (un-deformed) system volume. The application of stress on a body results in a modific at ion in the comparative positions of atoms within the body shown by the strain tensor:

$$\varepsilon_{ij} = \begin{bmatrix} \varepsilon_{11} & \varepsilon_{12} & \varepsilon_{13} \\ \varepsilon_{21} & \varepsilon\varepsilon_{22} & \varepsilon_{23} \\ \varepsilon_{31} & \varepsilon_{32} & \varepsilon_{33} \end{bmatrix} \tag{9}$$

The elastic stiffness coefficients, relating several components of stress and strain are defined by:

$$C_{lmnk} = \frac{\partial \sigma_{ln}}{\partial \sigma_{nk}}\Big|_{T,Z_{nk}} = \frac{1}{V_0}\frac{\partial^2 A}{\partial \varepsilon_{lm} \partial \varepsilon_{nk}}\Big|_{T,Z_{lm}Z_{nk}} \tag{10}$$

Here, A signifies the Helmholtz free energy. For minor deformation, the connection among the stress and strain may be shown in term of a general Hook's law:

$$\sigma_{lm} = C_{lmnk}\varepsilon_{nk} \tag{11}$$

To analyze the axial young's modulus (E_{11}) the atoms are displaced by $u_1=\varepsilon_{11}^0$. The average strain and stresses are:

$$\varepsilon_{11} = \varepsilon_{11}^0 \tag{12}$$

$$\sigma_{11} \neq 0 \ but \ \sigma_{ij} = 0 \tag{13}$$

Through MD simulation, the longitudinal Young's modulus E_{11} can be inferred as:

$$E_{11} = \frac{\sigma_{11}}{\varepsilon_{11}^0} \tag{14}$$

In another simulation run, the load was applied either in the transverse or shear direction. The internal stress tensor was then inferred from the methodically calculated virial and used to obtain approximations of the six columns of the elastic stiffness matrix. The engineering constants were calculated from elastic constants using the following relations given by Christenen [31].

$$E_{33} = E_{22} = C_{22} + \frac{C_{12}^2(-C_{22}+C_{23})+C_{23}(-C_{11}C_{23}+C_{12}^2)}{C_{11}C_{22}-C_{12}^2} \tag{15}$$

$$V_{12} = V_{13} = \frac{C_{12}}{C_{22}+C_{23}} \tag{16}$$

$$G_{23} = \frac{1}{2}(C_{22} + C_{23}) \tag{17}$$

$$K_{23} = \frac{1}{2}(C_{22} + C_{23}) \tag{18}$$

Carbonaceous Composite Materials
Materials Research Foundations **42** (2018) 57-92

Materials Research Forum LLC
doi: http://dx.doi.org/10.21741/9781945291975-3

If the strain is only applied in uniaxial direction, for example, in longitudinal direction, the constant E_{11} is given as:

$$E_{11} = \frac{\sigma_{11}}{\varepsilon_{11}} \tag{19}$$

Likewise the shear moduli can be calculated. This method was given by Parrinello-Rahman [10] and has been used by Griebel and Hamaekers [9].

3. Total potential energies and inter-atomic forces

The MD simulation method rests on the usage of appropriate inter-atomic energies and forces. In this investigation, we have used the Condensed-phase Optimized Molecular Potentials for Atomistic Simulation Studies (COMPASS) force field. This force field is associated with the continuous family of force fields (CFF91, PCFF, CFF, and COMPASS), which are strictly connected to second-generation force fields. COMPASS is the primary force field that has been parameterized and authenticated using compressed level possessions in accumulation to detect information for particles in isolation. Thus, this force field allows precise and concurrent estimation of physical, conformational, vibrational, and thermo-physical properties for an extensive-range of molecules in isolation and in condensed phases.

The COMPASS force field comprises of expressions for bonds (b), angles (θ), dihedral (φ), out-of-plane angles (χ) as well as cross-terms, and two non-bonded functions, a Coulombic function of electrostatic interactions and a 9-6 Lennard-Jones potential for van der Waals interactions.

$$E_{total} = E_b + E_\theta + E_\varphi + E_x + E_{b,b'} + E_{b,\theta} + E_{b,\varphi} + E_{\theta,\varphi}$$
$$+E_{\theta,\theta'} + E_{\theta,\theta',\varphi} + E_q + E_{vdW} \tag{20}$$

$$E_b = \sum_b [k_2(b - b_0)^2 + k_3(b - b_0)^3 + k_4(b - b_0)^4] \tag{21}$$

$$E_\theta = \sum_\theta [k_2(\theta - \theta_0)^2 + k_3(\theta - \theta_0)^3 + k_4(\theta - \theta_0)^4] \tag{22}$$

$$E_\varphi = \sum_\varphi [k_1(1 - cos\varphi) + k_2(1 - cos2\varphi) + k_3(1 - cos3\varphi)] \tag{23}$$

$$E_x = \sum_x k_2 \, x^2 \tag{24}$$

$$E_{b,b'} = \sum k(b - b_0)(b' - b_0') \tag{25}$$

$$E_{b,\theta} = \sum k(b - b_0)(\theta - \theta_0) \tag{26}$$

$$E_{b,\varphi} = \sum_{b,\varphi,} (b - b_0)[k_1(cos\varphi) + k_2(cos2\varphi) + k_3(cos3\varphi)] \tag{27}$$

$$E_{\theta,\varphi} = \sum_{\theta,\varphi} (\theta - \theta_0)[k_1(cos\varphi) + k_2(cos2\varphi) + k_3(cos3\varphi)] \tag{28}$$

$$E_{\theta,\theta'} = \sum_{\theta,\varphi} (\theta - \theta_0)(\theta - \theta_0') \tag{29}$$

$$E_{\theta,\theta',\varphi} = \sum_{\theta,\theta',\varphi} k(\theta - \theta_0)(\theta - \theta_0') \cos\varphi \tag{30}$$

$$E_q = \sum_{ij} \frac{q_i q_j}{r_{ij}} \tag{31}$$

$$E_{vdW} = \sum_{ij} \epsilon_{ij} \left[2\left(\frac{r_{ij}^0}{r_i}\right) - 3\left(\frac{r_{ij}^0}{r_i}\right) \right] \tag{32}$$

Where,
$k_1, k_2, k_3, and k_4$ = Force constant determined experimentally
b, θ= bond angle and bond strength after stretching
b_0, θ_0 = Equilibrium bond length and bond angle respectively
φ= bond torsion angle
X= out of plane inversion angle

$E_{b,b'}$, $E_{b,\theta}$, $E_{b,\varphi}$, $E_{\theta,\varphi}$, $E_{\theta,\theta'}$, $E_{\theta,\theta',\varphi}$= cross relations demonstrating the energy due to interface between bond stretch-bond stretch, bond stretch-bond bend, bond stretch-bond torsion, bond bend-bond torsion, bond bend-bond bend and bond bend-bond bend-bond torsion respectively.

4.   Stiffness of SWCNTs

The stiffness of capped and uncapped SWCNTs has been simulated using the Materials Studio 8.1 software. Simple steps for predicting the stiffness of capped and uncapped SWCNTs have been discussed below.

Carbonaceous Composite Materials   Materials Research Forum LLC
Materials Research Foundations **42** (2018) 57-92   doi: http://dx.doi.org/10.21741/9781945291975-3

4.1   Modeling of SWCNTs

The primary step was to model the SWCNTs using the "Build" tool in Materials Studio. We can build SWCNTs with dissimilar chirality (n, m). In this study, we have built capped and uncapped armchair (n, n) SWCNTs. Here, the integer 'n' controls the total size of the CNT. The minimum value for 'n' is 1. Models of capped and uncapped SWCNTs have been shown in Figures 5-6.

4.2   Geometry optimization

The second step in MD simulation was the optimization or minimization of the model that we're going to investigate. It is necessary to improve the structure after it has been sketched because sketching generates the molecules in higher energy configuration and beginning our simulation without optimizing our structure may lead to wrong results. There are different kinds of optimization methods present in Material Studio viz., steepest descent method, conjugate gradient and Newton-Raphson method. In the steepest descent method, the line search direction was well-defined along the direction of the local downhill gradient. These entire line search forms a new direction which was always perpendicular to the earlier gradient. This ineffective performance was characteristic of steepest descents, particularly on energy surfaces ensuring constricted valleys. Convergence was slow near the minimum as the gradient approaches zero, but the process was tremendously healthy even for structures that are far from being equilibrated.

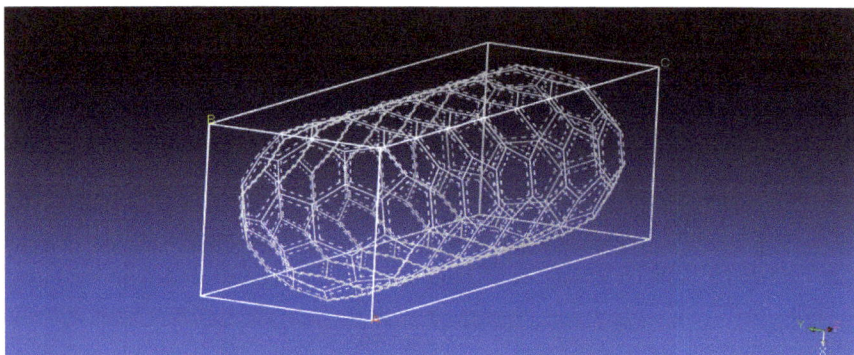

Figure 5 Model of armchair (5,5) capped SWCNT.

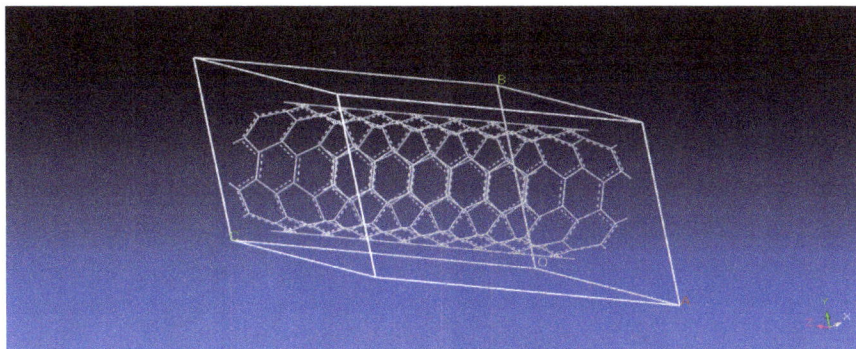

Figure 6 Model of armchair (5,5) uncapped SWCNT.

This method was generally used for generating low energy structure. It does not depend on what the function was or where the process has begun. So this method of optimization was used when the configurations were far from the minimum and gradients are very large. In this study, the smart algorithm has been used which is a cascade of all the overhead stated approaches. The parameters that were used in the optimization of models have been listed in Table 1.

4.3   **Dynamics**

After geometry optimization or when an optimized structure has been achieved, the next step was dynamics simulation. Classical equation of motion was modified in this step to agree with the result of temperature and pressure of the model. The constructed model was put into an ensemble with constant number of atoms, volume and temperature

Table 1 Geometry optimization parameters for capped and uncapped SWCNTs.

S.No	Parameters	Value
1	Algorithm	Smart
2	Quality convergence tolerance	Fine
3	Energy convergence tolerance	10^{-4} kcal/mol
4	Force convergence tolerance	0.005 kcal/mol/Å
5	Displacement convergence tolerance	5×10^{-5} Å
6	Maximum no. of iterations	5000

Table 2 Dynamics run parameters for capped and uncapped SWCNTs.

S.No	Parameters	Value
1	Ensemble	NVT
2	Initial velocity	Random
3	Temperature	300 K
4	Time step	1 fs
5	Total simulation time	50 ps
6	No. of steps	5000
7	Frame output every	5000
8	Thermostat	Anderson
9	Collision ratio	1
10	Energy deviation	5×10^{12} kcal/mol
11	Repulsive cutoff	6 Å

(NVT) simulation at a temperature of 298 °K for 5ps with a time step of simulation 1 fs. In the NVT approach the structure was thermally stabilized. After that Anderson thermostat was provided in the simulation and a temperature was maintained at 298 °K. The main purpose of the dynamics run was a trajectory file that has data for the atomic configuration, velocities and other information that were recorded in a period of time steps. The different parameters used in these steps have been listed in Table 2.

4.4   Mechanical properties

Table 3 Mechanical properties simulation parameters for capped and uncapped SWCNTs.

S.No	Parameters	Values
1	Number of strains	6
2	Maximum strain	0.003
3	Pre-optimize structure	Yes
4	Algorithm	Smart
5	Maximum number of iterations	500
6	Forcefield	Compass
7	Repulsive cutoff	6 Å

Carbonaceous Composite Materials   Materials Research Forum LLC
Materials Research Foundations **42** (2018) 57-92   doi: http://dx.doi.org/10.21741/9781945291975-3

After the dynamics run the next step was to calculate the mechanical properties. For the calculation of mechanical properties of capped and uncapped SWCNTs, the "Forcite"module has been used. With the help of Forcite module, elastic modulus of single structure or series of structures can be calculated. External stresses balanced the internal forces if the structure was in equilibrium. For the calculation of mechanical properties of the capped and uncapped SWCNTs the parameters that were used have been listed in Table 3.

5.   Results and discussion

In this section the results obtained for the mechanical properties of capped and uncapped SWCNTs have been discussed in detail. Several models of capped and uncapped (5,5) armchair SWCNT of different length but having the same diameter were constructed. Two capped armchair (5,5) SWCNTs having lengths of 12 and 34 Å have been shown in Figures 7 and 8 respectively. The variation in temperature with time for capped armchair (5,5) SWCNT has been shown in Figure 9. The dynamics run was performed for 50 ps using Forcite module of Material Studio 8.1. The structures of uncapped armchair (5,5) SWCNTs with lengths of 12 and 34 Å have been shown in Figure 10 and 11.

Variation in Young's modulus (E_{11}) for capped and uncapped armchair (5,5) for different lengths has been shown in Figure 12. Gao and Li [14] have explored the dependence of Young's modulus of capped and uncapped SWCNT for different length using MD simulation and observed that Young's Modulus for capped and uncapped SWCNT increases quickly when the length was less than 20 Å and after that becomes almost constant. It has been observed that Young's modulus for capped and uncapped armchair (5,5) SWCNT increases with an increase in the length of CNT.

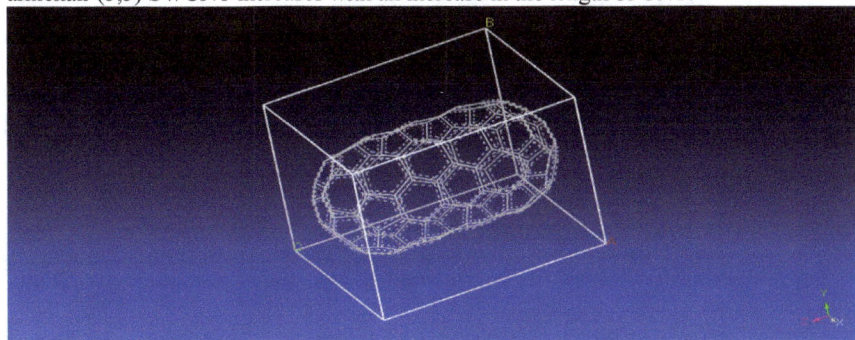

Figure 7. A capped armchair (5,5) of length 12 Å.

Carbonaceous Composite Materials   Materials Research Forum LLC
Materials Research Foundations **42** (2018) 57-92   doi: http://dx.doi.org/10.21741/9781945291975-3

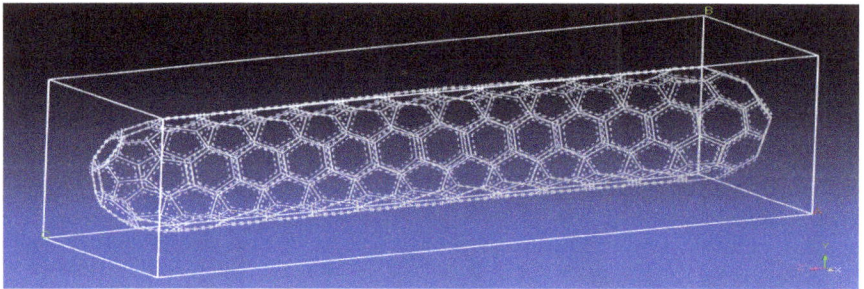

Figure 8. A capped armchair (5,5) of length 34 Å.

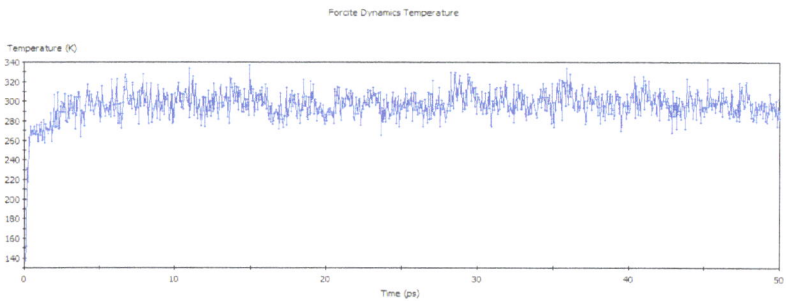

Figure 9. Dynamics run showing the variation of temperature with time for capped (5,5) SWCNT.

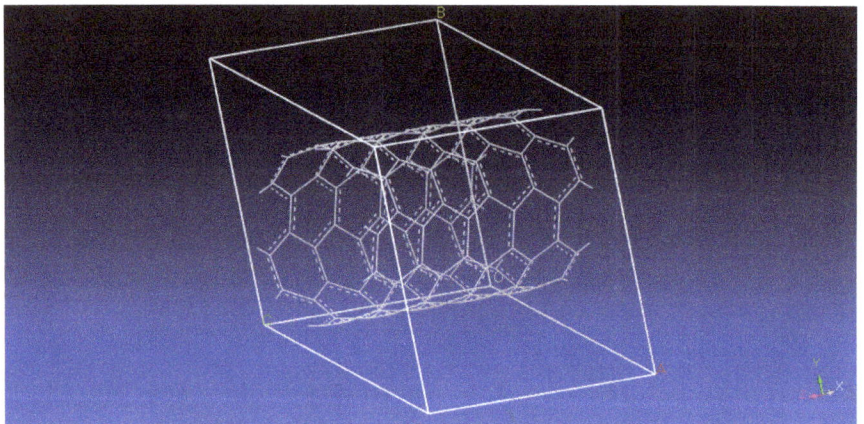

Figure 10. An uncapped armchair (5,5) of length 12 Å.

Carbonaceous Composite Materials   Materials Research Forum LLC
Materials Research Foundations **42** (2018) 57-92   doi: http://dx.doi.org/10.21741/9781945291975-3

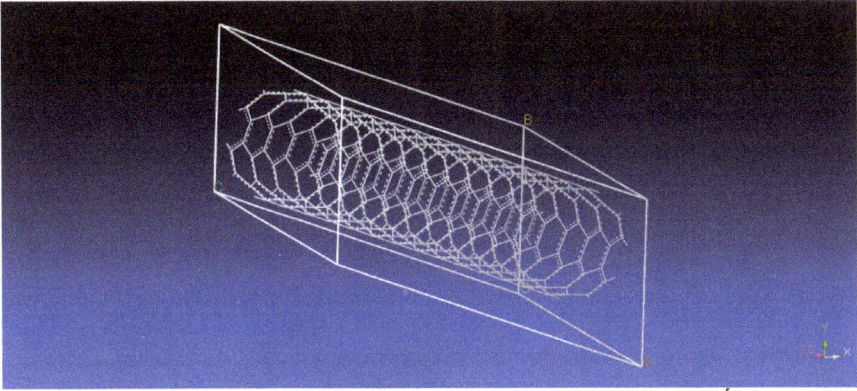

Figure 11. An uncapped armchair (5,5) SWCNT of length 34 Å.

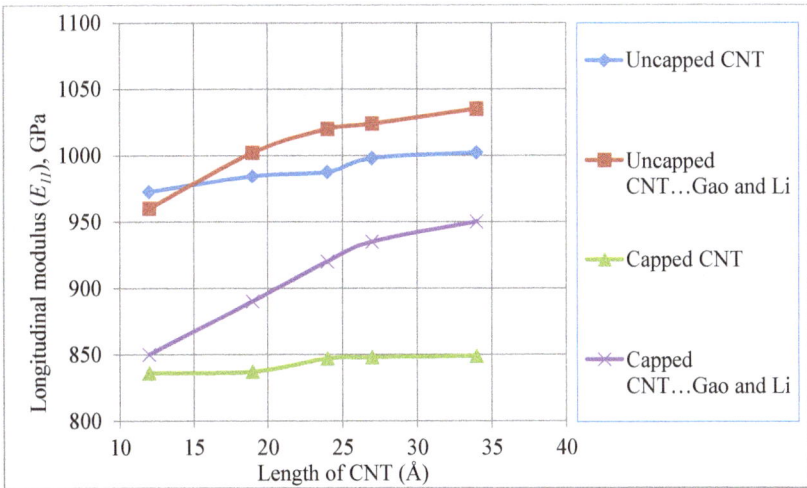

Figure 12. Variation of Young's modulus for capped and uncapped SWCNT for different length.

Carbonaceous Composite Materials   Materials Research Forum LLC
Materials Research Foundations **42** (2018) 57-92   doi: http://dx.doi.org/10.21741/9781945291975-3

Table 4 Young's modulus of capped and uncapped SWCNTs for different lengths.

S. No	Length (Å)	No. of Atoms for uncapped SWCNT	No. of Atoms for capped SWCNT	Young's modulus for capped (5,5) SWCNT (GPa)	Young's modulus for uncapped (5,5) SWCNT (GPa)
1	12	100	110	835.89	972.64
2	19	160	170	836.90	984.23
3	24	200	210	846.81	987.60
4	27	220	230	847.86	998.02
5	34	280	290	848.72	1002.24
Average				843.23	988.94

Table 5 Bulk modulus and shear modulus of capped and uncapped SWCNTs for different lengths.

S.No	Length (Å)	Bulk modulus of uncapped SWCNT (Gpa) (Voigt)	Bulk modulus of capped SWCNT (Gpa) (Voigt)	Shear modulus of uncapped SWCNT (Gpa) (Voigt)	Shear modulus of capped SWCNT (Gpa) (Voigt)
1	12	106.86	34.94	59.06	3.68
2	19	107.25	42.59	59.10	8.68
3	24	111.95	51.20	59.47	12.12
4	27	112.052	58.142	58.89	12.90
5	34	112.59	62.95	59.50	13.19
Average		110.14	49.96	59.204	10.11

The average Young's modulus, E_{11} for capped (5,5) armchair SWCNT was 843.23 GPa whereas for the uncapped (5,5) armchair SWCNT, E_{11} was 988.94 GPa. Figure 12 shows the variation in the Young's modulus of capped and uncapped armchair (5,5) SWCNT for the different lengths. The results of the present study differ slightly from those given by Gao and Li [14] because of the length and simulation technique. Gao and Li [14] used molecular structural mechanics for calculation of Young's modulus of capped SWCNT and the present study was based on MD simulation. The present study concludes that Young's modulus of capped and uncapped armchair (5,5) SWCNT increase with the increase in length.

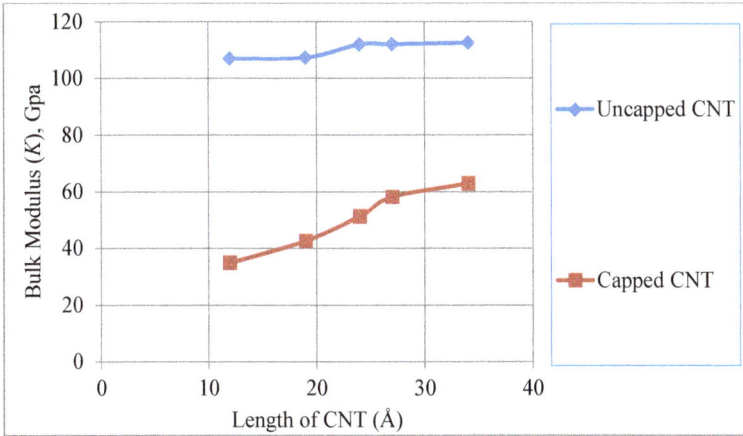

Figure 13. Variation in Bulk modulus of capped and uncapped SWCNT for different length.

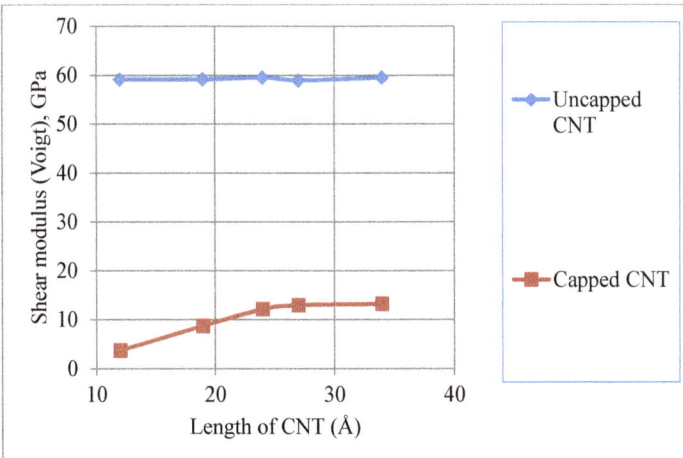

Figure 14. Variation in Shear modulus of capped and uncapped SWCNT for different lengths.

Table 4 display the values of Young's modulus (E_{11}) for capped and uncapped SWCNT for different length. Table 5 shows the values of Bulk modulus (K) and Shear modulus (G) for capped and uncapped SWCNT for different lengths. The present study was also compared with studies of Gao and Li [14]. The present study concludes that Young's modulus of capped and uncapped armchair (5,5) SWCNT increase with increase in length.

Variation of Bulk modulus (K) for capped and uncapped armchair (5,5) for different lengths has been displayed in Figure 13. The average Bulk modulus (K) of uncapped armchair (5,5) SWCNT was 110.14 GPa and average Bulk modulus (K) of capped armchair (5,5) SWCNT was 49.36 GPa. Similarly, Figure 14 displayed a variation in shear modulus (Voigt) of capped and uncapped SWCNT for different lengths. Average shear modulus of uncapped armchair (5,5) SWCNT is 59.204 GPa and average shear modulus for capped armchair (5,5) SWCNT was 10.11 GPa. It was concluded that the bulk modulus for capped and uncapped SWCNT increased gradually with an increase in length. Shear modulus (G) for uncapped SWCNT remains same with increase in the length of CNT whereas for capped SWCNT, the value of shear modulus was very less as compared to the uncapped CNT.

6.   Polymer/CNT Composites

In this section we have focused on the modeling and simulation of capped and uncapped armchair (5,5) SWCNT reinforced polymer composite using MD approach. MD simulation of polymer/CNT composite, composed of capped and uncapped armchair (5,5) CNT in amorphous polymer matrix poly methyl metha-acrylate (PMMA) has been performed for different volume fraction. Comparison and verification of the results obtained for the mechanical properties of capped and uncapped (5,5) on the basis of the present study has been made with those given by Han and Elliott [17]. The Cerius software was used by Elliot and Han [17] for determining the mechanical properties of uncapped armchair (10,10) in amorphous polymer matrix PMMA. MD simulation has been performed using Material Studio 8.1. Elastic moduli have been calculated after applying the energy minimization method of MD simulations. Periodic boundary conditions were applied along the tube axis and also in the transverse direction.

6.1   Molecular model of polymer matrix

With the help of "Amorphous" module of Material Studio 8.1, single chain of PMMA with 10 repeat units was constructed. Figure 15 displays methyl metha acrylate (MMA) monomer. These units were inserted in a periodic box by giving an initial density of

Carbonaceous Composite Materials   Materials Research Forum LLC
Materials Research Foundations **42** (2018) 57-92   doi: http://dx.doi.org/10.21741/9781945291975-3

0.1g/cc using COMPASS force field factors. Next step was to pack the polymer around the capped and uncapped (5,5) armchair SWCNT. The system was put in to ensemble with a constant number of atoms, volume and temperature (NVT) simulation with a pressure of 10 atm, volume 9023.272 Å^{3}and temperature 298.00 K for 50 ps with simulation time step of 1 fs. This step was required to compress the assembly gradually and to produce a primary amorphous matrix with accurate density and low residual stresses. The density of the absolute matrix was 1.203 g/cm^3which was very near to the investigated value of 1.19 g/cm^3. The total number of atoms in the case of capped CNT was 1538 atoms and for uncapped was 1528 atoms. MD simulation was performed for roughly 50000 steps so that thetemperature achieved a persistent value and after that constant minimization has been carried out with strain 0.003.

6.2   Elastic moduli of polymer

With the help of the guidelines provided in section 6.1, the elastic modulus of theoretical isotropic amorphous PMMA matrix from the current study was obtained as 2.73 GPA which matched well with the experimental range of 2.24-3.8 GPA.

6.3   PMMA/CNT composite system

A capped armchair (5,5) SWCNT was positioned in the center of the periodic simulation cell. PMMA molecules with different number of repeat units were located arbitrarily around the tube in non-overlapping places. The total number of atoms in a cell varies from 7780-15380. A capped armchair (5,5) SWCNT wrapped helically in PMMA in vacuo after equilibrium has been shown in Figure 16. Similar procedure was followed for the uncapped armchair (5,5) SWCNT as in section 6.1. An uncapped armchair (5,5) SWCNT wrapped helically in PMMA in vacuo after equilibrium has been shown in Figure 17.

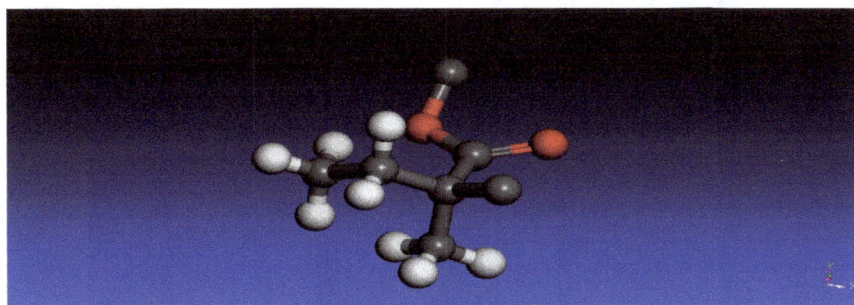

Figure 15. MMA monomer.

Carbonaceous Composite Materials   Materials Research Forum LLC
Materials Research Foundations **42** (2018) 57-92   doi: http://dx.doi.org/10.21741/9781945291975-3

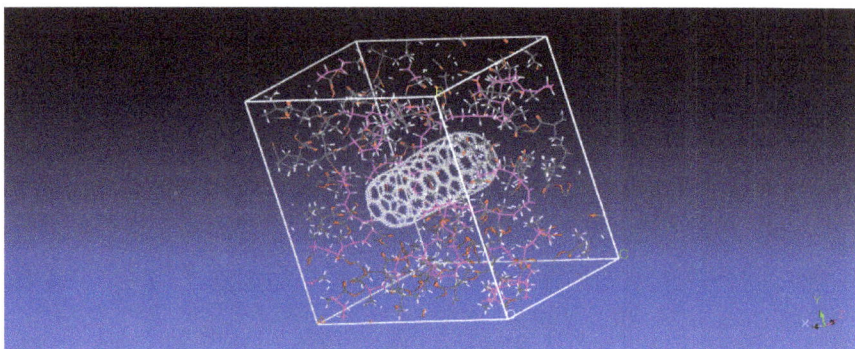

Figure 16. A capped (5,5) SWCNT wrapped helically with PMMA.

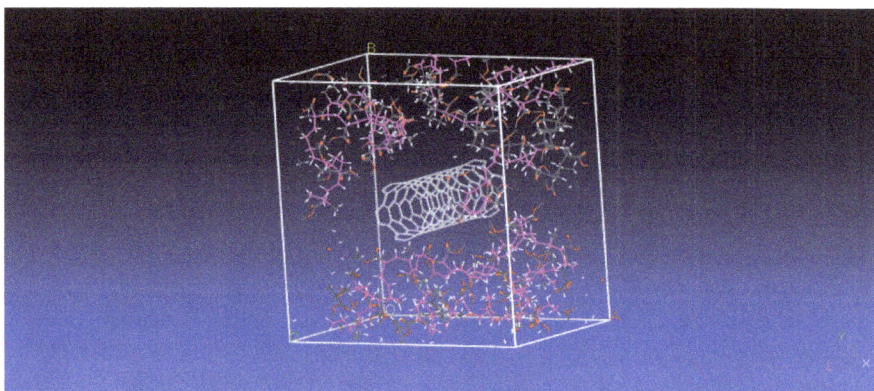

Figure 17. Uncapped (5,5) SWCNT wrapped helically with PMMA.

Table 6 Summary of results for longitudinal Young's modulus of PMMA/CNT composites.

S.No	Volume Fraction (%)	MD result of present study on (5,5) SWCNT		Elliot and Han (2007) Study on (10,10) SWCNT
		Longitudinal modulus of capped (5,5) PMMA/CNT composites (E_{11}), GPa	Longitudinal modulus of uncapped (5,5) PMMA/CNT composites (E_{11}), GPa	Longitudinal modulus of uncapped (10,10) PMMA/CNT composites (E_{11}), GPa
1	0	2.73	2.73	2.80
2	4	35.27	42.13	39.32
3	8	75.15	94.74	86.02
4	12	107.62	145.18	140.90
5	14	128.33	173.21	165.41
6	16	152.21	205.17	185.25

Table 6 shows the summary of results for longitudinal Young's modulus (E_{11}) of PMMA/CNT composite and Table 7 shows a summary of results for transverse Young's modulus (E_{22}) of PMMA/CNT composite. Figure 18 shows the graphical representation of the variation of E_{11} of capped and uncapped armchair (5,5) PMMA/CNT composite for different volume fraction. The comparison of study for uncapped armchair (5,5) has been done with the previous study for uncapped armchair (10,10) proposed by Han and Elliott [17]. The results of the present study were found to be in agreement with those of Han and Elliott [17]. The small difference between the present study and Han and Elliott [17] study might be because the present study was based on MD simulation of armchair (5,5) SWCNT and Han and Elliott [17] study was based on armchair (10,10 SWCNT).

E_{11} of uncapped (5,5) PMMA/CNT system, increased approximately 74 times when volume fraction of CNT was increased from V_f=0 to V_f=16. For the same range of V_f, Han and Elliott [17] observed an increase in E_{11} by approximately 65 times. Also on the basis of the present study, there was a large deviation in the values of E_{11} of capped CNT as compared to uncapped CNT. The difference in the result may be because Young's

Materials Research Forum LLC

doi: http://dx.doi.org/10.21741/9781945291975-3

modulus of uncapped (5,5) SWCNT is 984.23 GPa and Young's modulus of capped (5,5) SWCNT is 836.90 GPa for the same length as used in polymer/CNT composites. Table 8 shows the summary of results for Poisson's ratio of PMMA/CNT composites.

Table 7 Summary of results for transverse Young's modulus of PMMA/CNT composites.

		MD result of present study on (5,5) SWCNT		Elliot and Han(2007) Study on (10,10) SWCNT
S.No	Volume Fraction (%)	Transverse modulus of capped (5,5) PMMA/CNT composites (E_{22}), GPa	Transverse modulus of uncapped (5,5) PMMA/CNT composites (E_{22}), GPa	Transverse modulus of uncapped (10,10) PMMA/CNT composites (E_{22}), GPa
1	0	2.73	2.73	2.80
2	4	2.74	3.18	3.94
3	8	3.14	4.61	4.97
4	12	3.56	7.01	7.12
5	14	4.73	7.98	8.42
6	16	5.72	8.21	8.60

Figure 19 shows the graphical representation of the variation of E_{22} of capped and uncapped armchair (5,5) PMMA/CNT composite for different volume fraction. Table 9 shows the percentage increment of longitudinal modulus for capped and uncapped PMMA/CNT composites. It was concluded that longitudinal modulus of uncapped PMMA/CNT system increased approximately 34 times when volume fraction was increased from $V_f = 0$ to $V_f = 0.08$ and increases approximately 2 times when volume fraction was increased from $V_f = 0.08$ to $V_f = 0.12$. Similarly for capped PMMA/CNT system, longitudinal modulus increased approximately 27 times when volume fraction was increased from $V_f = 0$ to $V_f = 0.08$ and increased approximately 2 times when volume fraction increses from $V_f = 0.08$ to $V_f = 0.12$. Overall, it was concluded that longitudinal modulus of uncapped (5,5) PMMA/CNT system, increased approximately 74 times when volume fraction of SWCNT was increased from $V_f=0$ to $V_f=16$ %

Carbonaceous Composite Materials   Materials Research Forum LLC
Materials Research Foundations **42** (2018) 57-92   doi: http://dx.doi.org/10.21741/9781945291975-3

whereas the longitudinal modulus of capped (5,5) PMMA/CNT system, increased approximately 55 times when volume fraction of CNT was increased from V_f=0 to V_f=16 %.

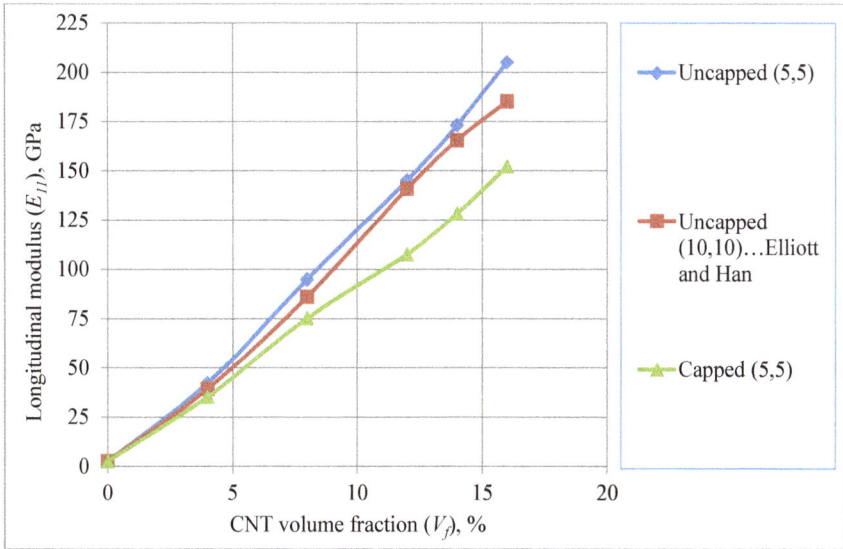

Figure 18 Comparison of results for longitudinal modulus of PMMA/CNT system.

Table 8 Summary of results for Poisson's ratio of PMMA/CNT composites.

S.No	Volume Fraction(V_f)	Poisson's ratio for capped (5,5) PMMA-CNT composites	Poisson's ratio for uncapped (5,5) PMMA-CNT composites
1	0	0.29	0.29
2	4	0.31	0.34
3	8	0.31	0.35
4	12	0.30	0.34
5	14	0.32	0.41
6	16	0.32	0.43

Figure 20 shows the graphical representation of the results of Poisson's ratio of capped (5,5) and uncapped (5,5) SWCNT/PMMA composite. It was concluded that uncapped

Carbonaceous Composite Materials   Materials Research Forum LLC
Materials Research Foundations **42** (2018) 57-92   doi: http://dx.doi.org/10.21741/9781945291975-3

SWCNT/PMMA composite has greater poison's ratio value as compared to capped SWCNT/PMMA system. Also on the basis of the present study it was concluded that Poisson's ratio of capped and uncapped SWCNT increased with V_f.

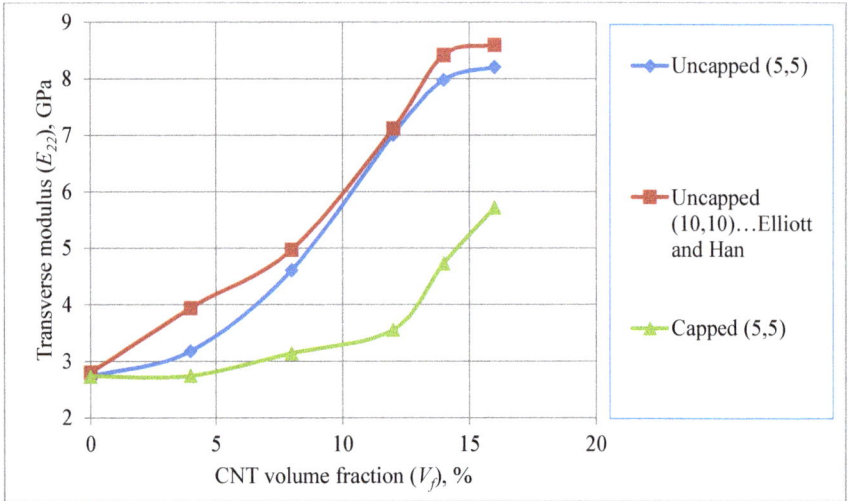

Figure 19. Comparison of results for transverse modulus of PMMA/CNT system.

Table 9 Percentage increment for longitudinal modulus of capped and uncapped SWCNT.

Volume Fraction (%)	Longitudinal modulus of capped PMMA-CNT composites	Longitudinal modulus of uncapped PMMA-CNT composites	Increment (%) for capped PMMA-CNT composites	Increment (%) for uncapped PMMA-CNT composites
0	2.73	2.73	-	-
4	35.21	42.13	11.89	14.43
8	75.15	94.74	11.34	12.48
12	107.62	145.18	43.20	53.24
14	128.33	173.21	19.24	19.30
16	152.21	205.11	18.60	18.45

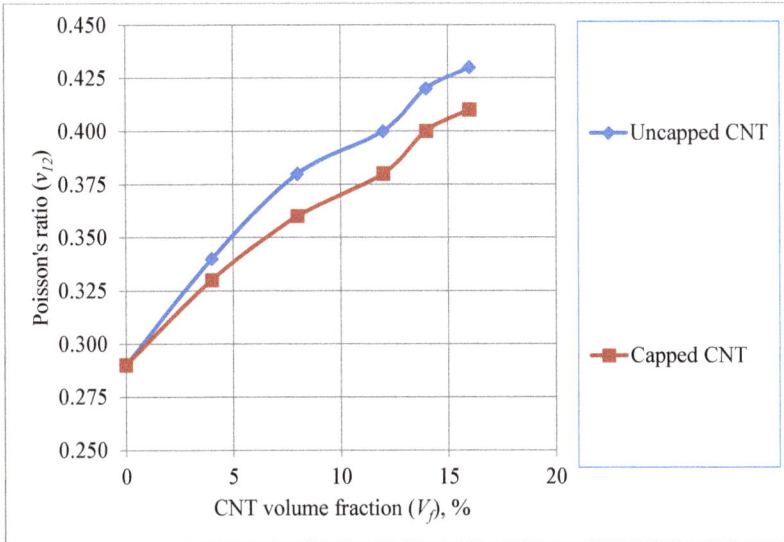

Figure 20 Comparison of results for Poisson's ratio of PMMA/CNT system.

7.   Conclusion

With the help of MD simulation, Young's modulus of capped and uncapped armchair (5,5) SWCNT has been investigated. It was observed that Young's modulus of uncapped armchair (5,5) SWCNT was 988.94 GPa and Young's modulus of capped armchair (5,5) SWCNT was 834.23 GPa.

Main findings of the present study have been highlighted below:

(i) Young's modulus of capped and uncapped SWCNT increases with increase in length of CNT. The shear modulus and bulk modulus also increase with increase in length of CNTs.

(ii) Uncapped SWCNTs have higher Young's modulus as compared to capped SWCNTs for the same length of CNTs.

(iii) Both capped and uncapped SWCNT are effective as mechanical reinforcement for polymer matrices, especially in the longitudinal direction. Reinforcement does not cause any significant change in properties in transverse direction.

(iv)For a fixed tensile load, capped and uncapped SWCNT should be associated parallel to the loading direction to accomplish largest tensile modulus.

(v) Presence of pentagonal and heptagonal defects in the end caps of capped CNTs can be the probable reason for lower values of E_{11}, E_{22} and poisson's ratio in comparison to the uncapped CNTs.

(vi) With increase in V_f from 0-12%, E_{11} increases rapidly after which the rate of increase slows down. It has been argued that proper dispersion at lower compositions, stronger π-π interaction between CNTs and base matrix, combination of large aspect ratio and high surface to volume ratio of CNTs and improved load transfer capability of CNTs assisted in improvement of properties but for higher compositions curvy and slippery nature of CNTs did not assist in further improvement of mechanical properties.

References

[1] H.W. Kroto, J.R.Heath, S.C.O'Brien, R.F. Curl, R.E. Smalley, C60: Buckminsterfullerene, Nature, 318 (1985) 162-163. https://doi.org/10.1038/318162a0

[2] P.R. Wallace, The band theory of graphite, Phys. Rev., 71 (1947) 622-634. https://doi.org/10.1103/PhysRev.71.622

[3] R. E. Smalley, Formation and properties of C60 and the fullerenes, National Institute of Standards and Technology (1990) Dec. 6-7.

[4] M. S. Dresselhaus, Oral presentation at fullerene workshop, University of Pennsylvania (1991).

[5] S.Iijima, Helical microtubules of graphitic carbon, Nature354 (1991)56-58.

[6] R.Bacon, Growth, structure, and properties of graphite whiskers, J. Appl. Phys. 31 (1960) 283-290. https://doi.org/10.1063/1.1735559

[7] M.Endo, A.Oberlin, T. Koyama, Filamentous growth of carbon through benzene decomposition, J. Cryst Growth. 32 (1976) 335-349. https://doi.org/10.1016/0022-0248(76)90115-9

[8] L.G. Zhou and S.Q. Shi, Molecular dynamic simulation on tensile mechanical properties of single-walled carbon nanotubes with and without hydrogen, Comput. Mater. Sci.23 (2002) 166-174. https://doi.org/10.1016/S0927-0256(01)00233-6

[9] M.Griebeland J.Hamaekers, Molecular dynamics simulations of the elastic moduli of polymer-carbon nanotube composites,Comput. Methods Appl. Mech. Eng.193 (2003) 1773-1788. https://doi.org/10.1016/j.cma.2003.12.025

[10]   M.Parrinello and A.Rahman, Crystal structure and pair potentials: A Molecular dynamics study, Phys. Rev. Lett. 45 (1980) 1196-1199. https://doi.org/10.1103/PhysRevLett.45.1196

[11]   K.Mylvaganam and L.C. Zhang, Important issues in a molecular dynamics simulation for characterizing the mechanical properties of carbon nanotubes, Carbon 42 (2004) 2025-2032. https://doi.org/10.1016/j.carbon.2004.04.004

[12]   X. Zhou,E. Shin,K. Wang, C.Bakis, Interfacial damping characteristics of carbonnanotube-based composites, Compos.Sci. Technol.64 (2004) 2425-2437. https://doi.org/10.1016/j.compscitech.2004.06.001

[13]   W.X.Bao, C.C. Zhu, W.Z. Cui, Simulation of Young's modulus of single walled carbon nanotubes by molecular dynamics, Physica B. 352 (2004) 156-163. https://doi.org/10.1016/j.physb.2004.07.005

[14]   X.L. Gao and K. Li,A shear-lag model of carbon nanotube-reinforced polymer composites, Int. J. Solids Struct.42 (2005)1649-1667. https://doi.org/10.1016/j.ijsolstr.2004.08.020

[15]   C.Deng, D. Wang, X. Zhang, Damping characteristics of carbon nanotube reinforced aluminum composite, Mater. Lett.61 (2006) 3229-3231. https://doi.org/10.1016/j.matlet.2006.11.073

[16]   A. Adnan, C.T. Sun, H. Mahfuz, A molecular dynamics simulation study to investigate the effect of filler size on elastic properties of polymer nanocomposites, Compos. Sci. Technol. 67 (2007) 348-356. https://doi.org/10.1016/j.compscitech.2006.09.015

[17]   Y. Han and J.Elliott, Molecular dynamics simulations of the elastic properties of polymer/carbon nanotube composites, Comput.Mater. Sci. 39 (2006) 315-323. https://doi.org/10.1016/j.commatsci.2006.06.011

[18]   V.V.Mokashi, D.Qian, Y. Liu, A study on the tensile response and fracture in carbon nanotube-based composites using molecular mechanics, Compos. Sci. Technol. 67 (2007) 530-540. https://doi.org/10.1016/j.compscitech.2006.08.014

[19]   J. Cho and C. Sun, A molecular dynamics simulation study of inclusion size effect on polymeric nanocomposites, Comput. Mater. Sci.41 (2007) 54-62. https://doi.org/10.1016/j.commatsci.2007.03.001

[20]   L.H. Gan and J.Q. Zhao, Theoretical investigation of (5,5), (9,0) and (10,10) closed SWCNTs, Physica *E:*Low Dimens. Syst. Nanostruct.41 (2009) 1249-1252. https://doi.org/10.1016/j.physe.2009.02.014

[21] M.D.Ganji, A. Fereidoon, M.Jahanshahi, M.G.Ahangari (2012),Elastic properties of SWCNTs with curved morphology: Density functional tight binding based treatment, Solid State Commun.152 (2012) 1526-1530. https://doi.org/10.1016/j.ssc.2012.06.005

[22] A.Fereidoon, M.Jahanshahi, M.G. Ahangari, M.D. Ganji, Density functional theory investigation of the mechanical properties of single-walled carbon nanotubes, Comput. Mater. Sci.53(2012) 377-381. https://doi.org/10.1016/j.commatsci.2011.08.007

[23] F.Gao, J.Qu, M. Yao,Electrical resistance at carbon nanotube/copper interfaces: Capped versus open-end carbon nanotube, Mater. Lett.82 (2012) 184-187. https://doi.org/10.1016/j.matlet.2012.05.095

[24] S.Haghighatpanah and K. Bolton, Molecular-level computational studies of single wall carbon nanotube–polyethylene composites, Comput. Mater. Sci. 69 (2013) 443–454. https://doi.org/10.1016/j.commatsci.2012.12.012

[25] B.Jeong and H. Kim, Molecular dynamics simulations of the failure behaviors of closed carbon nanotubes fully filled with C60 fullerenes, Comput. Mater. Sci. 77 (2013) 7–12. https://doi.org/10.1016/j.commatsci.2013.04.024

[26] M.Mahboob andM.Z. Islam,Molecular dynamics simulations of defective CNT-polyethylene composite systems, Comput. Mater. Sci. 79 (2013) 223-229. https://doi.org/10.1016/j.commatsci.2013.05.042

[27] X.Shao, H.Luo, J.Cai, C.Dong,First-principles study of single atom adsorption on capped single-walled carbon nanotubes, Int. J. Hydrogen Energy39 (2014) 10161-10168. https://doi.org/10.1016/j.ijhydene.2014.04.157

[28] S.Sharma, R.Chandra, P.Kumar, N.Kumar,Effect of Stone-Wales and vacancy defects on elastic moduli of carbon nanotubes and their composites using molecular dynamics simulation, Comput. Mater. Sci. 86 (2014) 1-8. https://doi.org/10.1016/j.commatsci.2014.01.035

[29] S.Sharma, R.Chandra, P.Kumar, N.Kumar,Molecular dynamics simulation of polymer/carbon nanotube composites, Acta Mech. Solida Sin. 28 (2015) 409-419. https://doi.org/10.1016/S0894-9166(15)30026-4

[30] R.J. Swenson, Comments on virial theorems for bounded systems, Am. J. Phys.51 (1983) 940-942. https://doi.org/10.1119/1.13390

[31] R. Christensen, Mechanics of composite materials, Krieger Publishing Company, Malbar, Fl, (1991) 74.

Carbonaceous Composite Materials
Materials Research Foundations **42** (2018) 93-110

Materials Research Forum LLC
doi: http://dx.doi.org/10.21741/9781945291975-4

Chapter 4

Fullerenes and its Composites

P. Senthil Kumar[*1], G. Janet Joshiba[1] and Abishek Sankaranarayan[2]

[1]Department of Chemical Engineering, SSN College of Engineering, Chennai 603 110, India

[2]School of Geography and the Environment, University of Oxford, South Parks Road, Oxford OX13QY, United Kingdom

senthilkumarp@ssn.edu.in*

Abstract

The evolution in the nanotechnology has created an immense enthusiasm for utilization of composites of fullerenes in various industrial applications. Fullerene is one of the greatest achievement and remarkable advancement in material sciences. In recent times they have pulled the significant consideration of diverse industrial domains. Due to their remarkable physicochemical feature, they serve as an important material in the manufacture of various gadgets, medicines, and materials. They show an extensive variety of unique mechanical and electrical properties. The smallest size and the unique structure of the fullerene make it exhibit extensive electric, magnetic, optical, structural, mechanical and chemical characteristics. This chapter deals with the characteristic and function of fullerenes and their composites.

Keywords

Fullerenes, Composites, Physicochemical Feature, Unique Structure

Contents

1.   Introduction

Carbon is an outstanding element with a tremendous scope of interesting properties. The carbon nanostructures such as graphene, fullerene, graphene, diamond stone, nanotube and carbon onions are widely used in various applications of mechanical and biomedical applications because of their unique potential and flexibility [1]. A composite material is synthesized by mixing two or more different components of varying qualities which will bring about an enhanced framework with unrivaled attributes than that of its individual segments by themselves. The nanocomposites are acknowledged for its unique features and it is often preferred for several processes in industrial sectors. It is made up of blending of two different compounds such as natural polymer material and some inert materials. The nanocomposites are synthesized by reinforcing the polymer material on the inorganic material at an optimized atmospheric condition. The polymer with multiple layers is preferred for the reinforcement of the polymer with the composites. Initially, graphite and clay are the two substances which consist of several layers and are considered to be the material for reinforcement.

Graphite is a material with several layers and it comprises of carbon molecules reinforced covalently in a hexagonal course of action in the sheets powered by van der Waals force acting between progressive layers. As the van der Waals forces are moderately frail, several particles can easily interchelate between the distinct layers and can easily reinforce with the graphite layers delivering graphite intercalation mixes. Due to the absence of active functional groups in the graphite intercalate with natural materials straightforwardly and it is found inefficient in the reinforcement of composites with graphite layers. Later the elongated graphite material is used for the manufacture of polymer/graphite reinforced composites. These elongated graphite materials consisted of inexhaustible pores going from 10 to 100 nm and it is comprised of substantial interlayer partitioning the diverse layers of extended graphite. Several particles and ions of different sizes can easily interchelated inside the elongated graphite resulting in the production of graphite reinforced composites. Subsequently, in 1985 a new class of substance came into existence and named as fullerenes [2].

Carbonaceous Composite Materials   Materials Research Forum LLC
Materials Research Foundations **42** (2018) 93-110   doi: http://dx.doi.org/10.21741/9781945291975-4

2.   Fullerenes

Fullerene is a significant classification of an allotrope of carbon and it is also knowns as Buckminster Fullerenes. It is composed of an array of carbon molecules arranged in various forms such as a hollow sphere, ellipsoid, and tube. It was unearthed in 1985 by a research group including Richard Smalley, Robert Curl and Sir Harry Kroto for which they were awarded the noble prize in 1996. As these materials had earmarks of geodesic dome of Buckminster Fuller it is named as Buckminster fullerenes. After the remarkable discovery of the fullerenes many other types of fullerenes composed of carbon atoms ranging from 18 to several 100 atoms were discovered and among the various types of fullerenes the bucky ball structured fullerene composed of 60 carbon atoms is prominent. As the synthesis of bucky ball is very simple it is highly preferred than other types of fullerene and it is popularly known as C_{60} atom [3]. The structure of fullerene material is distinguished into different structures such as endohedral, exohedral and on-site fullerene. In the endohedral fullerene, the dopant material is imprisoned inside the circle. In addition, the central part of the endohedral fullerene makes them more unique and more helpful than other fullerene structures [1]. The structure of fullerenes is found to be in three major forms such as elliptical, spherical and tubular shaped and the entire body is fully made up of carbon atoms [3]. Due to the unusual physical and chemical properties of the fullerenes, it is widely used in diverse fields of science. The disclosure and accessibility of fullerenes have invigorated their broad research in a wide assortment of fields such as physical science, science, material science, science and so forth. It has pulled in the significant consideration of everyone because of its compatibility with different atmospheric conditions [4]. The elite materials synthesized using the fullerene material possess a very good antibacterial and anticancerous properties [5]. Fullerenes possessed some of the unique features which were absent in diamond and graphite material. Due to its uniqueness, it is used as an important component in electrical devices, polymer reinforced composites, medical equipment, fuels and optical device [6]. In modern applications, the fullerenes are used efficiently as lubricants in industrial sectors [2].

2.1   Types of fullerenes

After the discovery of fullerenes many structural variations have been made in the structure of fullerenes and the fullerene has evolved into diverse structural variations and some of the famous types of fullerenes are listed below:

Carbonaceous Composite Materials   Materials Research Forum LLC
Materials Research Foundations **42** (2018) 93-110   doi: http://dx.doi.org/10.21741/9781945291975-4

2.1.1 Nanotubes

Nanotubes are thoroughly made from carbon and the length of the nanotube can vary from few nanometers to several millimeters. They possess small dimension and are hollow in shape. They can be of single-walled or multi-walled nanotubes with open and closed ends. They are especially used in aeronautical and electrical industries.

2.1.2 Mega tubes

These tubes possess larger dimensions and its thickness depends on the purpose for which it is used. In all other characteristics it is similar to nanotubes and it is used in the transport industries.

2.1.3 Bucky ball clusters

This type of fullerenes is the tiniest of all fullerenes and its clusters are ball-shaped. It occurs in nature and predominantly is found in mine extracts and flacky substance derived from combustion of organic matter. The littlest individual from these bunches incorporate C20 and the most widely recognized are C60.

2.1.4 Polymers

Polymers are entirely composed of carbon chains and they are formed by the macromolecules linked by covalent bonds. When polymer is subjected to high temperature and high pressure two dimensional and three-dimensional polymers are formed.

2.1.5 Nano onion

It is the same as the bucky ball structure and it consists of round particles in view of numerous carbon layers encompassing the center of the buckyball. It is mainly used as lubricant.

2.1.6 Linked "ball and chain" dimers

It consists of two bucky balls connected by a chain of carbon [7].

3.   Structure of fullerene

Carbon atoms under the influence of catenation reaction give rise to compounds of various shapes and sizes. Fullerene is the third allotrope of carbon and it is an enclosure containing only carbon elements. It is a group of confined mixes made from just carbon components comprising of 16 hexagonal rings and 12 pentagonal rings and resembles a geodesic arch. The complete Structure of fullerenes looks like a ball ex: a soccer ball. The run of the mill distance across of a fullerene particle is 0.7 nm [4].

3.1   Bucky ball structure

The Bucky ball structure is one of the popular structure of the fullerenes and it is commonly called as C60 structure. It resembles the structure of soccer ball and is composed of carbon atoms arranged in a spheroid shape with 60 vertices of 60 carbon atoms. The bucky ball fullerene is a mix of 20 hexagonal and 12 pentagonal rings arranged and fitted into a sphere. The distance between the C-C bond separation measured using the nuclear magnetic resonance was found to be 1.44 A° [3]. With the help of Sp2 hybridization, a single carbon atom is attached to three other carbon atoms. The bucky ball structure possesses higher symmetry when compared to other structures and it has a probability of nearly 120 symmetrical operations.

3.2   Cylindrical structure

Fullerenes in another way form a cage-like cylindrical structure and it is composed of a sheet of graphite rolled into a cylindrical structured material. This type of fullerene is similar as carbon nanotubes and they possess some unique properties which are absent in the primitive type of fullerene. These fullerenes are further reinforced with metal composites and it gives rise to several advancements in electronics. These fullerenes can be used as nanowires for serving communication purpose and it is very cheap widely used fullerenes than the primitive fullerenes [8].

4.   Synthesis

Initially, the fullerene material is synthesized using the vaporization technique of graphite with the help of inert gas and the synthesis was carried out at an optimized pressure condition [6]. Primitively, the fullerenes are synthesized with the help of chemical vapor deposition methods, arc-discharge method and by combustion processes. These traditional methods were found to be dissatisfying and many new approaches have been enunciated by researchers for improving the quality of the fullerenes produced. Many different synthetic techniques have been formed and implemented to improve the production and quality of the fullerene [9].

4.1   Arc discharge vaporization of graphite

The arc discharge vaporization of graphite is the ancient method followed to synthesize fullerene. It is initially discovered by Kratschmer and his crew in 1990 and this method is named as a Kratschmer-Huffman method. In the Kratschmer method, they found that when the carbon rods are subjected to heat in a helium gas environment it produced a minor quantity of fullerene materials. Later, the graphite electrodes are used to produce fullerenes. The graphite electrodes in the form of contact arcs are placed in the helium atmosphere of about 200 torr and heated to optimal temperature and pressure by passing direct electric current inside the graphite electrodes. As the current is passed through the electrodes the vaporization of the graphite electrode takes place and it is modified into a variety of soot. Then the soot is immersed into the nonpolar solvent and then the solvent is removed from the residue using drier. The bucky ball structured fullerene are found in the remains derived from the electrode. This method is predominantly used to produce C60 Fullerene [3].

4.2   Low – pressure Benzene/Oxygen diffusion flame method

Fullerenes can be synthesized using hydrocarbon flames at low pressure with the help of fuels. The derivatives of fullerene and their components are produced using the flame diffusion method. In the benzene diffusion flame method, the low-pressure benzene gas is diluted with the argon gas in the presence of oxygen. An optimal range of 12 to 40 torr of pressure was maintained in the combustion chamber. After burning of the benzene and argon mixture the residue containing fullerenes, soot and other derivatives of fullerenes are collected in a filter. Then the filter containing residues are immersed in the solvent and it is also subjected to a sonication process. The fullerenes are finally analyzed using high-performance liquid chromatography and the other structures of fullerenes are determined using a high transmission electron microscope [6].

4.3   Combustion process

The idea of producing fullerenes through flames came into existence in 1986 and the fullerene particles were identified and proven in premixed acetylene-oxygen and benzene-oxygen blazes in 1987. Fullerenes were found in residue created from the flame produced by burning low-pressure premixed mixture of benzene/oxygen/argon flares. This discovery remains as a remarkable achievement and it paved the way for the production of C60 material by the combustion process. From the combustion process, it is also inferred that the reactivity of fullerene completely depends on the pressure, temperature, residence time and weight ratio. From these examinations, the outturn of fullerenes from the collected residue seemed to be at the level of 0.003% to 9%. In this combustion process, two methodologies are followed such as premixed flame studies and diffusion flame studies [10].

Carbonaceous Composite Materials   Materials Research Forum LLC
Materials Research Foundations **42** (2018) 93-110   doi: http://dx.doi.org/10.21741/9781945291975-4

4.4   Laser ablation

The laser ablation is a prominent technology used for the synthesis of fullerenes. Initially, in 1996 this laser ablation technique is used to synthesize fullerene and then followed by the creation of carbon nanotubes especially single-walled carbon nanotubes. The laser ablation set up consists of a heater, laser source, and a quartz. A source of the laser is placed in the center of the graphite pole in the reactor. The sample is subjected to high temperature and vaporized with the help of argon gas and further subjected to the cooler and coated. The laser has the adequately high vitality to vaporize the graphite focus at the nuclear level, which is then utilized as the material for the union of SWNTs. The SWNT production depends on the temperature, reactant metals, and stream rate [11].

4.5   Chemical vapor deposition (CVD)

The chemical vapor deposition is the method involving pyrolysis of hydrocarbons and carbon feedstocks. The hydrocarbons such as ethylene, propylene, toluene, etc. are pyrolyzed under the CVD method. The carbon feedstocks such as polymers are weakened in the surge of idle gas in the heater framework over the surface of metal catalysts. The catalyst material might be strong, fluid, or gas and can be put inside the heater or nourished in ceaselessly from outside. Average temperature extends for the blend is 500 to 1200 °C [11]. The catalyst remains as the fundamental part of the chemical vapor deposition and it is of two types such as drifting and bolstered catalyst. Different sorts of catalyst backings such as Zeolite, Al_2O_3, $CaCO_3$, and MgO were used for the processing of chemical vapor deposition. The recovery of the catalyst supports from the final residue is difficult. Then again, water-dissolvable materials as impetus backings can be effortlessly isolated from the item. Chemical vapor deposition is an assuring technique because of its moderately minimal effort and conceivably high return [12].

4.6   Chemical synthesis of fullerene

The fullerenes are synthesized by various new techniques but the quantity of fullerene produced becomes a limitation in several synthesis processes. To increase the quality and quantity in the production of fullerene the chemical synthesis method is followed. The C60 structure of fullerene seems to be possessing higher exertion energy and it is preferred mostly than other types of fullerenes. The C60 synthesized by the chemical process should also possess the higher exertion energy same as the natural fullerenes. The thermal energy value of the fullerenes is found to be higher than 600 Kcal mol^{-1}. The chemically synthesized fullerene should also meet all the criteria and specifications same as the natural fullerene. The chemical synthesis method followed by Barth and Lawton for forming corannulene is a remarkable chemical synthesis method and it paved the way to produce C60 fullerenes. Many new approaches such as coupling reactions, bending reactions, dehydrogenation, retrosynthesis and hydrogen atom shifting are implemented towards the chemical synthesis of fullerenes. Further researches are focused on finding new methodologies for the synthesis of other types of fullerenes using the chemical synthesis method [13].

5.   Properties

5.1   Physical properties

Fullerenes are completely composed of carbon atoms and it is popularly known as bucky ball due to its structure. It is a black /brown colored powdery substance. The fullerenes materials are familiar for their Young's modulus and compressibility factor which was found to be at 15.9 Gpa and $6.9 \times 10^{-12} cm^2 dyn$. The bulk modulus of about 8.8 Gpa was found in the fullerene material. The boiling point of the fullerenes was found to be at 800 K and it possesses a refractive index of about 2.2. It is a completely odorless material and the density of fullerene was found to be at 1.65 g/cm [4]. Fullerenes are greatly solid atoms, ready to stand up to incredible weights they will ricochet back to their unique shape in the wake of being liable to more than 3,000 atmospheres. Hypothetical figuring recommend that a solitary C60 particle has a viable bulk modulus of 668 GPa when compacted to 75% its size. This property influences fullerenes to end up plainly harder than steel, precious stone, whose bulk moduli are 160 GPa and 442 Gpa, individually. An intriguing test demonstrates that Fullerenes can withstand crashes of up to 15,000 mph against stainless steel, simply bobbing back and keeping their shapes. This test looks like the high steadiness of the particle [2].

5.2   Size

The shape and structure of the fullerene is similiar to that of a soccer ball. The average size of the fullerene seemed to be at $7 \times 10^{-10} m$ [8].

5.3   Solubility

Solubility is one of the important criteria to be studied in a fullerene. The fullerene shows reduced solubility in polar solvents. The solubility of the fullerene is studied importantly for drug designing in medical application [13]. The extent of solubility of fullerene in solvents vary widely and the understanding of solubility helps in finding the appropriate solvent for purifying the fullerenes [4]. The solubility of fullerenes in water is found to be higher than other polar solvents and the fullerenes form complexes with calixerenes, phospholipids, micelles, liposomes and polyvinylpyrrolidone [14].The fullerenes are dissolvable in solvents such as benzene, toluene and chloroform [8].

5.4   Chemical properties

The fullerenes are sp_2 and sp_3 hybridized of which the sp2 carbons are in charge of the impressively point strain introduced inside the particle. C60 has a confined pi-electron framework, which keeps the atom from showing superaromaticity properties. It also undergoes reversible oxidation reactions and it has high electron affinity due to the existence of triply-degenerate low-lying LUMOs (lowest unoccupied molecular orbital) [2]. Initially, the fullerene materials were chemically dormant material and after many inspections, it is inferred that the fullerene materials involve them in many important chemical reactions and it possesses quiet good reactivity. Fullerenes remain as a secure electron acceptor for many organic and inorganic donor ions. The chemical reactivity of the fullerenes is due to the cleavage of the double bond. The fullerene is classified into two types based on the presence of functional groups such as exohedral and endohedral fullerenes. The fullerenes participate actively in different kinds of chemical reactions such as nucleophilic addition, electrophilic addition, hydroxylation, hydrogenation, oxidation, cycloaddition and hydroarylation reaction [4].

5.5   Optical properties

The optical properties of the fullerenes are studied with the help of solutions. The straight assimilation range of fullerene demonstrates a solid ingestion in the UV region and an exceptionally feeble assimilation in the visible range. The intensity of optical absorption is enhanced by various mechanisms such as nonlinear refraction, thermal effects, nonlinear scattering and nonlinear absorption. Immaculate C60 is one of an important atom possessing high symmetry and it also contains p-electrons delocalized along the entire 3-D structure. These properties make fullerene a fascinating material in the nonlinear optical field and it is also responsible for a wide assortment of phenomenal physical properties going from optical limiting to superconductivity and photoconductivity. The optical properties of the fullerene seem to be strong even if the symmetrical arrangement of fullerenes is disturbed and broken [15]. Detached pi electrons in fullerenes are known to give incredibly vast nonlinear optical reactions. Fullerenes have demonstrated specific guarantees in optical constraining, what's more, force subordinate refractive list. Furthermore, the exchange of electrons from the encased atom(s) to the Fullerene upgrades the third-arrange nonlinear optical impact by requests of extent contrasted with discharge confine fullerenes [3].

5.6   Mechanical properties

Nanocomposites like fullerenes have been widely utilized as a part of different designing fields for as far back as the olden times. The fullerenes reinforced with other material shows high solidness, low thickness, high quality to weight proportion and high solidness to weight proportion, high durability with enhanced crawl protection and wear protection. Joining high hardened nanoparticles to the low modulus polymer network, enhances the heap conveying limit of the fullerene. The consolidate between the constituents fundamentally influences the mechanical properties of nano-based composites. The expectation of compelling mechanical properties of composites depends on the suppositions of flawless holding. This supposition is insufficient for the portrayal of the composite with blemished debonds. To examine composite material properties with the nearness of debonding, one needs to choose test or hypothetical methodologies [16]. The bond length of a material generally gives information about its strength and the two bond lengths of the fullerenes are seemed to be 1.46 A$^{\circ}$ and 1.40 A$^{\circ}$ [17].

5.7   Vibrational properties

The vibrational properties of a material are studied using Raman spectroscopy and IR spectroscopy. The Buckminster fullerene is the prominent class of fullerene which exhibit clear isotopes. The vibrational properties are seemed to be dissatisfying according to the Raman spectroscopy and it possesses nearly 46 modes of vibrations [18]. Based on the extent of coupling between particles strong fullerene is classified into two major classifications such as intramolecular and intermolecular modes. The intramolecular modes are frequently called as atomic modes because of their low frequencies. The intermolecular modes are also called as cross section modes and it is additionally subdivided into acoustic, vibrational and optic modes. The frequencies for the intermolecular modes are low, mirroring the powerless van der Waals bonds between fullerene particles [19].

5.8   Electrical properties

The conduct of the fullerenes in electric fields pulled in ahead of schedule much consideration because of the potential utilization of this class of particles as proficient nonlinear optical gadgets [20]. From the beginning, carbon has been a good conductor of electricity and the conductivity and the insulating property of the carbon fully depend on the material. The thermal activation mechanism and the higher activation energy are the two important parameters which decide the conductivity of immaculate fullerenes. The polymer-reinforced fullerenes possess good conductivity than the pristine fullerenes [21]. The fullerenes in its original form remain as a bad conductor of electricity and when the pristine fullerene is reinforced with a polymer or other membranes the conductivity increases [22].

5.9 Magnetic properties

Fullerenes are delegated sweet-smelling mixes despite the fact that the term aromaticity forces a few prerequisites, not which are all completely met by fullerenes. Cyclic delocalization of electrons in fragrant mixes straightforwardly influences the attractive properties by fluctuating the diamagnetic powerlessness: an outer attractive field actuates a diamagnetic ring current. This offers to ascend to an attractive field inside the sweet-smelling ring; the bearing of this field is inverse to that of the outer attractive field. Conversely, the outer attractive field is improved outside the sweet-smelling ring. In this manner, the estimation of diamagnetic weakness is bigger in sweet-smelling hydrocarbons than in other unsaturated mixes. Initially, when fullerenes were discovered it was thought that due to its spheroidal shape it possesses an extraordinary magnetic particle. The amount of paramagnetic current passing through the benzene ring will be equal to the amount of current passing through the pentagon rings. In the fullerenes, the hexagons are responsible for the diamagnetic behavior of the particle and the pentagons are responsible for the paramagnetic behavior of the material. The fullerene materials possess an extraordinary magnetic property due to its unique structure and it depends on the structure of the fullerenes [23].

5.10 Lubricating properties

Graphite is likewise esteemed in modern applications for its self-greasing up and dry greasing up. The free interlamellar coupling between the several sheets in the structure is highly responsible for the lubricating properties of the fullerenes. Especially the C60 atom of the fullerene have exceptional grease impacts and they have been regarded as an oil because of its round shape. Researchers have found that the graphite interchelating compounds derived from the graphite can be used as superlubricants. The blending of the graphite with fullerenes will result in synergistic impacts as a lubricant [2].

6. Composites of fullerenes

Composite materials are supplanting a few customary materials considering their high quality and high firmness to weight proportions. These materials improve the quality of the fiber and also the high longitudinal modulus of the fiber constituent. In the present decade, composite materials are further reinforced by nanocomposite material for bringing out an extraordinary feature. The portrayal of nanocomposites materials is accomplished by utilizing two methodologies such as computational science strategies and mechanics approach. The principle constituents of buckminster fullerene strengthened composites are fullerene and epoxy grid. The coating of fullerene in the epoxy material is typically of nonuniform nature. The fullerene is glorified as an empty circle with uniform thickness [24].

Polycarbonate (PC) is one of the most grounded general polymers accessible today and is broadly utilized as a part of requesting aviation applications. The thermal and the mechanical properties of the polycarbonate is very alluring Promote from a mechanical building viewpoint. Due to its uniqueness, the polymer reinforced composites are used in aeronautics. Then again, fullerenes have pulled in much consideration as a captivating nanocarbon material in material science. The circular p-conjugated structure of fullerenes with extensive pyramidalization edge is less demanding to synthetically functionalize than CNT and in this manner brings about considerably higher solvency to numerous natural and fluid solvents. Such functionalization eases to set up the nanocomposites as well as gives an enhanced noticeable light transmittance [25].

7.   Applications

Fullerenes are said to have various applications in diverse domains of science. The applications of the fullerenes are as follows.

7.1   Fullerenes as wires

Late trials have archived electron transport through single atoms. Under certain trial conditions, atomic conduction through a solitary particle instead of through a group of atoms is ensured. This wonder is conceivable due to the high electron partiality of fullerenes. In the event that a sub-atomic PC is ever to be fabricated, at that point, it will require atomic wires with a specific end goal to associate with its different segments. The fullerene atoms get energized and electrons move from the Porphyrin wire towards the fullerenes at the point when a wellspring of UV light is connected to the framework. These electrons leave gaps in the Porphyrin through which electrical current can spill out of one cathode to the next.

7.2   Medicinal applications

- The fullerene is used to inhibit the reactivity of the enzyme and serves as an enzyme inhibitor. It was discovered by Tokuyama et al. in 1993. He discovered an arrangement of fullerenes (5a) for photoinduced chemical restraint.

- Fullerene 5a was found to hinder cysteine proteinases and serine proteinases at the point when presented to low-vitality light. Dendrimer 4p containing 18 carboxyl gatherings is the additionally encouraging today.
- The enthusiasm of researchers in water-dissolvable fullerene mixes is specifically identified with their natural movement. The combination and the utilization of this compound in prescription were protected in the U.S. (C-sixty Corporation); at the display, these tranquilize experiences clinical trials as a promising solution for treatment of AIDS.
- The fullerene has good antiviral activity in the case of HIV virus. The fullerene material is checked for antiviral activity against HIV virus with the help of examining the fullerene with cells infected with HIV and free cells. The fullerene remains as molecular scissors because of its usage to cleave DNA. When the mixture of 5a and 5c is incubated with supercoiled pBR322 it was found to cut DNA at the point particularly at 182 base-match sections at guanine buildups upon presentation to light.
- Fullerene by addition into phospholipidic bilayers could create layer disturbances and consequently have antibacterial activity.
- Fullerenes are astronganticancerous agent which is responding promptly and at a high rate with free radicals, which are regularly the reason for cell harm or death.
- Fullerenes used to protect the neurological system from damage and they control the neurological harm of sicknesses for disease such as Alzheimer's illness and Lou Gehrig's sickness (ALS).
- Fullerenes are known to act like a "radical wipe," as they can wipe up and kill at least 20 free radicals for every fullerene atom. They have appeared execution 100 times more compelling than current driving cancer prevention agents, for example, Vitamin E.
- Fullerene is exceptionally dissolvable in almond oil and along these lines, it can be utilized for a screening test for visual tissue danger demonstrating no unfavorable impact [3,4,8,14].

7.3   Fullerenes in organo photovoltaics

The fullerenes are used in oregano photovoltaics. The fullerene goes about as then-sort semiconductor (electron acceptor). Then-sort is utilized as a part of conjunction with a p-sort polymer (electron contributor), regularly a polythiophene. The record effectiveness for a mass heterojunction polymer sunlight based cell is higher due to the fullerene/polymer mix. This mass heterojunction is the combination of dynamic layers of polymer and semiconductor. Due to their solubility, it is used in photovoltaics. The most regularly utilized subsidiary in photovoltaics is C60, yet C70 has been appeared to have a 25% higher power change proficiency than C60. In November of 2005, a record cell productivity of 4.4% utilizing a fullerene subordinate and outlining the significance of the attributes of the dynamic layer on execution was published [8].

7.4   Fullerenes as hydrogen gas storage

Fullerenes have the capability to easily hydrogenate and dehydrogenate because of its remarkable atomic structure. At the point when fullerenes are hydrogenated, the C=C twofold bonds progress toward becoming CC single bonds and C-H bonds. The H-C bond is lighter than the C=C bond and so the H-C easily disintegrates with the increase of temperature and the shade of the hydrogenated fullerene changes from dark to darker, at that point to red, orange, furthermore, light yellow with expanding hydrogen content. Fullerenes with up to 6.1% hydrogen content have been produced tentatively. A potential utilization of fullerene hydrides is in hydrogen gas stockpiling gadgets for electric vehicles that would utilize a power module. At present accessible hydrogen stockpiling advances like packed gas or capacity as metal hydrides are conceivably risky and additionally have low hydrogen stockpiling densities.

7.5   Fullerenes as sensors

Fullerenes are used as sensors due to its good electron accepting properties. Fullerene-based interdigitated capacitors (IDCs) as of late have been produced to investigate sensor applications. This novel strong state sensor configuration depends on the electron tolerating properties of fullerene films and the progressions that happen when planar atoms communicate with the film surface. Fullerene science gives a high level of selectivity and the IDC configuration gives high affectability. The strong state compound sensor's little size, straightforwardness, reproducibility, and minimal effort make them an appealing contender for fullerene applications advancement. Investigations of IDC arrangements with fullerene films are ready to detect water in isopropanol with a determination of 40 ppm. These outcomes show the plausibility of utilizing fullerenes as specific dielectric films for IC substance sensors [3].

Conclusion

Fullerenes, the third type of carbon, have turned out to be essential atoms in science and innovation. Due to their exceptionally commonsense properties, fullerenes are a key subject on nanotechnology and mechanical research these days. The applications displayed in this paper are conceivable employments of the particles because of at least one of their exceptional properties. Fullerenes are utilized as a part of the present business as of now, for the most part in beauty care products, where they assume a critical part as cancer prevention agents. Fullerenes possess a unique structure and properties. This chapter has clearly explained about the structural properties of the fullerene. The bucky ball structure is the familiar structure with various remarkable properties. The fullerene has a distinct electrical, magnetic, lubricating, mechanical and vibrating property. The fullerene reinforced with polymer and ceramics has an enhanced functionality. The fullerenes with its composites are used in various places such as medicinal, electrical and several other domains.

References

[1]   E. Kantar, Superconductivity-like phenomena in an ferrimagneticendohedral fullerene with diluted magnetic surface, in: A. Pinczuk, Solid State Communications, Elsevier Ltd, New York, 2017, pp. 31-37. https://doi.org/10.1007/s10853-009-4187-z

[2]   S. Yoshimoto, J. Amano, K. Miura, Synthesis of a fullerene/expanded graphite composite and its lubricating properties, J Mater Sci. 45 (2010) 1955–1962.

[3]   E. Ulloa, Fullerenes and their Applications in Science and Technology. Introduction to Nanotechnology. Spring 2013.

[4]   S. B. Singh, A. Singh, The Third Allotrope of Carbon: Fullerene and update, Int J Chemtech Res. 5 (2013) 167-17.

[5]   S. Duri, A. L. Harkins, A. J. Frazier, C. D. Tran, Composites Containing Fullerenes and Polysaccharides: Green and Facile Synthesis, Biocompatibility, and Antimicrobial Activity, ACS Sustainable Chem. Eng., 5 (2017) 5408–5417. https://doi.org/10.1021/acssuschemeng.7b00715

[6]   P. Hebgen, A. Goel, J. B. Howard, L. C. Rainey, J. B. V. Sande, Synthesis of fullerenes and fullerenic nanostructures in a low- pressure benzene/ oxygen diffusion flame, Proceedings of the Combustion Institute, 28 (2000) 1397–1404. https://doi.org/10.1016/S0082-0784(00)80355-0

[7]   M. Safdar, Fullerene: Its definition, types and scope, 2010, Biotech articles, Retrieved from: https://www.biotecharticles.com/Nanotechnology-Article/Fullerene-Its-Definition-Types-and-Scope-469.

[8]   B.C. Yadav, R. Kumar, Structure, properties and applications of fullerenes, Int J Nanotechnol Appl. 2 (2008) 15–24.

[9]   M. Mojica, J. A. Alonso, F. Mendeza, Synthesis of fullerenes, J. Phys. Org. Chem. 26 (2013) 526–539. https://doi.org/10.1002/poc.3121

[10] A. Goel, P. Hebgen, J. B. V. Sande, J. B. Howard, Combustion synthesis of
      fullerenes and fullerenic nanostructures, Carbon, 40 (2002), Pages 177-182.
      https://doi.org/10.1016/S0008-6223(01)00170-1

[11] K. Koziol, B. O. Boskovic, N. Yahya, Synthesis of Carbon Nanostructures by
      CVD Method, in: N. Yahya, N. Yahya (ed.), Carbon and Oxide Nanostructures,
      Adv Struct Mater 5, Springer, Verlag Berlin Heidelberg, 2010, 23-49.

[12] Y. Yang, X. Liu, Y. Han, W. Ren, B. Xu, Ferromagnetic Property and Synthesis of
      Onion-Like Fullerenes by Chemical Vapor Deposition Using Fe and Co Catalysts
      Supported on NaCl, J Nanomater, 2011 (2011).
      https://doi.org/10.1155/2011/720937

[13] L. T. Scott, Methods for the Chemical Synthesis of Fullerenes, Angew. Chem. Int.
      Ed. 43 (2004) 4994 – 5007. https://doi.org/10.1002/anie.200400661

[14] A. W. Jensen, S. R. Wilson, D. I. Schuster, Biological Applications of Fullerenes,
      Bioorganic Med. Chem. 4 (1996) 767-779. https://doi.org/10.1016/0968-
      0896(96)00081-8

[15] G. Brusatin and R. Signorini, Linear and nonlinear optical properties of fullerenes
      in solid state materials, J. Mater. Chem. 12 (2002) 1964–1977.
      https://doi.org/10.1039/b202399g

[16] P. Prasanthi, G. S. Rao and B. U. Gowd, Mechanical Behavior of Fullerene
      Reinforced Fiber Composites with Interface Defects through Homogenization
      Approach and Finite Element Method, ISSN: 2005-4238 IJAST, 78 (2015) 67-82.

[17] A. Zettl and J. Cunnings, Elastic properties of fullerenes, in: M. Levy, H. Bass, R.
      Stern, Handbook of Elastic Properties of Solids, Liquids, and Gases, Academic
      press, 2001. https://doi.org/10.1016/B978-012445760-7/50037-X

[18] H. Kuzmany, J. Winter, Vibrational properties of fullerenes and fullerides, in: W.
      Andreon, The Physics of Fullerene-Based and Fullerene-Related Material,
      springer, 2000.

[19] M.S. Dresselhaus, G. Dresselhaus, Fundamental Properties of Fullerenes, in: P. C.
      Eklund, A. M. Rao, Fullerene Polymers and Fullerene Polymer Composites,
      Springer, Verlag Berlin Heidelberg.

[20] D. Jonsson, P. Norman, K. Ruud, H. A. Gren, T. Helgaker, Electric and magnetic
      properties of fullerenes, J. Chem. Phys.109 (1998).
      https://doi.org/10.1063/1.476593

[21] T.L. Makarova, B. Sundqvist, P. Scharff, M.E. Gaevski, E. Olsson, V.A. Davydov,
      A.V. Rakhmanina, L.S. Kashevarova, Electrical properties of two-dimensional
      fullerene matrices, Carbon 39 (2001) 2203–2209. https://doi.org/10.1016/S0008-
      6223(01)00036-7

[22] F. Gardea, D. C. Lagoudas, Characterization of electrical and thermal properties of carbon nanotube/epoxy composites, Composites: Part B. 56 (2014) 611–620. https://doi.org/10.1016/j.compositesb.2013.08.032

[23] T. L. Makarova, Magnetic Properties of Carbon Structures, Semiconductors, 38 (2004) 615–638. https://doi.org/10.1134/1.1766362

[24] P. Prasanthi, G. S. Rao, B. U. Gowd, Effectiveness of Buckminster Fullerene Reinforcement on Mechanical Properties of FRP Composites, Procedia Materials Science, 3rd International Conference on Materials Processing and Characterisation (ICMPC 2014), 6 (2014) 1243 – 1252.

[25] T. Saotome, K. Kokubo, S. Shirakawa, T. Oshima, H.T. Hahn, Polymer nanocomposites reinforced with C60 fullerene: effect of hydroxylation, J. Compos. Mater.45(25) 2595–2601. https://doi.org/10.1177/0021998311416682

Carbonaceous Composite Materials   Materials Research Forum LLC
Materials Research Foundations **42** (2018) 111-142   doi: http://dx.doi.org/10.21741/9781945291975-5

Chapter 5

Graphene Oxide Composites and their Potential Applications

Deepika Jamwal[1, 2] Surinder Kumar Mehta[1], Dolly Rana[2*], Akash Katoch[3*]

[1]Department of Chemistry and Centre of Advanced Studies in Chemistry, Panjab University, Chandigarh 160014, India

[2]School of Chemistry, Faculty of Basic Sciences, Shoolini University, Solan, H.P. 173212, India

[3]Centre for Nanoscience and Nanotechnology, Panjab University, Chandigarh 160014, India

ranadolly079@gmail.com*, katochakash16@gmail.com*

Abstract

Graphene oxide (GO) is the oxidized form of graphene which makes it hydrophilic in nature i.e., water soluble and significantly altered their various properties such as high surface area, mechanical stability, tunable electrical and optical properties. GO is easily manufactured via chemical treatment through oxidation and exfoliation by ultra-sonication of graphite with low-cost production than the other carbon related materials. GO is capable to form mono-layer sheet through different functional groups (hydroxyl, epoxy and carboxyl) which is present on the surface of GO on many substrates, making it an admirable candidate to coordinate with other materials such as biopolymer or polymers, metal or metal oxide, metal sulfide, magnetic materials, etc. which are significantly used in various potential applications including super capacitor, photocatalysis, removal of heavy metal ion, water purification, sensors, batteries, biomedical applications (antibacterial activity, cancer cell detection, etc.).

Keywords

Graphene, Graphene Oxide, Photocatalysis, Water Purification, Biomedical Applications

Contents

1.   Supercapacitors or electrochemical capacitors

Supercapacitors or electrochemical capacitors are well known for energy storage device applications, presenting higher power density ($\sim$10000 W.kg^{-1}), long cycle life over repeated charge-discharge (>100 000 cycles) and low cost of maintenance, which make them one of the most promising candidates for next-generation devices [1, 2]. Based on their energy storage mechanism, electrochemical capacitors can be mainly divided into two ways: In a first way, the energy is stored physically through the adsorption of ions on the electrodes surface which is known as electrochemical double-layer capacitors and in the second approach, energy storage is endorsed by rapid redox reactions ensuing among the active material used as electrode and the electrolyte which is known as pseudocapacitors [3].

The usage of transition metal oxides (MnO$_2$, RuO$_2$, NiO, etc.) as an electroactive material have shown better energy density up to 1000 F^{3}g^{-1} [4-6], although suffered from poor electrical conductivity, while the densely packed structure system limits their applications in the development of high-performance supercapacitors [7]. It is well recognized that one of the finest ways to resolve these difficulties is to mix the transition metal oxide with carbon materials (carbon nanotube or carbon nanofibers, graphene, conducting polymer) [8-11]. In the last few years, composites of transition metal oxide and graphene (including rGO) have been proven to show superior system for obtaining better supercapacitor performance, which actually arises from the improved electrical conductivity of composites and the additional electrical double-layer capacitance contribution from graphene [12].

Graphene have extremely large specific surface area and constitutes high electrical conductivity compare to other materials. In addition to this, excellent chemical or electrochemical or mechanical stability with higher flexibility remarkably count it as an ideal material for supercapacitors [13]. Moreover they have good dispersion in a wide range of solvents, particularly in water which is extremely desirable for solution processing and further functionalization or interaction with composites of polymer and inorganic materials [14]. Graphene act as active materials for energy storage devices which storing electrostatic charges on the electrode double-layer [15]. Due to the strong π-π interaction among the graphene sheets, which allow easying incorporation of various other materials such as conductive polymers, heteroatoms, carbon-based materials, and inorganic materials [16]. Graphene is a 2D material and could be an interesting substrate to improve the charge-storage capacity of transition metal oxide. Using graphene as a substrate for transition metal oxide which can improve the charge transfer through metal oxide materials, and likely the easier and quick ion diffusion between metal oxide/electrolytes results in the high specific capacitance [17]. However, due to the strong van der Waals interactions, graphene sheet oxide usually suffer from restacking of the neighboring components in the reduction process due to which the entire surface area of rGO cannot be completely utilized, and the specific capacitance could be lowered to a significant extent [18]. To address this issue, the current strategy is to introduce spacers in between graphene sheets to avoid their restacking. The introduction of metal or metal oxide, carbon-based materials (CNT), conductive polymers and their binary or ternary combination, etc. were reported as effective spacers between graphene layers to improve capacitance. Thus, the mixture of these aforementioned materials can produce an advanced composite system which possesses useful properties for electrochemical supercapacitors [9].

The molybdenum based metal-organic porous composites frameworks prepared by simple mixing of MoO_3 with rGO sheets and has been utilized as the electrode material for the preparation of entire symmetric supercapacitor devices those mainly belong to or termed in solid-state, flexible category. The rGO film wrapped MoO_3 porous composites shows that the hierarchical porous structure decreases the diffusion length of electrolytic ions as well as delivers extra lively surface region for the redox reaction, leading to an improved electrochemical presentation. The structure of rGO/MoO_3 porous composites electrode offered electrically conductive systems for charge transport and also avoid the agglomeration of MoO_3 which results in extended cycle life in supercapacitors which are essentially required for practical applications in energy-storage devices [19].

In another study, rGO/MoS_2 composite was synthesized by growing the layered MoS_2 on rGO sheets. The fabricated composite showed relatively superior performance towards charge storage with high specific capacitance and high energy density. Specific capacitance was depending on the optimum amount of $MoCl_5$ precursors in the reaction condition and found high specific capacitance (265 at 10 mV s^{-1}) and the reason for this behavior is due to both the faradic and non-faradic processes of the MoS_2 layers combined with graphene sheets, The supercapacitor electrode also showed higher cyclic stability with 92% of the specific capacitance retained after 1000 cycles [20]. A similar approach was used for fabrication of the graphene/Fe_2O_3 composite hydrogels and was investigated as high-performance anode materials for supercapacitors. Here, single-crystalline self-assemble Fe_2O_3 particles were directly grown on flexible graphene sheet to form interconnected porous structures with high specific surface area, which intensely facilitate charge and ion transport in the complete electrode. This porous structure of electrode displayed extremely high specific capacitance of 908 F.g^1 at 2 A.g^{-1} within the potential range from 1.05 to 0.3 V and an excellent retention capability about 69% retention at 50 A.g^{-1}. Additionally, the cycling performance is clearly much better for the graphene/Fe_2O_3 composite hydrogels than that for bare Fe_2O_3 [21].

In the past few years, some efforts have been made to develop ternary composites, in order to improve the performance of supercapacitors. GO/Polyaniline/MnO_2 ternary composites were proposed where MnO_2 nanorods were incorporated to intercalate PANI/solution-exfoliated GO nanosheets. GO/Polyaniline/MnO_2 composites with different content of MnO_2 were calculated as electrode materials for supercapacitors. Results found that the porosity of electrode (surface area 91.37 and 73.65 m^2/g) was increased as increased the mass percentage of MnO_2 from 46 wt. % to 70 wt. % in the electrode, respectively. It has been found that the MnO_2 nanorods enable the charge transport in energy storage application. The electrode with 70 wt. % MnO_2 exhibits a highest specific capacitance of 512 F.g^1 at 0.25 A.g^{-1} current density. Polyaniline conductive coating on electrode exhibited high electrochemical stability in cycling, with, 97% capacitance retention observed after 5100 cycles [22].

2. Lithium-ion batteries

In Lithium-ion batteries, carbon-based materials such as graphite, graphene including its derivatives and composites with other materials have shown their potential as a lithium-accepting anode and Li$^+$ ions continuously travel between a lithium-releasing cathode. Generally, the capacity of the battery is defined by the amount of Li$^+$ ions accommodated per gram of material where a layered lithium-metal-oxide usually used as cathode material [23, 24]. In carbon-based materials, single-layered graphene can host Li$^+$ ions by two times more as compared to conventional graphites and other carbon-based materials [25]. Lithium storage by graphene-based anodes is strongly depended on the structure, morphology, composition, fabrication method of electrode ionic diffusion kinetics, electrical conductivity, as well as surface characteristics of the electrode material [26]. In recent time, chemically modified graphene with transition metal oxide (CuO, NiO, SnO₂, TiO₂, VO₂ etc.) has been an attractive choice for synthesizing composites with the purpose of improving their battery capacities [27-31]. In most reported studies, rGO is the excellent material for Li⁺ ions storage and showed an extremely high-capacity value during the Li⁺ insertion, which is higher than the single-layer graphene [32, 33]. Owing to excellent electrical conductivity, tunable porosity and higher mechanical strength of 3D graphene-based materials as an electrode for Li⁺ ions batteries have been attracting increasing attention in the field of energy storage [26].

Graphene/SnO₂ composite was synthesized via a one-step hydrothermal method where microwaves were used for the commencing of the reaction system and the size of SnO₂ nanoparticle ~ 3.5 nm was deposited on graphene sheets. The quick charge and discharge capacities at a current density of 100 mA g⁻¹ were 2213 and 1402 mA h g⁻¹ with Coulomb efficiencies of 63.35%. The discharge specific capacity remains at 1005 mA h g⁻¹ after 100 cycles at a current density of 700 mA g⁻¹. Moreover, at a high current density of 1000 mA g⁻¹, the first discharge and charge capacities were 1502 and 876 mA h g⁻¹, and the discharge specific capacities remain 1057 and 677 mA h g⁻¹ after 420 and 1000 cycles, respectively. The study suggested that the SnO₂/graphene composite have stable cyclic performance and high reversible capacity for energy storage device applications [34].

In recent times, silicon has widely been explored as a high-capacity anode material because of its known capacity value which is on the higher side compared to the other materials (4200 mAh g⁻¹), nearly ten times higher than that of graphite (372 mAh g⁻¹). However, silicon's performance as an anode material is stalled due to the large volume changes (4300%) and low intrinsic charge transport on the upsurge of lithium and discharge from silicon, leading to intensive pulverization of silicon nanoparticles and fast capacity decrement. In addition, for the synthesis of graphene/ nanosized silicon composites generally required high temperature which repeatedly results in the conglomeration of silicon nanoparticles. In order to solve aforementioned issues as well as to improve the electrochemical performance of silicon anodes, the use of carbon-based materials can be a solution, because it gives an opportunity to limit the extensive variation in the volume of silicon in order to retain the electrode veracity as well as to improve the electrical conductivity [35-37]. A study reported silicon nanoparticles intercalated uniformly into graphene sheets results in the formation of a hierarchical micro and nanostructures. Here, the graphene act as the backbone of the system by controlling the charge transport as well as act as an elastic buffer for uniformly intercalated the silicon nanoparticles. The uniform intercalation silicon nanoparticles into graphene were done by combine processes including freeze-drying and thermal reduction process. Composite electrode exhibit remarkably improved cycling performance (1153mA h g-1 after 100 cycles) and rate performance in comparison with pure Si nanoparticles [38].

Pyrite iron sulphide is an interesting cathode or anode material for lithium batteries due to its abundance in nature and its cost-effectiveness. As per theoretical study, it has a high theoretical capacity of 890 mA h g^{-1} supposing four electron transfer, therefore has an potential to be used as electrode material, however, reduction of pyrite Fe_2S with lithium forms polysulphide Li_2S_x ($2 < x < 8$), and have eased to dissolution in the liquid electrolytes and disrupt the electrical exchange between the active material and current collector, which leads to a poor cycle performance and hinders its real-time use in lithium batteries. In order to resolve this problem, some current studies showed that covering sulfur over graphene sheets can significantly lower the dissolution of polysulphide into the electrolyte and enhances the cycling stability and rate performance [39]. For example, rGO/FeS$_2$ composite prepared via the one-step solvothermal method and used as an anode in lithium-ion batteries. Pyrite FeS_2 microspheres wrapped by reduced graphene oxide and showed high specific capacitance of 970 mA h g^{-1} at a current density of 890 mA g^{-1} after 300 cycles. Moreover, this anodic electrode displays a remarkable capacity of 380 mA h g $^{-1}$ inevitably at high current densities of 8900 mA g^{-1} over 2000 cycles with long cycling life [40]. The use of sulfur in lithium-sulphur batteries is the effective way when used as a lithium-accepting anode, the resultant theoretical energy density is 2600 Wh kg^{-1}. The usage of sulfur is advantageous in term of easily available in the earth crust, cost-effective and has multielectron transfer redox chemistry. However, there are still many difficulties; for example, the shuttle of polysulfide induces rapid capacity degradation and poor cycling stability. These issues were resolved by introducing the conductive porous frameworks which reduce the sulfur composite based cathodes by means of large charge transport activity as well as enough room for the volume expansion. In addition, also restrict the transportation of polysulfides by the confining the polysulfides with the help of hierarchical pores and resilient interfacial coupling. In another study, a distinctive lithium-sulfur based battery structure was proposed and was recommended with an ultrathin graphene oxide (GO) membrane for obtaining maximum stability [41]. The GO surface was modified with the electronegative oxygen atoms and functionalized with the carboxyl groups acted as ion-hopping sites of positively charged lithium species (Li^+) and stopped the transference of negatively charged species because of the electrostatic interactions. These aforementioned activities extensively hampered the transfer of polysulfides through the GO membrane. Therefore, the GO membrane can be recognized as an efficient perm-selective separator system in lithium-sulfur batteries.

The addition of GO effectively lowered the cyclic capacity decay rate from 0.49 to 0.23% per cycle. It was suggested that the GO membrane likely to restrict the transfer of polysulfides through the GO membrane pores, and was also beneficial for anti-self-discharge properties.

Currently, sodium ion batteries have fascinated much consideration of research curiosity due to the sodium being present in huge abundance and can easily be extract from the earth crust and its cost-effectiveness as compared to Li [42]. Significantly, sodium has similar physical and chemical properties as in Li elements. Sodium ion batteries have shown their credential for a large-scale storage system. The stored energy can be sufficiently utilized for renewable energy applications such as solar and wind where energy is generated as compared to lithium-ion battery [43]. However, compared with lithium ions batteries, sodium ion batteries have less worthy specific capacity, rate capability and good reversibility. On the other hand, Li-ion batteries have a higher energy density, and capability to accumulate a bigger quantity of energy in small dimensions, but due to the disadvantages, such as higher cost, low cyclic stability life and safety concerns limit their universal application. The huge size of sodium ions (1.02 Å), which is nearly twice to the lithium-ions (0.59 Å), that leads to slower Na^+ diffusion efficiency, larger volumetric change and severe pulverization of the lithium-accepting anode material. It may also result in a less stable solid-electrolyte interface layer, therefore, consequences is a massive failure in the cyclic stability of anode materials [44]. Several attempts have been dedicated to developing the anode material that would give sodium-ion batteries a longer life. Many carbon-based materials have been intensively studied as anode materials for sodium-ion batteries [45].

In a recent year focus has been given to metal chalcogenides/ carbon-based hybrids for obtaining better performance. rGO/MoS_2 composites with the optimal loading of rGO (0.4 g) were reported as a potential electrode material for rechargeable sodium-ion batteries than that of pure MoS_2, as they exhibit a maximum reversible specific capacity of about 305 m Ahg^{-1} at a current density of 100m Ag^{-1} after 50 cycles with excellent rate performance. The better performance was attributed to the porous conductive network structure that will accommodate the volume expansion, the increased specific surface area and the lowered charger transfer resistance due to the incorporation of rGO [46]. C-$MoSe_2$/rGO composite was prepared with both high porosity and the large surface area in which $MoSe_2$ nanosheets were covered by a carbon layer and also intensely rooted on the interconnected rGO network. The prepared composite was used as an anode in sodium ion batteries and remarkably enhanced the sodium ion storage capacity, with a high specific capacity of 445 $mAhg^{-1}$ at 200 mAg^{-1} after 350 cycles and 228 $mAhg^{-1}$ even at 4 Ag^{-1} and these values are much better than those of C-$MoSe_2$ nanosheets [47].

The 3D MoS_2 nanoflowers/rGO composite was successfully synthesized through a simple ultrasonic exfoliation technique. Further, it was tested as anode material for sodium-ion batteries and demonstrates a high reversible specific capacity of 575 mAhg^{-1} at 100 mAg^{-1} in between 0.01 V-2.6 V and 218 mAhg^{-1} at 50 mAg^{-1} when discharged in a potential range of 0.4 V-2.6 V. Notably this material showed exceptional reversible Na-storage capacity and decent cycling stability [48]. Sb_2S_3/sulfur doped graphene composite was reported as superior materials than the other antimony based materials with superior and a stable capacity retention of 83% for 900 cycles with high capacities and excellent rate performances. Resultant composites have a robust architecture which gives outstanding cycling stability that resolves the long-term cycling stability issue for sodium ion batteries [49]

3. Glucose sensors

Graphene-based composites have remarkably been used for glucose estimation in human blood samples of diabetes mellitus affected patients. Graphene-based materials are well known for non-enzymatic glucose sensors which are found to be an upfront system with superior stability and sensitivity with higher selectivity efficiently differentiate extent of glucose level in human blood samples [50, 51]. However, several methods have been used to prepare glucose sensors which rely on the glucose oxidase enzyme. But one of the major issues is immobilization of the glucose oxidase enzyme being a protein nature on the surface of the electrode. Moreover, their practical application limits due to their dependency or sensitivity towards temperature, pH, and humidity, results in the lacks in overall system stability [52, 53]. To overcome these problems, numerous efforts have been focused to realize non-enzymatic glucose sensors which are founded to overcome the aforementioned concern over immobolization of the glucose over the surface of the electrode by mean of the direct catalytic oxidation of glucose on the surface of the electrode. Various catalytic materials including different metal oxides (CuO, NiO, Co_3O_4) [54-56] and metal nanomaterials (Au, Ag, Ni, Pt, Pd) [57-61] are utilized on the electrode surface as oxidation materials for oxidation of glucose in non-enzymatic glucose sensors. The efficiency of non-enzymatic glucose sensors can be influenced by the extent of oxidation capacity of the material for glucose. The material should have important properties including a higher specific surface area with better electron transport property, non-toxicity, and chemical stability with better mechanical flexibility. However, most of the metal oxide and metal nanomaterials are a poor conductor and easily drop their performance due to the tendency of nanoparticles towards aggregation and the adverse influence of the chemisorbed chemical species which limits their use in sensors development. Therefore, in order to overcome this limitation, these materials are

Carbonaceous Composite Materials   Materials Research Forum LLC
Materials Research Foundations **42** (2018) 111-142   doi: http://dx.doi.org/10.21741/9781945291975-5

dispersed with proper conducting support (such as activated carbon, carbon nanotube-based materials and graphene-based materials), which would be an effective way to reduce the aggregation between the nanoparticle and provides conducting pathway [62-64].

Graphene-based materials with two-dimensional structure exhibit a high density of defect sites. The defect sites act as active sites and improve electron transport between graphene to the chemical and biological species [65]. Due to these fascinating features graphene is an excellent material system suitable for detection of different types of biological and chemical species which includes the glucose, dopamine, ascorbic acid, H_2O_2, uric acid, protein, DNA, nicotinamide adenine dinucleotide, cholesterol, histidine, organosulfate pesticides etc. [66]. The synergic effect between graphene and the upcoming moieties are mainly supported by the ease of graphene towards functionalization by chemical or biological species having covalent or non-covalent bond making it easier to use as composites with another matrix system of polymer, metal nanoparticles and metal oxide nanoparticles. As mentioned, the two-dimensional structure is also contributive towards dispersion of abovementioned nanomaterials results in better catalytic activity and detection ability.

The composites of graphene oxide decorated with silver nanoparticles were prepared by the anodic dissolution of silver in the aqueous dispersion of graphene oxide and resultant composites showed enzyme-less amperometric glucose sensing for real blood samples. The results showed that the detection limit for glucose was 4μM and very much selective in the existence of uric and ascorbic acid, which is considered as interfering molecules for glucose detection. The presence of graphene with silver nanoparticles results in a better sensors response in comparison to the pure silver nanoparticles [67].

In another study, the reduced graphene oxide/copper nanoparticles were prepared on a glass/Ti/Au electrode via using an electrophoretic deposition technique. The mixture of graphene oxide and copper sulphate were prepared in the form of colloidal suspension and their capability towards the detection of glucose in alkaline medium, respectively. Interestingly, the efficiency of Au electrode modified with reduced graphene oxide/copper nanoparticles exhibited superb electrocatalytic performance for glucose oxidation with a sensitivity of 447.65 μA mM^{-1} cm^{-2} as compared to Au electrode modified only with reduced graphene oxide. The resultant sensor showed detection limit is 3.4 mM and displayed negligible inference while crossed investigated in the existence of various oxidizable and carbohydrate molecules (dopamine, uric acid, ascorbic acid fructose, lactose and galactose) which specifies the superior selectivity of electrode [66].

Chang *et al.*, adopted a facile and clean method to synthesized nanocomposites of graphene and platinum nanoclusters. During the synthetic step, polyvinyl pyrrolidone added as a dispersing agent, which directed to improved dispersity as well as controlled the size of platinum nanoclusters on graphene support and further, evaluated their electrocatalytic activity towards the oxidation of glucose in neutral media. The results showed platinum nanoclusters/graphene exhibited a fast response time (~3 seconds) with a detection limit of 1.21 μAcm^{-2} mM^{-1}. Moreover, the influence of the oxidation of unwanted chemical species can be restricted to a certain extent by selecting the suitable detection potential [68].

The copper oxide/polypyrrole nanofiber/reduced graphene oxide nanocomposite based sensor was used for the detection of glucose. The sensors presented remarkable reproducibility, stability, selectivity properties and limit of detection is 3µM which is higher than the non-enzymatic glucose sensors based on copper oxide, polypyrrole and reduced graphene oxide [69].

Lu *et al.*, fabricated a composite film of nickel oxide hollow spheres/reduced graphene oxide /nafion modified glassy carbon electrode and magnificently utilized for the selective detection of glucose. The composite fits better in term of reflecting better catalytic activity with selectivity. Moreover, the long-term stability towards oxidation of glucose as well as a higher rate of reproducibility showed its potential towards sensing applications. The limit of detection is 0.03 µM, which is higher than the bare glassy carbon electrode, reduced graphene oxide/nafion modified glassy carbon electrode, and nickel oxide hollow spheres/nafion modified glassy carbon electrode. Furthermore, the composite was effectively capable of determining the glucose in real serum specimen under the required limits [70].

4. H_2O_2 sensors

The development of H_2O_2 sensor is needed and highly considerable because of their immense consumption in the field of food, paper, biomedical and chemical industries. The H_2O_2 is effective oxidizing agents and their direct and indirect participation in our daily life because of its direct involvement in charge transportation in the biological systems, its availability in the numerous cleaning goods or in the form of by-product in various chemical reactions. So, tremendous efforts have been devoted for the development of reliable, subtle, fast, and low cost effective H_2O_2 sensors. Several techniques such as surface plasma resonance, fluorescence, colorimetry, chemiluminescence, chromatography, electrochemical method etc. have been used for detection of H_2O_2.

Carbonaceous Composite Materials   Materials Research Forum LLC
Materials Research Foundations **42** (2018) 111-142   doi: http://dx.doi.org/10.21741/9781945291975-5

Among them, the electrochemical method is advantageous because of its better detection capability, cost towards the lower side, ease of operation and their prospect for real-time analysis. However, the enzyme-based H_2O_2 sensor is acute for the environment, because of which sensors stability is compromised in term of fluctuation in sensor response and rate of reproducibility. To overcome these challenges, many efforts have been made without using enzymes. Various kinds of nanomaterials (metallic and metal oxides nanoparticles) with high surface area and superb catalytic capability are immensely used to modify electrode to accomplish the non-enzymatic based H_2O_2 sensors. Recently, the combination of graphene-based materials produces a synergistic effect which results in higher catalytic activity, enhanced conductivity and improved stability of the metallic and metal oxides nanomaterials. As an amperometric sensor for detection of H_2O_2, silver nanoparticles decorated graphene oxide via electrochemical reduction method on glassy carbon electrode by an amperometry method and utilized. The modified electrode exhibit high sensitivity and selectivity with a low detection limit (0.085 mM) [71].

On the other hand, silver nanoparticles decorated 3D-graphene composite synthesized through the hydrothermal process. Interestingly, the composite was straightaway fabricated over freestanding sensing electrode for non-enzyme H_2O_2 detection in phosphate buffered solutions. The results revealed that the composite exhibited fast amperometric detection, highly selective, low detection limitation (14.9 µM) for H_2O_2 detection was attributed to the synergistic effect of silver nanoparticles for showing high electrocatalytic activity and 3D-graphene for the large surface area and better charge transport [72].

Woo *et al.*, designed a graphene/multiwalled carbon nanotube composite modified electrode for obtaining a combination of the large surface area with excellent charge transport activity for H_2O_2 detection. Graphene/multiwalled carbon nanotube composite results synergistic effect, which promotes the electro catalytic reduction of H_2O_2 and the limit of detection is estimated, is 9.4×10^{-6} mol L^{-1} [73].

In another study, paper-like 2D reduced graphene oxide-supported Cu_2O and 3D-reduced graphene oxide aerogel-supported Cu_2O composites were synthesized by hydrothermal and filtration process. The results demonstrated that the 3D-reduced graphene oxide aerogel-supported Cu_2O composite showed superior electrocatalytic activity towards the reduction of H_2O_2 results in the excellent detection capability for H_2O_2 with a detection limit as low as of value 0.37 µM, respectively. Meanwhile, the paper-like 2D reduced graphene oxide/Cu_2O composite also showed worthy response towards H_2O_2 detection with high sensitivity and selectivity. Both composites correspondingly displayed the real-time detection of H_2O_2 in human serum, which reflects the potential of the fabricated H_2O_2 sensor, respectively [74].

The non-enzymatic H_2O_2 sensor was developed using a rigid chain liquid crystalline polymer and reduced graphene composite. Polymer offered superb electrochemical performance and amended electrode fabricated with composite displayed superior electrocatalytic activity towards the reduction of H_2O_2 with a sensitivity 117.142 mA mM^{-1} cm^{-2} and detection limit is 1.253 µM, respectively [75].

In continuation, the one-step hydrothermal method was used to successfully synthesize silver nanoparticles/reduced graphene oxide/multiwalled carbon nanotube composite without reducing agent and composite showed superb electrocatalytic activity for the reduction of H_2O_2 with a quick response time of ~ 3 s. The electrocatalytic activity was highly influenced by the amount of silver ammonia solution in the prepared nanocomposites. Meanwhile, the maximum electrocatalytic activity response was witnessed for the nanocomposite with 6:1 volume ratios of multiwalled carbon nanotube-graphene oxide (3:1, v/v) to Ag $(NH_3)_2OH$ (0.04 M). The limit of detection was estimated to be 0.9 µM [76]. Another work was reported on the preparation of reduced graphene oxide /zinc oxide composite over glassy carbon electrode. The green synthesis approach was applied where simultaneous electrodeposition of ZnO was performed with electrochemical reduction of graphene oxide. The composite shows superb electrocatalytic activity towards H_2O_2 with a detection limit of 0.02 µM and fast response time of less than 5s [77].

Venegasa *et al.*, modify glassy carbon electrodes by preparing the cobalt doped stannates/reduced graphene oxide composites on it. The fabricated electrodes were tested for the amperometric evaluation of H_2O_2. Notably, the concentration of cobalt in the stannates compound as well as on the content of reduced graphene oxide present in the composite influences the catalytic activity of the composite towards the oxidation of H_2O_2. The pure cobalt stannate composite with a ratio of 8:1 (stannates: reduced graphene oxide) showed the best catalytic activity towards H_2O_2 and a detection limit of 0.31 µM [78]. Fabricated cuprous oxide/reduced graphene oxide nanocomposites via three approaches including physical adsorption, in situ reduction and one-pot synthesis, respectively. The resulted composites have different morphologies and exhibited considerably enhanced performance for the catalytic reduction of H_2O_2 than the bare Cu_2O. Among all three kinds of cuprous oxide/reduced graphene oxide nanocomposites, the nanocomposites synthesized via simple physical adsorption approach showed higher sensitivity (19.5 µA/mM) with detection limit is 21.7 µM and better stability as compared to the other two composites [79].

Carbonaceous Composite Materials   Materials Research Forum LLC
Materials Research Foundations **42** (2018) 111-142   doi: http://dx.doi.org/10.21741/9781945291975-5

5.   Photodegradation of organic pollutants

Recently, Combination of graphene oxide (GO and RGO) and semiconductors (ZnO, Cu_2O, TiO_2) have widely been used as advanced photocatalyst because of their superior charge transfer interaction, magnetic and electronic interaction between graphene sheets and semiconductor materials. Graphene sheet has the capability to receive the photoexcited electron from the conduction band and suppress the recombination of electrons and holes. Moreover, the higher surface area of graphene sheets causes the semiconductor materials to disperse which accomplished to support the growth and avoid agglomeration of the nanoparticles [80-83].

Xu et al., the synthesized a network structure of RGO coated ZnO flower via a simple one-pot hydrothermal method and used as photocatalyst for degradation of methylene blue. The results found that increasing the contents of GO (up to 2wt %) into ZnO could enhance the photocatalytic activity and moreover, RGO/ZnO composites have enhanced photocatalytic efficiency compared to the pure ZnO flowers [84].

Azarang et al., used the sol-gel method to prepare ZnO nanoparticle decorated on rGO in a gelatin medium. The role of gelatin was to terminate the growth of the ZnO nanoparticles on GO and stabilize them. The resulting composites were used for photodegradation of methylene blue. The results found that ZnO/rGO exhibits high MB degradation efficiency (99.5%) in comparison to the ZnO nanoparticles (63%) [81]. In continuation, they synthesized ZnO/rGO nanocomposites using a sol-gel method in a starch medium. Starch was utilized to stop the growth of the ZnO nanoparticles on rGO and stabilize them. Further, ZnO/rGO nanocomposites used as photocatalyst for degradation of methylene blue and results of photoluminescence spectroscopy showed that introduction of rGO sheet into ZnO nanoparticles suppressed the electron-hole recombination of the composite which results in enhanced photocatalytic degradation of MB compared to bare ZnO nanoparticles [80].

Zou et al., prepared different Cu_2O/rGO composites with Cu_2O having different crystal facets ($\{1\ 1\ 1\}$, $\{1\ 1\ 0\}$ and $\{1\ 0\ 0\}$) with nearly same particle sizes and surface areas. Different composites were used as photocatalysts for degradation of MB and further, compared their photocatalytic efficiency and the result is shown in order of Cu_2O $\{111\}$ ZnO/rGO>Cu_2O $\{110\}$ ZnO/rGO>Cu_2O $\{100\}$ ZnO/RGO.

The UV-vis diffuse reflectance and photoluminescence spectra investigations revealed the improvement in the visible-light absorption and the faster charge-transfer rate by Cu_2O $\{111\}$ ZnO/rGO nanocomposite. FTIR, X-ray photoelectron spectroscopy and Raman analysis also suggest that because of the developed electronic structures and interfacial connections of $\{111\}$ ZnO/rGO, the Cu^+ species and oxygen defects were much easier to occur. In addition, the presence of *In-situ* electrons spin resonance speculated detection of more super-oxide radicals over {111} ZnO/rGO, which promoted organic pollutants degradation [82].

TiO_2/rGO nanocomposite synthesized via the hydrothermal method. TiO_2 have nano spindle structure which is featured by large exposed {001} facets and further evaluated their photocatalytic activity by the degradation of MB. The result showed the photocatalytic efficiency is much higher than that of pure TiO_2 nano spindle under the same conditions because of superior electron conductivity and enhanced absorbtivity of rGO. In addition, cycling use of the nanocomposites indicates that the stability is high [85].

The ternary nanocomposite of rGO/ZnS-Ag_2S was successfully synthesized using the hydrothermal method and accessed the efficient photocatalyst and good recycling ability for the degradation of RhB under simulated sunlight irradiation compared to pure ZnS. This was attributed to an efficient charge transfer from ZnS to Ag_2S and graphene sheets. Strong π-π bonding between graphene sheets provided better structural stability by means of the superior mechanical strength and porous structure which is likely to be an ideal requirement for that for immobilizing the many nanoparticles which can provide huge water/solid interfaces. This leads to promote the easy retrieval of the organic dyes due to a interlinked porous structure [86].

In another work, successful preparation of Ag/rGO composites hydrogels in the presence of polyethyleneimine was reported and the prepared composite was used as a potential catalyst for degradation of RhD and MB dyes under UV light. They were found Ag/rGO composites can create more number of electrons and holes which results in the formation of additional superoxide anions or peroxides species thereby enhanced the efficiency of Photocatalysts [87]. The same composite system was developed by Bhunia *et al.*, Ag/rGO composites using a simple and low-cost chemical route and has shown commendable efficiency towards the photocatalyst degradation of colorless organic pollutants (phenol, bisphenol A, and atrazine) under visible-light. They found that photocatalytic efficiency of Ag/rGO composites under visible light was pointedly better than the bare rGO or silver nanoparticle and be influenced by on the optimized content of Ag. Generally, Ag nanoparticles are capable of offering the induced excitation of silver plasmons under visible-light, on the other hand, the conductive rGO provides better charge separation which indeed required to promote the oxidative degradation of the organic pollutant [88].

GO/Ag@AgCl composites as Z-scheme photocatalyst was used for the degradation of MB. They found that GO and AgCl stimulated the catalytic activity and metallic Ag considerably lowers the recombination of electron-hole pairs from GO and AgCl under exposure to visible light. They suggested that the electron-hole recombination follows in two ways; electron-hole pairs of low energy level can undergo recombination by Ag, on the other hand, the electron-hole pairs of the high energy level are available for the two photochemical reactions that work parallel in the system. GO/Ag@AgCl composites show an enhanced photocatalytic activity for the degradation of MB than the Ag@AgCl-rGO while the content of the graphene was the same (15 wt%) [89].

In the past two decades, many researchers have focused on materials (peroxymonosulfate oxidant (PMS)) that are able to generate sulfate radical ($SO4^{-\cdot}$) under UV irradiation which plays a key role in the degradation process. PMS has a higher potential (1.82 V), are cost-effective and eco-friendly [90]. Yao $et\ al.$, prepared $rGO/ZnFe_2O_4$ composite and evaluated their catalytic activity for degradation of orange-II using peroxymonosulfate as an oxidant under visible light irradiation. The $rGO/ZnFe_2O_4$ composites displayed remarkable enhancement in visible-light photoactivity toward orange-II in comparison to the bare $ZnFe_2O_4$ [91].

The introduction of graphene oxide into silver orthophosphate was found beneficial to improve separation efficiency of photogenerated electron-hole pairs for increasing the overall photocatalytic activity [92, 93]. Wang $et\ al.$, prepared GO/ Ag_3PO_4 composites via one-step route and further evaluated their photocatalytic activity for the degradation of RhB and Bisphenol A. The outcome suggest that the overall catalytic activity and good stability depend on the introduction of GO (6wt%) and the main reason is to promote the charge separation, therefore catalytic degradation efficiency of Bisphenol A over GO/Ag_3PO_4 was higher than that of Ag_3PO_4 [94]. In another study, RGO/ Ag_3PO_4 composites were synthesized in situ deposition of Ag_3PO_4 nanoparticles on the surface of RGO sheets and evaluated their stability and photocatalytic activity for the degradation of RhB of composites. Results revealed that the amount of rGO up to 3 wt% in the composites significantly enhanced catalytic activity, which could be ascribed to the superb charge compliant and transport properties of rGO under visible-light irradiation [95]. In extension to this an excellent photocatalytic degradation of RhB, MB and methyl orange under visible light was reported using GO/ Ag_2CO_3 composites which were prepared via modest and operative precipitation method. The Ag_2CO_3 with 0.9 wt% GO concentration showed the photodegradation efficiency for organic dyes and it was 2 times higher than the bare Ag_2CO_3 crystal. The enhanced photocatalytic activity of GO/Ag_2CO_3 composites was ascribed to the mutual contribution of GO sheets with high surface area, improved absorption of organic dyes, optimum band gap, efficient charge separation and decent charge acceptor properties of GO [96].

Carbonaceous Composite Materials   Materials Research Forum LLC
Materials Research Foundations **42** (2018) 111-142   doi: http://dx.doi.org/10.21741/9781945291975-5

The degradation of the MB and phenol under visible-light irradiation by Ag_2CrO_4/GO composite prepared via a precipitation method which acts as a Z-scheme photocatalyst. Experimental results shown that, composite with 1.0 wt% GO content displays the photocatalytic activity for MB degradation as fast as within 15 min, and was 3.5 times that of bare Ag_2CrO_4 and could be achieved by the creation of Z-scheme Ag_2CrO_4/GO heterojunctions that own better efficiency towards charge separation and transfer along with the resilient oxidation and reduction capability [97].

Several researchers have focused their attention towards the use of $BiVO_4$ materials with different crystal structures as well as various three-dimensional morphologies as photocatalysts for the degradation of organic pollutants under visible-light irradiation. Nevertheless, pure $BiVO_4$ have limitation including low adsorptive ability and difficulty in passage of photo-generated electron and holes, respectively [98]. In this concern, some researchers have been reported some publications on the incorporation of pure $BiVO_4$ with graphene oxide materials and they found as a reliable way to develop the photocatalytic efficiency for degradation of organic pollutants under visible light irradiation [99]. Dong *et al.*, synthesized three-dimensional acicular sheaf shaped $BiVO_4$ architectures and incorporated with rGO which resulted in $BiVO_4$/rGO composites. Composites displayed outstanding photocatalytic activity towards the degradation of RhB and stability during prolonged exposure to natural sunlight. Additionally, results revealed that the photocatalytic activity of the composites improved with respect to the rGO content ranging from 0.25 wt% to 1 wt% in the composites. The reason behind enhancement was the higher light harvesting efficiency and reduction in rate of electrons and holes recombination [100].

Table 1 Photocatalytic degradation of Graphene oxide and reduced graphene oxide-based composites used for the degradation of organic pollutants.

S.No.	GO & RGO composites	Organic pollutants	Light source	Photocatalytic efficiency	Conc. of photocatalyst & organic pollutant	Method	Ref.
1	ZnO/ (rGO)	MB	UV irradiation produced by a 500 W, high-pressure Hg lamp	92.5% in 120 min	10 mg of the the resulting material was dispersed in 30 mL of the MB aqueous solution (10 mg/L).	Sol–gel	[80]
2	ZnO/ (rGO)	MB	UV irradiation produced by a 500 W high-pressure Hg lamp.	99.5% in 180 min	10 mg of the obtained material was dispersed in 30 ml of the MB aqueous solution (10 mg/L).	Sol–gel	[81]
3	Cu_2O-rGO	MB	Visible light was provided by the 400 W metal halide lamp	72% in 2 hrs.	15 mg of sample was added to 45 mL MB (10 mg/L)solution	---	[82]
4	TiO_2/rGO	MB	Light irradiation by a 300W xenon lamp.	---	10 mg of the as-prepared samples were suspended in 60 ml of MB aqueous solutions (10 mg/L).	Hydro-thermal	[85]
6	rGO / ZnS-Ag_2S	RhB	Solar simulator equipped with an AM 1.5G filter and 150 W Xe lamp.	54.08% in 120 min.	100 mg of photocatalyst was suspended in a 100 mL aqueous solution of RhB	Hydro-thermal	[86]
7	Ag/rGO	Phenol, BPA, and Atrazine	250 W Hg vapor lamp was used for	---	22 mg of rGO–Ag composite was	---	[88]

			visible light		dispersed in the 50 mL of pollutant solution with a conc. of 100 mg/L		
8	Ag/rGO	RhB and MB	UV light high pressure mercury lamp (365 nm, 100 W)	100%70 min, 100% 30 min	rGO/PEI/Ag hydrogel catalyst in 100 mL of RhB (4 mg/L) and MB solution (10 mg/L).	Reduction approach	[87]
9	GO/Ag@ AgCl	MB	Visible light irritation the LED lamp	---	30 mg were dispersed in 50 mL 2.5 x10^{-5}M MB solution under	---	[89]
10	rGO/ ZnFe$_2$O$_4$	Orange II	Visible irradiation	100 % of Orange II was degraded by the ZnFe$_2$O$_4$–rGO within 150 min.	ZnFe$_2$O$_4$–rGO (20 mg) powder was added into Orange II (20 mg/L, 100 ml)	---	[91]
11	GO/ Ag$_3$PO$_4$	RhB, BPA	Visible light irradiation under 300 W Xe lamp	100% in 12 min and 87%	0.075 g photocatalysts were added into 75 ml RhB (10 mg/L). BPA (20mg/L) over GO/Ag$_3$PO$_4$ composites	---	[94]
12	rGO/Ag$_3$ PO$_4$)	RhB	A 500 W xenon lamp	92% in 20 min	50 mg of photocatalysts was added to 100 mL of RhB solution	---	[95]
13	GO/Ag$_2$CO$_3$	RhB, MB and MO	Under visible light 300 W Xe lamp	100% in 30 min	Dye solutions (30 mL, 1 × 10^{-5} M) containing 100 mg of photocatalyst	Precipita-tion method	[96]

| 14 | GO/Ag$_2$CrO^{4-} | MB, Phenol | Visible-light irradiation A300 W simulated solar Xe arc lamp | 100% MB in 15 min and 90% Phenol in 60 min | 20 mg of the photocatalyst was dispersed in 100 mL of MB aqueous solution (1 × 10^{-5}mol L^{-1}) | Precipita-tion method | [97] |
| 16 | rGO /BiVO$_4$ | RhB | Sunlight irradiation. | 98.5%in 10 hrs | As prepared photocatalyst (0.10 g) were added to 200 mL RhB aqueous solution (5 mg/L) | --- | [100] |

GO=Graphene oxide, RGO=Reduced graphene oxide, MB=Methylene blue, BPA= Bisphenol A, RhB= Rhodamine B, MO=Methyl Orange

6. Cancer therapy

In recent times, graphene-based composites have been receiving increased attention in biomedical applications including drug or gene delivery, tissue engineering, biosensing [101]. The use of graphene-based nanocomposites as drug carriers for cancer therapy has been grown rapidly in the past few years. Graphene material shows favorable biocompatibility with low toxicity, admirable physical properties, a surface availability for further bio or chemical modification and development of multifunctional activity. Graphene possessing delocalized π electrons and high surface area (2600 m^2 g^{-1}) which offers interaction with various biomolecules or aromatic ring-containing anticancer drugs and provide efficient loading capacity and target delivery than other materials which can improve drug efficiency without increasing the dose of the chemotherapeutic agent [102]. Graphene oxide is mostly used in biomedical application because it is soluble in water. But in physiological solution have lots of salts and proteins which creates some issues such as, the formation of protein corona on graphene oxide surface, adverse effects occur which disturb its biodistribution or interaction with the immune system causing toxicity. However, to overcome these issues, it is desired to functionalize the graphene sheets to impart with aqueous solubility, biocompatibility, suitable sizes are necessary to suitably interface with biological systems in vitro or in vivo [103].

Sun *et al.*, synthesized nanographene oxide/ polyethylene glycol polymers at nanoscale of size ranges from 10-50 nm and explore the biological applications. The result presented high stability in physiological solutions and delocalized π electrons system of graphene surface can be employed for the active involvement of aromatic anticancer drugs such as doxorubicin and SN 38 via π-π stacking, which is insoluble in the water. The higher surface area of graphene provides high drug loading efficiency on nanographene oxide/ polyethylene glycol polymers. The terminals of polyethylene glycol were accessible for the formation of links between targeting antibodies, which eased the drug delivery to certain cancerous cell [104].

Yang et al., checked the performances of nanographene sheets with polyethylene glycol coating done via a fluorescent labeling method for photothermal therapy by intravenous direction and fluorescence imaging reveals amazingly large tumor acceptance of nanographene sheets in numerous xenograft tumor mouse models in vivo behavior. Moreover, polyethylene glycol/ nanographene sheets show superior permeability and retention effect of cancer lumps with the unnoticeable signal of toxicity [105].

Zhang et al., employed folic acid conjugated nanoscale graphene oxide as a nanocarrier to load two anticancer drugs (doxorubicin and camptothecin) via π-π stacking and hydrophobic interactions. Result confirmed that simultaneously two drugs loaded folic acid conjugated nanoscale graphene oxide shows explicit aiming to MCF-7 cells and shown low cytotoxicity in comparison to folic acid conjugated nanoscale graphene oxide loaded with either doxorubicin or camptothecin only [106].

Feng et al., designed polyethylene glycol/polyethylenimine conjugated nanographene oxide via amide bonds which is physiologically stable with ultra-small size. The results demonstrated that the prepared system shows enhanced gene transfection activity with reduced cytotoxicity as compared to bare polyethylenimine and polyethylenimine conjugated nanographene oxide without polyethylene glycol. Cellular uptake of polyethylene glycol/polyethylenimine conjugated nanographene oxide is enhanced under a low power NIR laser irradiation and because of slight photon based thermal heating increases the cell membrane transportation without any considerable cell damage which causes an enhancement in plasmid DNA transfection activities persuaded by the NIR laser were retained using polyethylene glycol/polyethylenimine conjugated nanographene oxide as the light-responsive gene carrier. Moreover, the system also able to deliver siRNA into cells under the control of NIR light irradiation [107].

Shi et al., synthesized graphene oxide/silver nanocomposite via chemical deposition of silver nanoparticles over graphene oxide via a hydrothermal method. An anti-cancerous drug (doxorubicin) was conjugate with graphene oxide/silver nanocomposite by ester bonds which shows a huge drug loading capacity (wt. 82.0%, weight ratio of doxorubicin: graphene oxide/silver), active tumor targeting ability and superb solidity in physiological solutions then graphene oxide/silver nanocomposite/doxorubicin was actively functionalized by DSPE-PEG2000-NGR. Results found that the release profiles of doxorubicin from graphene oxide/silver nanocomposite/doxorubicin/NGR revealed strong dependences on near-infrared laser and the surface plasmon resonance effect of silver nanoparticles. Graphene oxide/silver nanocomposite/doxorubicin/NGR displayed enhanced antitumor efficiency deprived of toxic involvement to organs because of about 8.4 times higher doxorubicin validation of tumor and about 1.7-fold higher doxorubicin released in tumor with a near-infrared laser than the other tissues [108]. Some *et al.*, synthesized curcumin/graphene composites and practiced them as nanovectors for the transportation of the hydrophobic anticancer drug curcumin that relies on the pH reliance. The degree of drug-loading was enhanced by increasing the number of functional groups having oxygen group in the derivatives of graphene. The results confirmed a mutual contribution of curcumin and graphene on the demise of the cancer cell (HCT 116) together in-vitro and in-vivo investigations. Curcumin/graphene quantum dots consisted of the curcumin nanoparticles on higher side showed realistic anticancer activity in comparison to the other composite systems having curcumin with similar dose [109].

Conclusion

This chapter summarizes mainly graphene oxide composites. Graphene oxide has excellent chemical or electrochemical activities, wide specific area, rapid response towards pH, temperature, etc., which made graphene-based composites widely applied to the fields of sensors, biomedicine, energy storage devices, waste water treatment, etc. The applications of graphene-based composites have tremendous applications and are not restricted to the above-discussed areas, so it is feasible to introduce novel functional materials into graphene-based composites with embellished properties in different fields through the existing and advanced techniques.

Acknowledgment:

Author Deepika Jamwal acknowledges financial support from National Post-Doctoral Fellowship (PDF/2017/001869), from Department of Science and Technology, SERB, India and Panjab University, Chandigarh.

List of Abbreviations

BPA	Bisphenol A
CNT	Carbon Nano Tubes
DNA	Deoxyribonucleic Acid
FTIR	Fourier-transform Infrared Spectroscopy

GO	Graphene Oxide
MB	Methylene Blue
MO	Methyl Orange
NIR laser	Near Infra-Red Laser
PMS	Peroxymonosulfate
rGO	Reduced Graphene Oxide
RhB	Rhodamine B
RNA	Ribonucleic Acid
UV-Vis	Ultra Violet-Visible

References

[1] L.F. Chen, Z.Y. Yu, J.J. Wang, Q.X. Li, Z.Q. Tan, Y.W. Zhu, S.H. Yu, Metal-like fluorine-doped β-FeOOH nanorods grown on carbon cloth for scalable high-performance supercapacitors, Nano Energy 11 (2015) 119-128. https://doi.org/10.1016/j.nanoen.2014.10.005

[2] P. Yang, W. Mai, Flexible solid-state electrochemical supercapacitors, Nano Energy 8 (2014) 274-290. https://doi.org/10.1016/j.nanoen.2014.05.022

[3] R. Raccichini, A. Varzi, S. Passerini, B Scrosati, The role of graphene for electrochemical energy storage, Nat. Mater. 14 (2015) 271-279. https://doi.org/10.1038/nmat4170

[4] Z.S. Wu, W. Ren, D.W. Wang, F. Li, B. Liu, H.M. Cheng, High-energy MnO2 nanowire/graphene and graphene asymmetric electrochemical capacitors, ACS Nano 4 (2010) 5835-5842. https://doi.org/10.1021/nn101754k

[5] J. Zhang, J. Ma, L.L. Zhang, P. Guo, J. Jiang, X. S. Zhao, Template synthesis of tubular ruthenium oxides for super capacitor applications, J. Phys. Chem. C. 114 (2010) 13608–13613. https://doi.org/10.1021/jp105146c

[6] D.W. Wang, F. Li, H.M. Cheng, Hierarchical porous nickel oxide and carbon as electrode materials for asymmetric super capacitor, J. Power Sources. 185 (2008) 1563–1568. https://doi.org/10.1016/j.jpowsour.2008.08.032

[7] X. Dong, W.Shen, J. Gu, L. Xiong, Y. Zhu, H. Li, J. Shi, MnO2-embedded-in-mesoporous-carbon-wall structure for use as electrochemical capacitors, J. Phys. Chem. B. 110 (2006) 6015-6019. https://doi.org/10.1021/jp056754n

[8] S. Ma, K. Ahn, E. Lee, K. Oh, K. Kim, Synthesis and characterization of manganese dioxide spontaneously coated on carbon nanotubes, Carbon. 45 (2007) 375–382. https://doi.org/10.1016/j.carbon.2006.09.006

[9]   O.C. Compton, S.T. Nguyen, Graphene oxide, highly reduced graphene oxide, and graphene: Versatile building blocks for carbon-based materials, Small. 6 (2010) 711–723. https://doi.org/10.1002/smll.200901934

[10] Z.S. Wua, G. Zhoua, L.C. Yina, W. Rena, F. Lia, H.M. Cheng, Graphene/metal oxide composite electrode materials for energy storage, Nano Energy. 1 (2012) 107–131. https://doi.org/10.1016/j.nanoen.2011.11.001

[11] L. Adamczyk, P.J. Kulesza, K. Miecznikowski, B. Palys, M. Chojak, D. Krawczyk, Nanostructured and advanced materials for fuel cells, J. Electrochem. Soc. 152 (2005) E98. https://doi.org/10.1149/1.1859710

[12] K. Qingqing, J. Wang, Graphene based materials for super capacitor electrodes: A review, J Materiomics. 2 (2016) 37-54. https://doi.org/10.1016/j.jmat.2016.01.001

[13] C. Lee, X. Wei, J. Kysar, J. Hone, Measurement of the elastic properties and intrinsic strength of monolayer grapheme, Science. 321 (2008) 385–388. https://doi.org/10.1126/science.1157996

[14] O. Compton, S. Nguyen, Graphene oxide, highly reduced graphene oxide, and graphene: versatile building blocks for carbon-based materials, Small. 6 (2010) 711–723. https://doi.org/10.1002/smll.200901934

[15] S. Chen, J. Duan, Y. Tang, S. Z. Qiao, Hybrid hydrogels of porous grapheme and nickel hydroxide as advanced super capacitor materials, J. Chem. Eur. 19 (2013) 7118. https://doi.org/10.1002/chem.201300157

[16] V. Singh, D. Joung, L. Zhai, S. Das, S. Khondaker, S. Seal, Graphene based materials: Past, present and future, Prog, Mater. Sci. 56 (2011) 1178–1271. https://doi.org/10.1016/j.pmatsci.2011.03.003

[17] J. Ji, L.L. Zhang, H. Ji, Y. Li, X. Zhao, X. Bai, X. Fan, F. Zhang, R.S. Ruoff, Nanoporous $Ni(OH)_2$ thin film on 3D ultrathin-graphite foam for asymmetric super capacitor, ACS Nano. 7 (2013) 6237–6243. https://doi.org/10.1021/nn4021955

[18] C. Yang, L. Dong, Z. Chen, H. Lu, High-performance all-solid-state super capacitor based on the assembly of graphene and manganese (ii) phosphate nano sheets, J. Phys. Chem. C. 118 (2014) 18884−18891. https://doi.org/10.1021/jp504741u

[19] X. Cao, B. Zheng, W. Shi, J. Yang, Z. Fan, Z. Luo, X. Rui, B. Chen, Q. Ya, H. Zhang, Reduced Graphene Oxide-Wrapped MoO3 Composites prepared by using metal–organic frameworks as precursor for all-solid-state flexible super capacitors Adv. Mater. 27 (2017) 4695-4701. https://doi.org/10.1002/adma.201501310

[20] E.G.D.S. Firmiano, A.C. Rabelo , C.J. Dalmaschio, A.N. Pinheiro, E.C. Pereira, W.H. Schreiner, E.R. Leite, Super capacitor electrodes obtained by directly bonding 2d MoS2 on reduced graphene oxide, Adv. Energy Mater. 4 (2014) 1301380 (1-8).

[21] H. Wanga, Z. Xua, H. Yia, H.Weib, Z. Guob, X. Wanga, One-step preparation of single-crystalline Fe_2O_3 particles/graphene composite hydrogels as high performance anode materials for super capacitors, Nano Energy. 7 (2014) 86-96. https://doi.org/10.1016/j.nanoen.2014.04.009

[22] G. Han, Y. Liu, L. Zhang, E. Kan, S. Zhang, J. Tang, W. Tang, MnO2 Nanorods intercalating graphene oxide/polyaniline ternary composites for robust high-performance super capacitors, Sci. Rep. 4 (2014) 4824 (1-7).

[23] R. Raccichini, A. Varzi, S. Passerini, B. Scrosati, The role of graphene for electro chemical energy storage, Nat. Mater. 14 (2015) 271-279. https://doi.org/10.1038/nmat4170

[24] B. Scrosati, J. Garche, Lithium batteries: status, prospects and future, J. Power Sources. 195 (2010) 2419–2430. https://doi.org/10.1016/j.jpowsour.2009.11.048

[25] J.R. Dahn, T. Zheng, Y. Liu, J.S. Xue, Mechanisms for lithium insertion in carbonaceous materials, Science. 270 (1995) 590–593. https://doi.org/10.1126/science.270.5236.590

[26] B. Luo, L. Zhi, Design and construction of three dimensional graphene-based composites for lithium ion battery applications, Energy Environ. Sci. 8 (2015) 456-477. https://doi.org/10.1039/C4EE02578D

[27] Y. Mai, J.Y. Xian, D. Zhang, J. Tu, X. Wang, Y.Q. Qiao, C. Gu, CuO/graphene composite as anode materials for lithium-ion batteries, Electrochem. Acta. 56 (2011) 2306–2311. https://doi.org/10.1016/j.electacta.2010.11.036

[28] I.R.M. Kottegoda, N.H. Idris, L. Lu, J. Wang, H. Liu, Synthesis and characterization of graphene-nickel oxide nanostructures for fast charge discharge application, Electrochim. Acta. 56 (2011) 5815–5822. https://doi.org/10.1016/j.electacta.2011.03.143

[29] P. Lian, X. Zhou, S. Liang, Z. Li, W. Yang, H. Wang, High reversible capacity of SnO_2/grapheme nano composite as an anode material for lithium-ion batteries, Electrochim Acta. 56 (2011) 4532–4539. https://doi.org/10.1016/j.electacta.2011.01.126

[30] T. Hu, X. Sun, H. Sun, M. Yu, F. Lu, C. Liu, J. Lian, Flexible free-standing graphene-TiO_2 hybrid paper for use as lithium ion battery anode materials, Carbon. 51 (2013) 322–326. https://doi.org/10.1016/j.carbon.2012.08.059

[31] C. Nethravathia, B. Viswanathb, J. Michaela, M. Rajamatha, Hydrothermal synthesis of a monoclinic VO_2 nanotube-graphene hybrid for use as cathode material in lithium ion batteries, Carbon. 50 (2013) 4839–4846. https://doi.org/10.1016/j.carbon.2012.06.010

[32] C.O.A. Vargas, A. Caballero, J. Morales, Can the performance of grapheme nano sheets for lithium storage in Li-ion batteries be predicted? Nanoscale 4 (2012) 2083–2092. https://doi.org/10.1039/c2nr11936f

[33] D. Jamwal, D. Rana, P. Singh, D. Pathak, S. Kalia, P. Thakur, E. Torino, Well-defined quantum dots and broadening of optical phonon line from hydrothermal method, RSC Adv. 6 (2016) 102010-102014. https://doi.org/10.1039/C6RA19818J

[34] L. Liu, M. An, P. Yang, J. Zhang, Superior cycle performance and high reversible capacity of SnO_2/grapheme composite as an anode material for lithium-ion batteries, Sci. Rep. 5 (2015) 9055 (1-10).

[35] Y. Yu, L. Gu, C. Zhu, S. T. sukimoto, P.A.V. Aken, J. Maier, Reversible storage of lithium in silver-coated three dimensional macroporous silicon Adv. Mater. 22 (2010) 2247-2250. https://doi.org/10.1002/adma.200903755

[36] B. Hertzberg, A. Alexeev, G. Yushin, Deformation in Si-Li anode upon electro chemical alloying in nano confined space, J. Am. Chem. Soc., 132 (2010) 8548-8549. https://doi.org/10.1021/ja1031997

[37] Y. Yao, M. T. McDowell, I. Ryu, H. Wu, N. Liu, L. Hu,W. D. Nix, Y. Cui, Interconnected silicon hollow nano spheres for lithium-ion battery anodes with long cycle life, Nano Lett. 11 (2011) 2949-2945. https://doi.org/10.1021/nl201470j

[38] X. Zhou, Y.X. Yin, L.J. Wan, Y.G. Guo, Facile synthesis of silicon nanoparticles inserted into graphene sheets as improved anode materials for lithium-ion batteries, W. Chem. Commun. 48 (2012) 2198–2200. https://doi.org/10.1039/c2cc17061b

[39] X. Huang, X. Qi, F. Boey, H. Zhang, Graphene-based composites, Chem. Soc. Rev, 41 (2012) 666-686. https://doi.org/10.1039/C1CS15078B

[40] H. Xue, Y.W.Y. Denis, J. Qing, Xia Yang, J. Xu, Z. Li, M. Sun, W. Kang, Y. Tang, S. Lee, Pyrite $FeS2$ microspheres wrapped by reduced graphene oxide as high-performance lithium-ion battery anodes, J. Mater. Chem. A, 3 (2015) 7945–7949. https://doi.org/10.1039/C5TA00988J

[41] J.Q. Huang, T.Z, Zhuang, Q. Zhang, H.J. Peng, C.M. Chen, F. Wei, Perm selective graphene oxide membrane for highly stable and anti-self-discharge lithium sulfur batteries, 9 (2015) 3002–3011.

[42] D. Kundu, E. Talaie, V. Duffort, L.F. Nazar, The emerging chemistry of sodium ion batteries for electrochemical energy storage, Angew. Chem. 54 (2015) 3431−3448. https://doi.org/10.1002/anie.201410376

[43] M.D. Slater, D. Kim, E. Lee, C.S. Johnson, Sodium-ion batteries, Adv. Funct. Mater. 23 (2013) 947–958. https://doi.org/10.1002/adfm.201200691

[44] B.L. Ellis, L.F. Nazar, Sodium and sodium-ion energy storage batteries, Curr. Opin. Solid State Mater. Sci. 16 (2012) 168–177. https://doi.org/10.1016/j.cossms.2012.04.002

[45] W. Li, M. Li, K.R. Adair, X. Sun, Y. Yu, Carbon nanofiber-based nanostructures for lithium-ion and sodium-ion batteries, J. Mater. Chem. A. 5 (2017) 13882-13906. https://doi.org/10.1039/C7TA02153D

[46] W. Qin, T. Chen, L. Pan, L.Y. Niu, B.W. Hua, D.S. Li, J. Li, Z. Sun, MoS2-reduced graphene oxide composites via microwave assisted synthesis for sodium ion battery anode with improved capacity and cycling performance, Electro Chimica Acta. 153 (2015) 55–61. https://doi.org/10.1016/j.electacta.2014.11.034

[47] D. Xie, W. Tang, Y. Wang, X. Xia, Y. Zhong, D. Zhou, D. Wang, X. Wang, J. Tu, Facile fabrication of integrated three-dimensional C-MoSe$_2$/ reduced graphene oxide composite with enhanced performance for sodium storage, Nano Research. 9 (2016) 1618-1629. https://doi.org/10.1007/s12274-016-1056-3

[48] T.S. Sahu, S. Mitra, Exfoliated MoS2 sheets and reduced graphene oxide-an excellent and fast anode for sodium-ion battery, Sci. Rep. 5 (2015) 12571 (1-8).

[49] X.H. Xiong, G. Wang, Y. Lin, Y. Wang, X. Ou, F.H. Zheng, C.G. Yang, J.H. Wang, M. Liu, Enhancing sodium ion battery performance by strongly binding nanostructured Sb2S3 on sulfur-doped graphene sheets, ACS Nano. 10 (2016) 10953-10959. https://doi.org/10.1021/acsnano.6b05653

[50] H.D. Jang, S.K. Kim, H. Chang, J.W. Choi, J.X. Huang, Synthesis of grapheme based noble metal composites for glucose biosensor, Mater. Lett. 106 (2013) 277–280. https://doi.org/10.1016/j.matlet.2013.05.033

[51] T. Kuila, S. Bose, P. Khanra, A.K. Mishra, N.H. Kim, J.H. Lee, Recent advances in graphene-based biosensors, Biosens. Bioelectron. 26 (2011) 4637–4648. https://doi.org/10.1016/j.bios.2011.05.039

[52] R. Wilson, A.P.F. Turner, Glucose oxidase-an ideal enzyme, Biosens. Bioelectron. 7 (1992) 165−185. https://doi.org/10.1016/0956-5663(92)87013-F

[53] G. Bharath, R. Madhu, S.M. Chen, V. Vediyappan, A. Balamurugan, D. Mangalaraj, C. Viswanathan, N. Ponpandian, Enzymatic electrochemical glucose

biosensors by mesoporous 1D hydroxyapatite-on-2D reduced graphene oxide, J. Mater. Chem. B. 3 (2014) 1360-1370. https://doi.org/10.1039/C4TB01651C

[54] C.L. Sun, W.L. Cheng, T.K. Hsu, C.W. Chang, J.L. Chang, J.M. Zen, Ultrasensitive and highly stable nonenzymatic glucose sensor by a CuO/graphene-modified screen-printed carbon electrode integrated with flow-injection analysis, Electrochem. Commun. 30 (2013) 91-94. https://doi.org/10.1016/j.elecom.2013.02.015

[55] Y. Mu, D. Jia, Y. He, Y. Miao, H.-L. Wu, Nano nickel oxide modified non-enzymati cglucose sensors with enhanced sensitivity through an electrochemical process strategy at high potential, Biosens. Bioelectron. 26 (2011) 2948–2952. https://doi.org/10.1016/j.bios.2010.11.042

[56] Y. Ding, Y. Wang, L. Su, M. Bellagamba, H. Zhang, Y. Lei, Electrospun Co_3O_4 nanofibers for sensitive and selective glucose detection, Biosens. Bioelectron.26 (2010) 542–548. https://doi.org/10.1016/j.bios.2010.07.050

[57] G. Chang, H. Shu, K. Ji, M. Oyama, X. Liu, Y. He, Gold nanoparticles directly modified glassy carbon electrode for non-enzymatic detection of glucose, Applied Surface Science. 288 (2014) 524–529. https://doi.org/10.1016/j.apsusc.2013.10.064

[58] K.M. Samant, V.R. Chaudhari, S. Kapoor, S.K. Haram, Filling and coating of multiwalled carbon nanotubes with silver by DC electrophoresis, Carbon. 45 (2007) 2126-2129. https://doi.org/10.1016/j.carbon.2007.06.020

[59] X. Niu, M. Lan, H. Zhao, C. Chen, Highly Sensitive and Selective Non enzymatic detection of glucose using three-dimensional porous nickel nanostructures, Anal. Chem. 85 (2013) 3561–3569. https://doi.org/10.1021/ac3030976

[60] Chen, P. Holt-Hindle, Platinum-based nanostructured materials: synthesis, properties, and applications, Chemical Reviews. 110 (2010) 3767–3804. https://doi.org/10.1021/cr9003902

[61] X.M. Chen, Z.J. Lin, D.J. Chen, T.T. Jia, Z.M. Cai, X.R. Wang, X. Chen, G.N. Chen, M. Oyama, Nonenzymatic ampere metric sensing of glucose by using palladium nanoparticles supported on functional carbon nanotubes, Biosens. Bioelectron. 25 (2010) 1803–1808. https://doi.org/10.1016/j.bios.2009.12.035

[62] R. Madhu, V. Veeramani, S.M. Chen, A. Manikandan, A. Ya Lo, Y.L. Chueh, Honeycomb-like porous carbon–cobalt oxide nanocomposite for high-performance enzyme less glucose sensor and super capacitor applications, ACS Appl. Mater. Interfaces. 7 (2015) 15812−15820. https://doi.org/10.1021/acsami.5b04132

[63]   H.W. Chang,Y.C. Tsai, C.W. Cheng, C.Y. Lin, P.H. Wu, Preparation of platinum/carbon nanotubes in aqueous solution by femto second laser for non-enzymatic glucose determination, Sens. Actuators B 183 (2013) 34-39. https://doi.org/10.1016/j.snb.2013.03.115

[64]   F. Liu, Y. Piao, K.S. Choi, T.S. Seo, Fabrication of free-standing graphene composite films as electrochemical biosensors, Carbon. 50 (2012) 123–133. https://doi.org/10.1016/j.carbon.2011.07.061

[65]   T. Kuila, S. Bose, P. Khanra, A. K. Mishra, N. H. Kim, J. H. Lee, Recent advances in graphene-based biosensors, Biosens. Bioelectron. 26 (2011) 4637-4648. https://doi.org/10.1016/j.bios.2011.05.039

[66]   Q. Wang, Q. Wang, M. Li, S. Szuneritsand, R.Boukherrou, Preparation of reduced graphene oxide/Cunanoparticle composites through electrophoretic deposition: application for non-enzymatic glucose sensing, RSC Adv. 5 (2015) 15861-15869. https://doi.org/10.1039/C4RA14132F

[67]   A.C. Joshi, G.B. Markad, S.k. Haram Rudimentary, simple method for the decoration of graphene oxide with silver nanoparticles: Their application for the ampere metric detection of glucose in the human blood samples, Electrochimica. Acta. 161 (2015) 108–114. https://doi.org/10.1016/j.electacta.2015.02.077

[68]   G. Chang, H. Shu, Q. Huang, M. Oyama, K. Ji, X. Liu, Y. He, Synthesis of highly dispersed Pt nanoclusters anchored grapheme composites and their application for non-enzymatic glucose sensing, Electrochim. Acta. 157 (2015) 149–157. https://doi.org/10.1016/j.electacta.2015.01.085

[69]   P.M. Nia, W.P. Meng, F. Lorestani, M.R. Mahmoudian, Y. Alias, Electrodeposition of copper oxide/polypyrrole/reduced grapheme oxide as a non-enzymatic glucose biosensor, Sens. Actuators B: Chem. 209 (2015) 100–108. https://doi.org/10.1016/j.snb.2014.11.072

[70]   P. Lu, J. Yua,Y. Lei, S. Lu,C. Wang,D. Liu, Q. Guo, Synthesis and characterization of nickel oxide hollow spheres–reduced graphene oxide–nafioncomposite and its biosensing for glucose, Sens. Actuators B: Chem. 208 (2015) 90–98. https://doi.org/10.1016/j.snb.2014.10.140

[71]   P.M. Nia, F. Lorestani, W.P. Meng, Y. Alias, A novel non-enzymatic H2O2 sensor based on poly pyrrole nanofibers–silver nanoparticles decorated reduced grapheme oxide nanocomposites, Appl. Surf. Sci. 332 (2015) 648–656. https://doi.org/10.1016/j.apsusc.2015.01.189

[72] B. Zhan, C. Liu, H. Shi, C. Li, L. Wang, W. Huang, X. Dong. A hydrogen peroxide electrochemical sensor based on silver nanoparticles decorated three-dimensional graphene, Appl. Phys. Lett. 104 (2014) 243704 (1-5).

[73] S. Woo, Y.R. Kim, T.D. Chung, Y. Piao, H. Kim, Synthesis of a graphene–carbon nanotube composite and its electrochemical sensing of hydrogen peroxide, Electrochim. Acta. 59 (2012) 509– 514. https://doi.org/10.1016/j.electacta.2011.11.012

[74] C. Cheng, C. Zhang, X. Gao, Z. Zhuang, C. Du, W. Chen, 3D network and 2D paper of reduced graphene oxide/Cu$_2$O composite for electrochemical sensing of hydrogen peroxide Anal. Chem. 90 (2018) 1983–1991. https://doi.org/10.1021/acs.analchem.7b04070

[75] Y.Q. Yang, H.L. Xie, J.T.S. Tang, J. Yi, H.L. Zhang, Design and preparation of a non-enzymatic hydrogen peroxide sensor based on a novel rigid chain liquid crystalline polymer/reduced grapheme oxide composite, RSC Adv. 5 (2015) 63662-63668. https://doi.org/10.1039/C5RA10540D

[76] F. Lorestani, Z. Shahnavaz, P. Mn, Y. Alias, N.S.A. Manan, One-step hydrothermal green synthesis of silver nanoparticle-carbon nanotube reduced-graphene oxide composite and its application as hydrogen peroxide sensor, Sens. Actuators, B: Chem. 208 (2015) 389–398. https://doi.org/10.1016/j.snb.2014.11.074

[77] S. Palanisamy, S.M. Chen, R. Sarawathi, A novel non enzymatic hydrogen peroxide sensor based on reduced grapheme oxide/ZnO composite modified electrode, Sens. Actuators, B: Chem. 166– 167 (2012) 372-377.

[78] C.J. Venegasa, E. Yedinaka, J.F. Marcod, S. Bollob, D.R. Leóna, Co–doped stannates /reduced graphene composites: Effect of cobalt substitution on the electrochemical sensing of hydrogen peroxide, Sens. Actuators, B: Chem. 250 (2017) 412-419. https://doi.org/10.1016/j.snb.2017.04.154

[79] F. Xu, M. Deng, G. Li, S. Chen, L. Wang, Electrochemical behavior of cuprous oxide–reduced graphene oxide nanocomposites and their application in non-enzymatic hydrogen peroxide sensing, Electrochim. Acta. 88 (2013) 59-65. https://doi.org/10.1016/j.electacta.2012.10.070

[80] M. Azarang, A. Shuhaimi, R. Yousefic, S.P. Jahromia, One-pot sol–gel synthesis of reduced graphene oxide uniformly decorated zinc oxide nanoparticles in starch environment for highly efficient photo degradation of methylene blue, RSC Adv. 5 (2015) 21888-21896. https://doi.org/10.1039/C4RA16767H

[81] M. Azaranga, A. Shuhaimia, R. Yousefic, A.M. Golsheikha, M. Sookhakiana, Synthesis and characterization of ZnO NPs/reduced graphene oxide nanocomposite prepared in gelatin medium as highly efficient photo-degradation of MB, Ceram. Int. 40 (2014) 10217–10221. https://doi.org/10.1016/j.ceramint.2014.02.109

[82] W. Zou, L. Zhang, L. Liu, X. Wang, Jingfang Sun, S. Wu, Y. Deng, C. Tang, F. Gao, L. Dong, Engineering the Cu_2O–reduced graphene oxide interface to enhance photo catalytic degradation of organic pollutants under visible light, Appl. Cat. B: Env. 181 (2016) 495–503. https://doi.org/10.1016/j.apcatb.2015.08.017

[83] Y. Gao, M. Hu, B. Mi, Membrane surface modification with TiO_2–grapheme oxide for enhanced photo catalytic performance, J. Membr. Sci. 455 (2014) 349–356. https://doi.org/10.1016/j.memsci.2014.01.011

[84] S. Xu, L. Fu, T.S.H. Pham, A. Yub, F. Han, L. Chena Preparation of ZnO flower/reduced grapheme oxide composite with enhanced photo catalytic performance under sunlight, Ceram. Int. 41 (2015) 4007–4013. https://doi.org/10.1016/j.ceramint.2014.11.086

[85] H. Wu, J. Fan, E. Liu, X. Hu, Y. Ma, X. Fan, Y. Li, C. Tang, Facile hydrothermal synthesis of TiO_2 nano spindles-reduced graphene oxide composite with a enhanced photo catalytic activity, J. Alloys Compd. 623 (2015) 298–303. https://doi.org/10.1016/j.jallcom.2014.10.153

[86] D.A. Reddy, R. Ma, M.Y. Choi, T.K. Kim, Reduced graphene oxide wrapped $ZnS-Ag_2S$ ternary composites synthesized via hydrothermal method: Applications in photo catalyst degradation of organic pollutants, Appl. Surf. Sci.324 (2015) 725-735. https://doi.org/10.1016/j.apsusc.2014.11.026

[87] T. Jiao, H.Guo, Q. Zhang, Q. Peng, Y. Tang, X. Yan, B. Li, Reduced graphene oxide-based silver nanoparticle-containing composite hydrogel as highly efficient dye catalysts for wastewater treatment. Sci. Rep. 5 (2015) 11873. https://doi.org/10.1038/srep11873

[88] S.K. Bhunia, N.R. Jana, Reduced graphene oxide-silver nanoparticle composite as visible light photocatalyst for degradation of colorless endocrine disruptors, ACS Appl. Mater. Interfaces. 6 (2014) 20085−20092. https://doi.org/10.1021/am505677x

[89] Y.L. Min, G.Q. He, Q.J. Xu, Y.C. Chen, Self-assembled encapsulation of graphene oxide/Ag@AgCl as a Z-scheme photo catalytic system for pollutant removal, J. Mater. Chem. A. 2 (2014) 1294-1301. https://doi.org/10.1039/C3TA13687F

[90]   Y. Yao, C. Xu, S. Yu, D. Zhang, S.Wang, Facile synthesis of Mn_3O_4–reduced graphene oxide hybrids for catalytic decomposition of aqueous organics, Ind. Eng. Chem. Res. 52 (2013) 3637–3645. https://doi.org/10.1021/ie303220x

[91]   Y. Yao, J. Qin, Y. Cai, F. Wei, F. Lu, S. Wang, Facile synthesis of magnetic $ZnFe_2O_4$–reduced graphene oxide hybrid and its photo-Fenton-like behavior under visible irradiation, Environ. Sci. Pollut. Res. 21 (2014) 7296-7306. https://doi.org/10.1007/s11356-014-2645-x

[92]   G.D. Chen, M. Sun, Q. Wei, Y.F. Zhang, B.C. Zhu, B. Du, Ag_3PO_4/Graphene-oxide composite with remarkably enhanced visible-light-driven photocatalytic activity toward dyes in water, J. Hazard. Mater. 244 (2013) 86-93. https://doi.org/10.1016/j.jhazmat.2012.11.032

[93]   Q.H. Liang, Y. Shi, W.J. Ma, Z. Li, X.M. Yang, Enhanced photocatalytic activity and structural stability by hybridizing Ag_3PO_4 nanospheres with graphene oxide sheets, Phys. Chem. Chem. Phys. 14 (2012) 15657-15665. https://doi.org/10.1039/c2cp42465g

[94]   C. Wang, J. Zhu, X. Wu, H. Xu, Y. Song, J. Yan, Y. Song, H. Ji, K. Wang, H. Li, Photocatalytic degradation of bisphenol A and dye by graphene-oxide/ Ag_3PO_4 composite under visible light irradiation, Ceram. Int. 40 (2014) 8061-8070. https://doi.org/10.1016/j.ceramint.2013.12.159

[95]   B. Chai, J. Li, Q. Xu, Reduced graphene oxide grafted Ag_3PO_4 composites with efficient photocatalytic activity under visible-light irradiation, Ind. Eng. Chem. Res. 53 (2014) 8744-8752. https://doi.org/10.1021/ie4041065

[96]   C. Dong, K.-L. Wu, X.-W. Wei, X.-Z. Li, L. Liu, T.-H. Ding, J. Wanga, Y. Yea, Synthesis of graphene oxide-Ag_2CO_3 composites with improved photoactivity and anti-photocorrosion, Cryst. Eng. Comm. 16 (2014) 730-736. https://doi.org/10.1039/C3CE41755G

[97]   D. Xu, B. Cheng, S. Cao, J. Yu, Enhanced photocatalytic activity and stability of Z-scheme Ag_2CrO_4-GO composite photocatalysts for organic pollutant degradation, Appl. Catal. B: Environ. 164 (2015) 380-388. https://doi.org/10.1016/j.apcatb.2014.09.051

[98]   H.Q. Jiang, H. Endo, H. Natori, M. Nagai, K. Kobayashi, Fabrication and efficient photocatalytic degradation of methylene blue over $CuO/BiVO_4$ composite under visible-light irradiation, Mater. Res. Bull. 44 (2009) 700-706. https://doi.org/10.1016/j.materresbull.2008.06.007

[99]   Y. Yan, S. Sun, Y. Song, X. Yan, W. Guan, X. Liu, W. Shi, Microwave-assisted in situ synthesis of reduced graphene oxide-$BiVO_4$ composite photocatalysts and

Carbonaceous Composite Materials   Materials Research Forum LLC
Materials Research Foundations **42** (2018) 111-142   doi: http://dx.doi.org/10.21741/9781945291975-5

their enhanced photocatalytic performance for the degradation of ciprofloxacin, J. Hazard. Mater. 250 (2013) 106-114. https://doi.org/10.1016/j.jhazmat.2013.01.051

[100] S. Dong, Y. Cui, Y. Wang, Y. Li, L. Hu, J. Sun, J. Sun, Designing three-dimensional acicular sheaf shaped BiVO$_4$/reduced graphene oxide composites for efficient sunlight-driven photocatalytic degradation of dye wastewater, Chem. Eng. J. 249 (2014) 102-110 https://doi.org/10.1016/j.cej.2014.03.071

[101] M. Orecchioni, R. Cabizza, A. Bianco, L.G. Delogu, Graphene as cancer theranostic tool: progress and future challenges, Theranostics. 5 (2015) 710-723. https://doi.org/10.7150/thno.11387

[102] M. Nejabat, F. Charbgoo, M. Ramezani, Graphene as multifunctional delivery platform in cancer therapy, J. Biomed. Mater. Res. A. 105 (2017) 2355-2367. https://doi.org/10.1002/jbm.a.36080

[103] D. Li, M.B. Muller, S. Gilge, R.B. Kaner, G.G. Wallace, Processable aqueous dispersions of graphene nanosheets. Nat. Nanotechol. 3 (2008) 101-105. https://doi.org/10.1038/nnano.2007.451

[104] X. Sun, Z. Liu, Nano-graphene oxide for cellular imaging and drug delivery, Nano Res. 1 (2008) 203-212. https://doi.org/10.1007/s12274-008-8021-8

[105] K. Yang, S. Zhang, G. Zhang, X. Sun, S.-T. Lee, Z. Liu, Graphene in mice: ultrahigh in nivo tumor uptake and efficient photothermal therapy, Nano Lett. 10 (2010) 3318-3323. https://doi.org/10.1021/nl100996u

[106] L. Zhang, J. Xia, Q. Zhao, L. Liu, Z. Zhang, Functional graphene oxide as a nanocarrier for controlled loading and targeted delivery of mixed anticancer drugs, Small 6 (2010), 537-544. https://doi.org/10.1002/smll.200901680

[107] L. Feng, X. Yang, X. Shi, X, Tan, R. Peng , J. Wang, Z. Liu, Polyethylene glycol and polyethylenimine dual- functionalized nano-graphene oxide for photothermally enhanced gene delivery, Small 9 (2013) 1989-1997. https://doi.org/10.1002/smll.201202538

[108] J. Shi, L. Wang, J. Zhang, R. Ma, J. Gao, Y. Liu, C. Zhang, Z. Zhang, A tumor-targeting near-infrared laser-triggered drug delivery system based on GO@Ag nanoparticles for chemo-photothermal therapy and X-ray imaging, Biomaterials 35 (2014) 5847-5861. https://doi.org/10.1016/j.biomaterials.2014.03.042

[109] S. Some1, A.R. Gwon, E. Hwang, G.H. Bahn, Y. Yoon, Y. Kim, S.H. Kim, S. Bak, J. Yang, D.G. Jo, H. Lee, Cancer therapy using ultrahigh hydrophobic drug-loaded graphene derivatives, Sci. Rep. 4 (2014) 6314 (1-9).

Carbonaceous Composite Materials
Materials Research Foundations **42** (2018) 143-177

Materials Research Forum LLC
doi: http://dx.doi.org/10.21741/9781945291975-6

Chapter 6

Bioceramics, Carbonaceous Composite and its Biomedical Applications

Sheeba Nuzhat Khan[1], Fazal-Ur-Rehman[2,*],

[1]Department of Kulliyat, AKTC, India

[2]Department of Anatomy, JNMC, Aligarh Muslim University, Aligarh 202002, India

*fazalorth12@gmail.com

Abstract

Biomaterial is a material that interacts with human tissue and body fluids to treat, improve, or replace anatomical element(s) of the human body. Biological materials such as human bone allograft, are considered to be biomaterials and they are used in many cases in orthopedic surgery. Due to compatibility of carbonaceous materials with bone and other tissue and the similarity of the mechanical properties of carbon to bone, carbonaceous composite is used for orthopedic implants. Nowadays, to obtain the most desirable clinical performance of the implants the mechanically superior metals are combined with ceramics and polymers of excellent biocompatibility and biofunctionality. Among ceramic/ceramic, ceramic/polymer and ceramic/metal composites, ceramic/ceramic composites enjoy superiority due to their similarity with bone minerals, exhibiting biocompatibility and ability to be shaped into a definite size. Among bioceramics alumina, ziconia and carbon revealed their blood compatibility, no tissue reaction and nontoxicity to cells, but none of the above three-bioinert ceramics exhibited bonding with the bone. However, this bioactivity of the bioinert ceramics can be achieved by forming composites with bioactive ceramics. Bioglass and glass ceramics are nontoxic and chemically bond to bone, elicit osteoinductive property. Calcium phosphate ceramics are nontoxic to tissues, and have bioresorption and osteoinductive property.

Keywords

Ceramic, Polymer, Carbonaceous Composites, Metal Composites, Alumina, Tissues

Contents

Bioceramics, Carbonaceous Composite and its Biomedical Applications .143

1.   Introduction (Types of implant materials)

Implant materials for medical uses especially for orthopedic applications have advanced over the past decades from materials of industrial applications into the development of materials with features of specific biological responses within the biological environment. Orthopedic implants are the medical devices used to replace or provide fixation of bone or to replace articulating surfaces of a joint. They are made mainly from stainless steel and titanium alloys for strength and lined with polyethylene plastic to act as artificial cartilage. Among orthopedics implant few are cemented and others are uncemented. The uncemented are pressed to fit so that bone can grow into the implant for strength. In most joint implants one component is made from metal, and another from polyethylene. Except some pure metal implants most metal implants are made from a mixture of two or more metals. A good material which has a balance of desired features can be created by combining metals, to make alloys. The most common metal alloys used in orthopedic implants are stainless steel, cobalt-chromium alloys, and titanium alloys. Bone fixation implants such as metal plates, intramedullary rods, and screws are constructed of either titanium or stainless steel because they are biocompatible, have higher mechanical strength, and a bending modulus that is greater than cortical bone. However, due to their stiffness they absorb most of the stress in load-bearing applications (stress shielding), leads to weak tissue growth at the site of implant. Continued use of these implants sometimes leads to adverse reaction in a patient's body as the metal, although inert has the potential to leach into the patient's system, thus a second surgery is required for removal of the implant. The innovation of ceramic materials for skeletal repair and reconstruction has been one of the major advances in the development of medical materials in the last four decades; they are often referred as bioceramics. Generally, all metal implants are non-magnetic and high in density. It is important that the implants be compatible with magnetic resonance imaging (MRI) techniques and should be visible under X-ray imaging. Most of artificial implants are subjected to either static or repetitive load which requires an excellent combination of strength and ductility. This is the superior characteristic of metals over polymers and ceramics. In general implant materials can be divided into the following categories: metals, polymers, ceramics, and composites.

1. Metals and alloys :   316L stainless steel, CP-Ti, Ti-Al-V, Co-Cr-Mo, Cr-Ni-Cr-Mo, Ni-Ti
2. Ceramics and glasses: Alumina Zirconia, Calcium phosphates, Bioactive glasses, Carbons
3. Polymers: Polyethylene, Polypropylene, PMMA(polymethyl methacrylate), Silicones
4. Composites: PMMA-glass fillers

Biomaterials used in orthopaedic surgery are divided into two groups: metals and nonmetals.

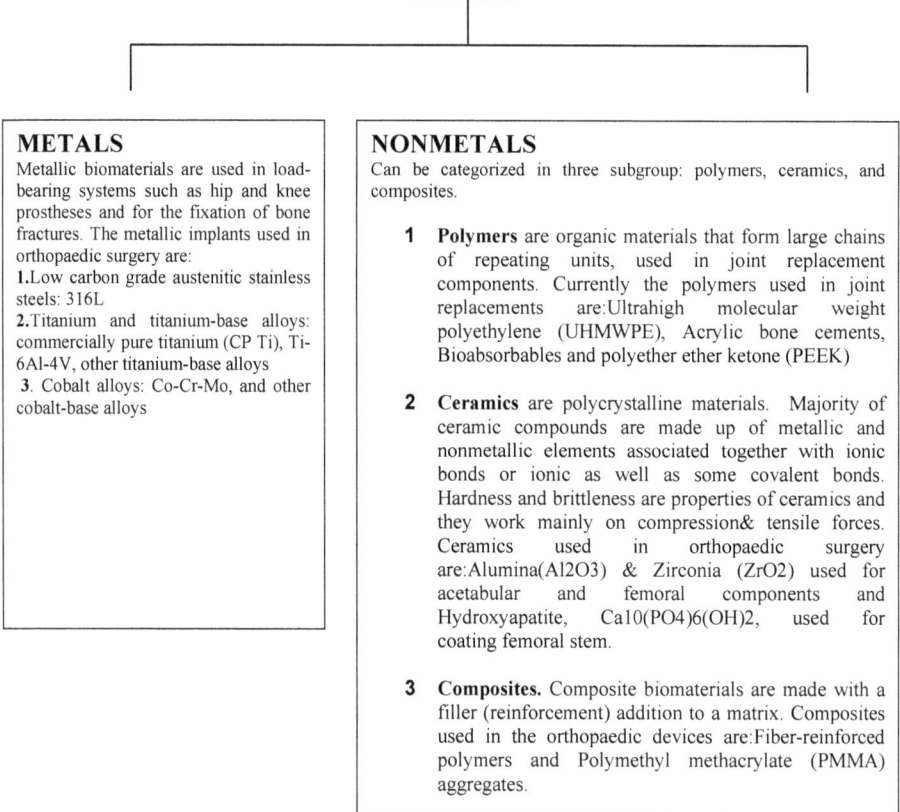

METALS

Metallic biomaterials are used in load-bearing systems such as hip and knee prostheses and for the fixation of bone fractures. The metallic implants used in orthopaedic surgery are:
1. Low carbon grade austenitic stainless steels: 316L
2. Titanium and titanium-base alloys: commercially pure titanium (CP Ti), Ti-6Al-4V, other titanium-base alloys
3. Cobalt alloys: Co-Cr-Mo, and other cobalt-base alloys

NONMETALS

Can be categorized in three subgroup: polymers, ceramics, and composites.

1 **Polymers** are organic materials that form large chains of repeating units, used in joint replacement components. Currently the polymers used in joint replacements are:Ultrahigh molecular weight polyethylene (UHMWPE), Acrylic bone cements, Bioabsorbables and polyether ether ketone (PEEK)

2 **Ceramics** are polycrystalline materials. Majority of ceramic compounds are made up of metallic and nonmetallic elements associated together with ionic bonds or ionic as well as some covalent bonds. Hardness and brittleness are properties of ceramics and they work mainly on compression& tensile forces. Ceramics used in orthopaedic surgery are:Alumina(Al_2O_3) & Zirconia (ZrO_2) used for acetabular and femoral components and Hydroxyapatite, $Ca10(PO_4)_6(OH)_2$, used for coating femoral stem.

3 **Composites.** Composite biomaterials are made with a filler (reinforcement) addition to a matrix. Composites used in the orthopaedic devices are:Fiber-reinforced polymers and Polymethyl methacrylate (PMMA) aggregates.

Carbonaceous Composite Materials   Materials Research Forum LLC
Materials Research Foundations **42** (2018) 143-177   doi: http://dx.doi.org/10.21741/9781945291975-6

The type of material used in biomedical depends on specific implant applications. Detail of the materials used in medical and dental implants are:

1.1   Stainless Steel

The majority of surgical equipment is made from stainless steel or a similar alloy. It is the metal of choice due to its hygienic abilities i.e. easy to sterilize and clean. Surgical stainless steel is a specific type of stainless steel, often used in implants that are intended to help repair fractures such as bone plates, screws, pins, and rods, because they are cheap, easily available and easy to process. Stainless steel is an alloy mainly made up of iron to which other metals such as chromium, nickel or molybdenum are added to make it more resistant to corrosion. There are many different types of stainless steel. The austenitic stainless steels, especially Types 316 and 316L, are most widely used for implant fabrication. They differ only in the content of carbon. These stainless steel used in orthopedic implants are designed to resist the normal chemicals found in the human body. In order to be austenitic at room temperature, stainless steel needs to contain a certain amount of austenite stabilizing elements such as Ni or Mn. The 316L contains 0.03 wt% C, 17–20 wt% Cr, 12–14 wt% Ni, 2–3 wt% Mo and minor amounts of nitrogen, manganese, phosphorus, silicon and sulphur. The new austenitic stainless steel having high Cr content (> 20%) with a high Nitrogen content (between 0.3 and 0.4%) and where Ni has been partially substituted by Mn is being used in joint prosthesis. Here nitrogen provides increase in the corrosion resistance and the mechanical properties i.e. yield stress and stabilize the austenitic phase. **Ni is substituted by Mn primarily to avoid Ni sensitivity.** The stainless steels are suitable to use only in temporary implant devices such as fractures plates, screws and hip nails. They are not suitable for permanent implant because of its poor fatigue strength and characteristic to undergo plastic deformation. If used for prolonged period even the 316L stainless steels may corrode inside the body under certain circumstances in a highly stressed and oxygen depleted region.

1.2   Cobalt-Chromium Alloys

While cobalt-chromium alloys contain mostly cobalt and chromium, others metals such as molybdenum and nickel are also added to increase their strength and corrosion resistance. There are basically two types of cobalt chromium alloys. One is the cobalt CoCrMo alloy, usually used to cast a dentistry product and the other is the CoNiCrMo alloy, these alloys are used in a variety of joint replacement implants, as well as some fracture repair implants. Prior to the use of titanium, cobalt-based alloy (Co-Cr-mo) and (Co-Cr-Ni) had largely replaced stainless steel as material for permanent implants. These alloys have more corrosion resistance due to the formation of a durable chromium-oxide surface layer i.e. passivation layer.

1.3   Titanium Alloys

Titanium alloys are the most flexible and the lightest of the orthopedic alloys. Titanium alloys mostly consist of titanium with varying degrees of other metals, such as aluminum and vanadium. Titanium is light and has good mechanical and chemical properties which are salient features for implant applications. Titanium alloy (Ti6Al4V) is most widely used to manufacture implants. The main alloying elements are aluminium (5.5 - 6.5%) and vanadium (3.5 - 4.5%). Although the strength of the titanium alloys may vary from lower than to equal of 316 stainless steel but when compared by specific strength (strength per density), the titanium alloys outperform any other implant material.

Physiological and Mechanical feature of titanium alloys: These materials are bioinert and when implanted into human bodies remain unchanged due to their excellent corrosion resistance. The human body recognizes these materials as foreign material, and tries to isolate them by encasing them in fibrous tissues. However, they do not produce any adverse reactions and are tolerated well by the human body. Some stainless steel may induce nickel hypersensitivity in surrounding tissues but titanium alloys do not induce any allergic reaction. The surface of titanium is often modified by hydroxyapatite coating; the commercially accepted technique for depositing such coatings is plasma spraying. The hydroxyapatite provides a bioactive surface, it actively participates in bone bonding so other mechanical fixation devices and bone cements are not required. The features of titanium and its alloys to be used in orthopedics and dental applications are mechanical properties such as strength, fatigue resistance and bend strength. Therefore, for load-bearing biomedical applications titanium alloys are used instead of materials such as hydroxyapatite which displays bioactive behavior. Other specific properties that make titanium alloys desirable biomaterial are density and elastic modulus. In terms of density, it has a significantly lower density (table 1) than other metallic biomaterials, so the implants are lighter than similar items fabricated out of stainless steel or cobalt chrome alloys.

Table1. Densities and young modulus of few ceramic biomaterials and cortical bone.

Material	Density [gm/cm^3]	Young Modulus [GPa]
Cortical Bone	~2.0 g.cm^{-3}	7-30
Cobalt-Chrome alloy	~8.5 g.cm^{-3}	230
316L Stainless Steel	8.0 g.cm^{-3}	200
CP Titanium	4.51 g.cm^{-3}	110
Ti6Al4V	4.40 g.cm^{-3}	106

Titanium alloys have lower elastic modulus as compared to the other metals. This is desirable because the bone hosting the biomaterial is less likely to atrophy and resorb. Femoral stem of total hip replacement prosthesis is made up of a titanium alloy. The white section is a hydroxyapatite coating to encourage bone bonding to the implant. The ball on top of the femoral stem is called the femoral head, is made up of zirconia ceramic and fits into the hip joint in the pelvis. The acetabular cup is also made from titanium alloy. It is coated in a porous alumina ceramic, to allow bone ingrowth for stabilisation. An ultra high molecular weight polyethylene (UHMWPE) liner fits inside the acetabular cup and provides the articulating surface for the femoral head. (Figure 1,2 & 3).

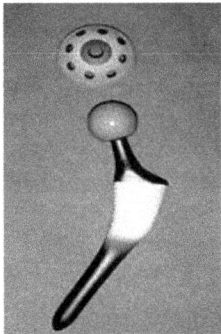

Figure 1. Implant components for a total hip replacement

Figure 2. A total knee replacement prosthesis components

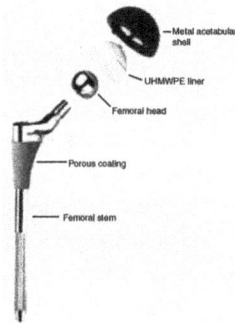

Figure3. Detailed components of Total Hip Replacement (THR) implant.

Figure 2 shows prototype of total knee replacement prosthesis. It consists of titanium alloy upper and lower structural components. A zirconia wear surface has been fabricated for the upper section. Similar to the hip prosthesis, this articulates against a UHMWPE insert on the lower section.

1.4 Commercially Pure Titanium (CP Ti)

Four grades of CP Ti are available according to oxygen content. CP Ti grade 4 contains the highest amount of oxygen, up to 0.4 per cent, and consequently has the highest tensile and yield strengths Ratner *et al*. 2004[1]. The most commonly used titanium alloys in orthopedics are commercially pure Ti (CP Ti), grade 4 (ASTM F67) and Ti_6Al_4V (ASTM F136).In dental implants CPTi with single phase alpha microstructure is used whereas Ti_6Al_4V, which has biphasic alpha-beta microstructure is used in orthopaedic applications. CP titanium may also be used in some implants (for example, to make fiber metal) in which a layer of metal fibers is bonded to the surface of an implant to allow the bone to grow in to the implant for a better grip.

1.5 Tantalum

Tantalum is a pure metal with excellent physical and biological characteristics. It is flexible, corrosion resistant, and biocompatible. Tantalum is gaining more attention as a new biomaterial. Elemental tantalum unites strength and corrosion resistance with excellent biocompatibility. Trabecular metal is made by the pyrolysis of thermosetting polymer foam, creating a low-density vitreous carbon skeleton, and then commercially pure Ta is deposited onto this interconnected scaffold using chemical vapor deposition/infiltration techniques that create a metallic strut configuration similar to trabecular bone. This explains tantalum's surgical applications such as cranioplasty plates and pacemaker leads [2]. Lack of sufficient mechanical strength for use as load-bearing metal implants is a major limitations of open cell porous scaffolds. Therefore, porous metals fabrication technologies which can ensure sufficient mechanical strength, net-shape fabrication capability, shape, pore size and high levels of purity for biomedical applications, become significantly important. Tantalum is often used as X-ray markers for stents because of its excellent X-ray visibility and low magnetic susceptibility.

1.6 Ultra High Molecular Weight Polyethylene (UHMWPE)

Sir John Charnley first used UHMWPE in 1962 and by 1970 it emerged as the dominant bearing material for total hip and knee replacements. UHMWPE has been used as a successful biomaterial for hip, knee, and spine implants over the last few decades. Highly crosslinked UHMWPE materials were introduced in 1998 and since then have rapidly become the standard of care for total hip replacements [3]. These new materials are crosslinked with gamma or electron beam radiation (50–105 kGy) and then thermally processed to improve their oxidation resistance [3]. Another important medical advancement for UHMWPE in the past decade has been the increase in use of fibers for sutures. UHMWPE is a type of plastic commonly used on the surface of one implant that is designed to come in contact with another implant, as in a joint replacement. In fact, a special type of medical-grade polyethylene was developed specifically for use in orthopedic implants. Abrasion, chemical resistance, and its compatibility with body tissue are basic properties for the medical use of UHMW-PE. Polyethylene is very durable when it comes into contact with other materials. When a metal implant moves on a polyethylene surface, as it does in most joint replacements, the contact is very smooth and the amount of wear is usually minimal. Polyethylene can be made even more resistance to wear, through a process called cross linking, which creates stronger bonds between the molecular chains of the polyethylene. The cross-linking processes can result in materials with lower mechanical properties than standard ultrahigh molecular weight polyethylene. The appropriate amount of cross linking depends on the type of implant. For example, the surface of a hip implant may require a different degree of cross linking than the surface of a knee implant. The radiation-cross linked remelted cups exhibited excellent resistance to oxidation. Because crosslinking can reduce the ultimate tensile strength, fatigue strength, and elongation to failure of ultra high molecular weight polyethylene, the optimal crosslinking dose provides a balance between these physical properties and the wear resistance of the implant and might substantially reduce the incidence of wear-induced osteolysis with total hip replacements. Osteolysis induced by ultra high molecular weight polyethylene wear debris is one of the primary factors limiting the lifespan of total hip replacements.

1.7   Ceramics

Ceramic materials are usually made by pressing and heating certain metal oxides (typically aluminum oxide and zirconium oxide) until they become hard and dense. These ceramic materials are strong, resistant to wear, and biocompatible. They are used mostly to make implant articulating surfaces that do not require flexibility, as in the surfaces of a hip joint.

1.8 Composite Materials

CLASSIFICATIONS
Based on the type of the Matrix Material

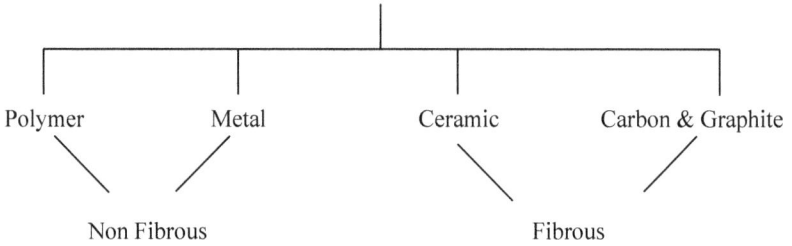

Polymer	Metal	Ceramic	Carbon & Graphite

Non Fibrous   Fibrous

Based on the form of the reinforcement

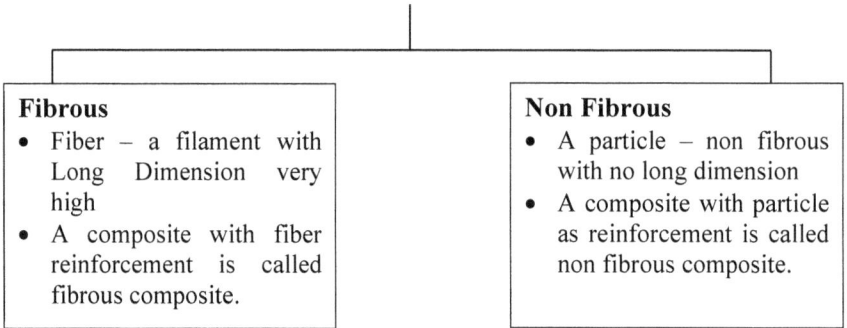

Fibrous
- Fiber – a filament with Long Dimension very high
- A composite with fiber reinforcement is called fibrous composite.

Non Fibrous
- A particle – non fibrous with no long dimension
- A composite with particle as reinforcement is called non fibrous composite.

One of the primary restrictions on clinical use of bioceramics is the uncertain lifetime under the complex stress states, slow crack growth, and cyclic fatigue that arise in many clinical applications. Composite materials are made by mixing two or more separate materials without creating a chemical bond between the materials. Metal alloys and ceramics are not considered to be composite materials because their ingredients are chemically bonded to create a new material. Bioactive and mechanically strong composite can be produced by combining of the bioinert and bioactive ceramics. On a larger scale two layers of these different materials can be combined to create a composite material with the desired characteristics, for example, hip implant stem may consist of layers of two different materials (alumina ceramic and hydroxyapatite composites) to get the desired combination of strength and flexibility. After introduction of ceramic particulate reinforcement there is a wide choice of material for implant applications like ceramic/ceramic, ceramic/polymer, and ceramic/metal composites. But the ceramic/polymer composites have been found to release toxic elements into the adjacent tissues and further have the limitations of organization into definite shape. The metal/ceramic composite faces problems of corrosion of metal and degradation of its ceramic coatings during its long time application. Ceramic/ceramic composites are more superior as they are similar to bone minerals, exhibiting biocompatibility and are able to be shaped into different sizes. The combinations of various absorbable and nonabsorbable polymers with calcium-based materials are well accepted eg. Polymer composites are used in orthopedic surgery whereas biocompatible polymers and polymer ceramic composites are used for tissue graft. These non-fibrous composites cannot be used as load-bearing skeletal implants due to their relatively poor mechanical properties.

1.9   Trabecular Metal

Trabecular Metal is made from tantalum over carbon. It is strong, flexible, and biocompatible.   Trabecular Metal material is structurally similar to the bone, approximates its physical and mechanical properties more closely than any other prosthetic materials and is the primary reason for wide acceptance.This unique, highly porous, trabecular structure is conducive to bone formation, enables rapid and extensive tissue infiltration and strong attachment[4,5].The Trabecular Metal Osteonecrosis Intervention Implant System (implant design) has been developed to intervene in stage I or II osteonecrosis of the femoral head and may be used with bone graft. The device has the potential to limit the progression of the disease, delaying or preventing the need for a hip replacement in many patients. The advantages of Trabecular Metal material and the implant design includes: (1)75-80% porosity (2) Cellular structure as of bone (3) Stiffness similar to bone (4) High coefficient of friction gives implant stability (5) Structural support to necrotic part of femoral head (6) Limit the progression of disease, delaying the need for total hip arthroplasty (7) Offers an alternative to grafting procedures

Physical and Mechanical Properties: Trabecular Metal allows greater bone in growth as compared to other conventional porous coatings because it is made up of interconnecting pores which leads to a structural biomaterial that is 70-80% porous, it also has higher interface shear strength. It can withstand physiological loads as it has high strength-to-weight ratio. The elastic modulus and compressive strength are similar to bone [5, 6]. The bone-like physical and mechanical properties of Trabecular Metal contribute to extensive bone infiltration. The pore size and high volume porosity of Trabecular Metal supports vascularization and rapid, secure soft tissue ingrowth. Its low stiffness helps physiological load transfer and minimizes stress shielding.

1.10   Bioabsorbable Materials

Bioabsorbable materials are absorbed by the body when their job is over. They are made from a biocompatible plastic that can be dissolved by normal body fluids. Many sutures used today in all types of surgery are bioabsorbable. These bioabsorbable materials may also be used in implants that reattach soft tissue to bone.

1.11   Silicone

Silicones are a general category of synthetic polymer, rubbery in nature and are very flexible. Silicones are made up of repeating silicon to oxygen bond and silicon atoms are also bonded with organic groups (methyl group). The most common Silicones are the polydimethylsiloxanes and trimethylsilyloxy. The most important silicone Implant is the Silicone breast implant. Due to its unique biocompatibility and biodurability, silicones have widespread application in health care. In orthopedics, it is most commonly used implant material that replaces the joints of hand and the foot, to replace the hand flexor tendon, to replace the thumb CMC (carpometacarpal) joint and the MCP(metacarpophalangeal) and PIP(proximal interphalangeal) joints.One of the most common problems with silicone implants is that they can leak or rupture. Others side effect are Infection or painful, swollen or inflamed implant site, fracture of the implant, loosening or dislocation of the prosthesis which requires revised surgery, bone restoration or over-production, allergic reaction(s) to prosthesis material(s) and untoward histological responses possibly involving macrophages and/or fibroblasts. Silicon inside the body is toxic and is a carcinogenic (cancer causing) substance, it is known to damage the immune system, kill cells and produce silicosis.

2.   Features of an ideal medical implants material

The ideal material or material combination should exhibit the following properties

- A biocompatible chemical composition to avoid adverse tissue reactions
- Excellent resistance to degradation (e.g., corrosion resistance for metals or resistance to biological degradation in polymers)
- Acceptable strength to sustain cyclic loading endured by the joint
- A low modulus to minimize bone resorption
- High wear resistance to minimize wear debris generation

Young's modulus, also known as the tensile modulus or elastic modulus, is a measure of the stiffness of an elastic material and is a quantity used to characterize materials.

3.   History of bioceramics origin

Earlier to 1925 in implant surgery pure metals were more often used as biomaterials. The 1930s marked the beginning of the era of better surgical techniques and also the first use of alloys such as Vitallium. In 1969 L. L. Hench and others discovered that various kinds of glasses and ceramics could bond to living bone [7,8]. In a scientific meeting Hench come to know that after an injury the bodies of soldiers would often reject the implant. Hench was very interested in this and wanted to know more about it and began to investigate materials that would be biocompatible. The final product was a new material which he called Bioglass. This work inspired a new field called bioceramics[9]. With the discovery of bioglass interest in bioceramics grew rapidly.The predecessor of bio-ceramics is traditional ceramic materials, a heat reflective material used for many purposes. A good example of this is ceramic coating of space shuttle that reflects solar radiation while in space, keeping the shuttle Columbia cool." Scientists started research on ceramics and FIR (Far infrared Rays) for application to the human body. Their goal was ceramics that could generate FIR at lower temperatures and at the correct wavelength so that it could align with FIR generated through the natural process of metabolism by one's body in the form of heat waves. Their work lead to the birth of Bioceramics.

4.   Bioceramics and its early uses

Bioceramics are subset of biomaterials [10,11]. Bioceramics and bioglasses are biocompatible ceramic materials i.e. being compatible with the human body environment[12]. Biocompatibility is a direct result of their chemical compositions which contain ions commonly found in the physiological environment (such as Ca^{2+}, K^+, Mg^{2+}, Na^+, etc.) and of other ions showing very limited toxicity to body tissues (such as Al^{3+} and Ti^{4+}). Bioceramics range in biocompatibility from inert nature in the body e.g.ceramic oxides to the other extreme of resorbable materials (which are eventually replaced by the materials which they were used to repair and reconstruction of diseased or damaged parts of the musculo-skeletal system). Bio-ceramic material is produced by combining various mineral oxides such as Silica, Aluminum, Magnesium, and more than 20 types of ceramics. These ingredients are heated at high temperatures (2900 F/1600C) to form the bio-ceramic material. Bioceramics can be single crystals (sapphire), polycrystalline (alumina or hydroxyapatite (HA)), glass (Bioglass), glass-ceramics (A/W glass-ceramic), or composites (stainless-steel-fiber-reinforced Bioglass or polyethylene-

Carbonaceous Composite Materials   Materials Research Forum LLC
Materials Research Foundations **42** (2018) 143-177   doi: http://dx.doi.org/10.21741/9781945291975-6

hydroxyapatite. Bioceramic materials must be long lasting, structural failure resistant, and corrosion resistant, also must have a low Young's modulus to prevent cracking of the material. Bone is a complex living tissue and structurally bone is a composite consisting of inorganic hydroxyapatite crystals (HA) and an organic mixture of mostly collagen fiber with small amounts of other proteins. The collagen fibers are aligned in bundles that contain apatite nano-crystals aligned along the fiber axis. Bone also contains smaller amounts of magnesium, fluoride and sodium. These minerals give bone its characteristic hardness and the ability to resist compression.This structural alignment is the foundation of the multi-level hierarchial structure of bone and contributes to bone's natural resistance to bending. So to be successful, a biocompatible synthetic bone composite must mimic this important orientation of components to achieve the mechanical properties of bone. Although the skeleton plays a vital role in the human body both in terms of support and locomotion and also the protectionof vital organs, it is susceptible to fractures as a result of injury and degenerative diseases which are often associated with ageing. Therefore there has always been a need, since the earliest time, for the repair of damaged hard tissue.The earliest attempts to replace hard tissue with biomaterials aimed to restore basic functions by repairing the defects caused by injury and disease and to elicit minimal biological response from the physiological environment.These materials are now largely classed as "Bioinert" and the absence of a toxic response would have been considered to be a successful outcome. In the 1920s de Jong[13] first observed the similarities betweenthe X-ray diffraction patterns of bone mineral and a calciumphosphate compound, hydroxyapatite. Later Posner and coworkers identified the crystallographic structure of bone mineral and hydroxyapatite [14, 15].

5.   Overview of bioceramics applications

In present time ceramics are commonly used in medical field for replacements of hips, knees, teeth, tendons, and ligaments and also for periodontal disease, maxillofacial reconstruction, augmentation and stabilization of the jaw bone, spinal fusion, and as bone fillers after tumor surgery. Carbon coatings are thrombo resistant, used for prosthetic heart valves. Therapeutic treatment of cancer has been achieved by localized delivery of radioactive isotopes via glass beads. A significant problem in the radiation treatment of cancer is the serious systemic side effects. Localization of the radiation at the site of the tumor decreases the radiation dosage required to kill the cancer cells and thereby minimizes side-effect toxicities. An innovative approach to the localized delivery of radioactive yttrium-90 (90Y) to treat liver cancer has been developed by using glass microspheres. An yttria-aluminosilicate glass, containing 89Y is made in the form of 25-μm microspheres. Prior to use in hepatic arterial infusion therapy, the microspheres are

Carbonaceous Composite Materials   Materials Research Forum LLC
Materials Research Foundations **42** (2018) 143-177   doi: http://dx.doi.org/10.21741/9781945291975-6

bombarded by neutrons which creates '90Y, a radioactive isotope which is a short half-life (64 h), short range (2.5 to 3 mm in the liver) β emitter. The microspheres are injected through a catheter placed in an artery, and the blood stream carries them to the liver where a high proportion goes to the cancerous part because of three times increased its blood supply. A localized dosage of up to 15000 rd can be delivered in this manner, whereas a maximum of 3000 rd of external radiation can be tolerated by the patient. Still another therapeutic application of bioceramics is delivery of various steroid hormones from aluminum calcium phosphate porous ceramics [16]. The advantage of this method is sustained delivery of a potentially toxic substance over long periods of time, again minimizing systemic side effects due to large dosages. Other medical uses of bioceramics are in pacemakers, kidney dialysis machines, and respirators. Applications of bioceramics are summarized as below in different areas of specialization.

Table 2: Biomedical Applications of Bioceramics

Medical Divisions of specialization	**Devices**	**Function**	**Biomaterial**
(A) For Implants in Orthopedic Surgery	Artificial total hip, knee, shoulder, elbow, wrist	Reconstruct arthritic or fractured joints	High-density alumina, metal bioglass coatings, stabilized zirconia.
	Bone plates, screws, wires	Repair fractures by holding the fragment in anatomical position. (stabilization and fixation of bone)i.e. osteosynthesis	Bioglass-metal fiber composite, Polysulfone-carbon fiber composite, PE-HA composite.
	Intramedullary nails	Align fractures for osteosynthesis	Bioglass-metal fiber composite, Polysulfone-carbon fiber composite
	Harrington rods	Correct abnormal spinal curvature	Bioglass-metal fiber composite, Polysulfone-carbon fiber composite
	Permanently implanted artificial limbs	Replace missing extremities	Bioglass-metal fiber composite, Polysulfone-carbon fiber composite
	Vertebrae Spacers and extensors	Correct congenital deformity	Al_2O_3

	Spinal fusion	Immobilize vertebrae to protect spinal cord	Bioglass
	Temporary bone space fillers		TCP, Calcium and phosphate salts
(B) For Dental Implants in Dentistry	Alveolar bone replacements, mandibular reconstruction	Restore the alveolar ridge to improve denture fit	Polytetra fluoro ethylene (PTFE) - carbon composite, Porous Al2O3, Bioglass, dense-apatite (HA), HA-autogenous bone composite.
	End osseous tooth replacement implants	Replace diseased, damaged or loosened teeth	Al_2O_3, Bioglass, dense hydroxyapatite, vitreous carbon
	Orthodontic anchors	Provide posts for stress application required to change deformities	Bioglass-coated Al2O3, Bioglass coated vitallium
	Periodontal pocket obliteration		HA, HA-PLA composite, TCP, Calcium and phosphate salts bioactive glasses
(C) Otorhinology	In ENT	Artificial eardrum	Al_2O_3,HA, Bioactive Glasses Bioactive Glass-ceramics
(D) Artificial tendon and ligament	Musculoskeletal System		PLA-carbon-fiber composite
(E) Artificial heart valves and stent	For cardiac uses	In cardio vascular	Pyrolytic carbon coatings
(F) Maxillofacial reconstruction			Al_2O_3 HA, HA-PLA composite Bioactive glasses

6.   Subdivision of bioceramics

Subdivision of bioceramics is based on their bioactivity. No material implanted in living tissues is inert; all materials elicit a response from living tissues. When a synthetic material is placed within the human body, tissue reacts towards the implant in a variety of ways depending on the material type. Materials elicit one of the four types of responses (Implant-Tissue Response), are

(a) If the material is toxic, the surrounding tissue dies.
(b) If the material is nontoxic and biologically inactive (nearly inert), a fibrous tissue of variable thickness forms.
(c) If the material is nontoxic and biologically active (bioactive), an interfacial bond forms.
(d) If the material is nontoxic and dissolves, the surrounding tissue replaces it.

These four types of response allow different types of attachment of prostheses to the musculo-skeletal system. The attachment mechanisms, with examples, are summarized below.

Table 3: Summary of Bioceramics-tissue attachment mechanism and their examples.

Type of bioceramic	Type of attachment mechanism	Examples
1	**Morphological fixation:** Dense, nonporous, nearly inert ceramics attach by bone growth into surface irregularities by cementing the device into the tissues, or by press fitting into a defect.	Al_2O_3 (single crystal polycrystalline)
2	**Biological fixation:** For porous inert implants bone ingrowth occurs, which mechanically attaches the bone to the material.	Al_2O_3 (porous polycrystalline) Hydroxyapatite-coated porous metals
3	**Bioactive fixation:** Dense, nonporous, surface-reactive ceramics, glasses, and glass-ceramics attach Hydroxyapatite directly by chemical bonding with the bone.	Bioactive glasses Bioactive glass-ceramics
4	**Bioresorbable:** Dense, nonporous (or porous), resorbable ceramics are designed to be slowly replaced by bone.	Calcium sulfate (plaster of Paris) Tricalcium phosphate Calcium phosphate salts

In general, there are three terms in which a biomaterial may be described, representing the tissues responses.

Carbonaceous Composite Materials   Materials Research Forum LLC
Materials Research Foundations **42** (2018) 143-177   doi: http://dx.doi.org/10.21741/9781945291975-6

Classification of Biocompatible Materials

Based on their effect on body tissues
- **Biotolerant** Example: Metal alloys, polymers
- **Bioinert** Example: Materials based on aluminum oxide, zirconium dioxide
- **Bioactive** Example: Materials based on calcium phosphates

Based on the effect of body tissues on materials
- **Biodegradable** Example: Metal alloys, polymers
- **Bioresistive** Example: Materials based on hydroxyapatite
- **Bioresorptive** Example: Materials bearing tricalcium phosphate or phosphate and silicophosphate glasses.

6.1   Bioinert

Are high strength ceramics, the term bioinert refers to any material that has minimal interaction with its surrounding tissue in the body, examples of these are stainless steel, titanium, alumina, partially stabilized zirconia. Generally a fibrous capsule might form around bioinert implants hence its biofunctionality relies on tissue integration through the implant. Examples with detailed are given below.

- Alumina (Al_2O_3)
- Zirconia (ZrO_2)
- Oxide ceramics
- Silica ceramics
- Carbon fiber
- Diamond-like carbon
- Ultra high molecular weight polyethylene (UHMWPE)

6.1.1   Alumina (Al_2O_3)

High density, high purity (> 9 9.5%) alumina (alpha-Al_2O_3) was the first bioceramic widely used clinically. An alumina ceramic has characteristics of high hardness and high abrasion resistance. The excellent wear and friction behavior of Al_2O_3 are associated with the surface energy and surface smoothness of this ceramic. There is only one thermodynamically stable phase, i.e. Al_2O_3 that has a hexagonal structure with aluminium ions at the octahedral interstitial sites. Although some dental implants are single-crystal sapphire, most alumina devices are very-fine-grained polycrystalline alpha-Al_2O_3. A very small amount of magnesia (<0.5%) is used as an aid to sintering and to limit grain growth during sintering.Alumina has been used in orthopedic surgery for nearly 20 years, motivated largely by (1) its excellent type 1 biocompatibility and very thin capsule formation which permits cementless fixation of prostheses and (2) its exceptionally low coefficients of friction and wear rate. Alumina on alumina load-bearing wearing surfaces, such as in hip prostheses, must have a very high degree of sphericity produced by grinding and polishing the two mating surfaces together. The alumina ball and socket in a hip prosthesis are polished together and used as a pair. The long-term coefficient of friction of an alumina-alumina joint decreases with time and approaches the values of a normal joint. This leads to wear of alumina on alumina articulating surfaces that are nearly 10 times lower than metal-polyethylene surfaces. Abrasion resistance, strength and chemical inertness of alumina have made it to be recognized as a ceramic for dental and bone implants. Other clinical applications of alumina include knee prostheses, bone screws, alveolar ridge and maxillofacial reconstruction, ossicular (middle ear) bone substitutes, keratoprostheses (corneal replacements), segmental bone replacements, plate & screw and posterior type dental implants.

6.1.2   Zirconia (ZrO_2)

Zirconia is of two types, partially stabilized zirconia (PSZ) and zirconia toughened alumina (ZTA). Zirconia (ZrO_2) is a ceramic material with adequate mechanical properties for manufacturing of medical devices. Zirconia stabilized with Y_2O_3 has the best properties for these applications. Zirconia ceramics have several advantages over other ceramic materials due to the transformation toughening mechanisms operating in their microstructure that can be manifested in components made out of them. In present day's main application of zirconia ceramics is in THR ball heads. When a stress occurs on a ZrO_2 surface, a crystalline modification opposes the propagation of cracks. Compression resistance of ZrO_2 is about 2000 MPa. Orthopedic research leads to this material being proposed for the manufacture of hip head prostheses. Prior to this, zirconia biocompatibility had been studied in vivo; no adverse responses were reported following the insertion of ZrO_2 samples into bone or muscle. In vitro experimentation showed absence of mutations and good viability of cells cultured on this material. Zirconia cores for fixed partial dentures (FPD) on anterior and posterior teeth and on implants are now available. Clinical evaluation of abutments and periodontal tissue must be performed prior to their use. Zirconia opacity is very useful in adverse clinical situations example, for masking of dischromic abutment teeth.

6.2   Bioactive

Bioactive material within the human body interacts with the surrounding bone and in few cases, even soft tissue. This occurs through a time dependent kinetic modification of the

Carbonaceous Composite Materials   Materials Research Forum LLC
Materials Research Foundations **42** (2018) 143-177   doi: http://dx.doi.org/10.21741/9781945291975-6

surface, triggered by their implantation within the living bone. An ion – exchange reaction between the bioactive implant and surrounding body fluids results in the formation of a biologically active carbonate apatite (CHAp) layer on the implant that is chemically and crystallographically equivalent to the mineral phase in bone. Examples of bioactive materials are

- Synthetic hydroxyapatite [$Ca_{10}(PO_4)_6(OH)_2$]
- Glass ceramic A-W
- Bioactive glass(e.g. 45S5 Bioglass)

A common characteristic of bioactive glasses and bioactive ceramics is a time-dependent, kinetic modification of the surface that occurs upon implantation [17, 18]. Bioactive Materials develop an adherent interface with tissues that resists substantial mechanical forces. In many cases the interfacial strength of adhesion is equivalent to or greater than the cohesive strength of the implant material or the tissue bonded to the bioactive implant. Bonding to bone was first demonstrated for a certain compositional range of bioactive glasses which contained SiO_2 Na_2O, CaO, and P_2O_5, in specific proportions. There were three key compositional features to these glasses that distinguished them from traditional Na_2O-CaO-SiO_2 glasses: (1) less than 60 mol% SiO_2, (2) high-Na_2O and high-CaO content, and (3) high-CaO/P_2O_5 ratio. These compositional features made the surface highly reactive when exposed to an aqueous medium.

6.2.1 Synthetic hydroxyapatite [$Ca_{10}(PO_4)_6(OH)_2$]

Medical uses of calcium phosphate-based bioceramics i.e. synthetic hydroxyapatite arises due to its similarity to the bone apatite, the major component of the inorganic phase of bone, which plays a key role in the calcification and resorption processes of bone. Most materials in the CaO – P_2O_5 – H_2O system easily satisfy the requirement of biocompatibility. The presence in the material of CO_3^{2-}, SiO_4^{4-}, Cl^-, F^-, Na^+, K^+, Mg^{2+}etc., i.e., all ions contained in the natural bone structure or in body tissues, do not disturb biocompatibility. The low strength of hydroxyapatite (HA, $Ca_{10}(PO_4)_6(OH)_2$)) have limited their scope for clinical applications and hence more research needs to be conducted to improve their mechanical properties. Scientist interest in this group of materials is for their use as a porous structure, as the bioactive phase in composites, as a bioactive coating on metallic implants, and as the bioactive matrix of composites.

Carbonaceous Composite Materials   Materials Research Forum LLC
Materials Research Foundations **42** (2018) 143-177   doi: http://dx.doi.org/10.21741/9781945291975-6

Ideal hydroxyapatite materials have to meet certain requirements:

1. They should be biocompatible with the living organism;
2. The material should have a certain level of bioresistivity or a certain biodegradation rate;(based on the treatment method used)
3. The material should possess biological activity, i.e., have an osteostimulating effect initiating the formation of bone tissue;
4. The materials should have certain strength;
5. The materials should withstand various types of sterilization and radiation (UHF, SHF, UV, x-ray, gamma radiation) without changing their properties of biocompatibility, bioresistivity, biodegradability, bioactivity, and strength parameters;
6. Porous materials should fulfill certain requirements e.g. type of porosity and the size of pores, whose dia. should be at least 100 nm,
7. Porous materials also ensure the desired interaction between the body tissue and the implanted material, i.e., the ingrowth of blood vessels and nerve fibers into the implant;
8. Osteoplastic materials should be easily amenable to mechanical treatment or other shape-correcting method used for surgery

6.2.2 Bioactive Composites

Bone is a natural composite material, having a complex structure in which several levels of organisation, from macro- to micro-scale, can be identified. Two levels of composite structure are considered when developing bone substitutes: first, the bone apatite reinforced collagen forming individual lamella at the nm to µm scale; second, osteon reinforced interstitial bone at the µm to mm scale. It is the apatite-collagen composite at the microscopic level that provides the basis for producing bioactive ceramic-polymer composites as analogue biomaterials for bone replacement [19]. Mechanical properties of bones have been well documented, which serve as the benchmark upon which the mechanical performance of bone analogue materials is evaluated. As an anisotropic material, cortical bone has a range of associated properties rather than a set of unique values. In order to overcome the problem of modulus-mismatch between existing implant materials and bone and promote the formation of a secure bond between the implant and host tissue, the concept of analogue biomaterials was introduced by Bonfield et al in the 1980s[20] since then, varieties of bioactive composite materials have been produced and investigated[19]. Due to the presence of particulate bioactive bioceramics such as HA, Bioglass and A-W glass-ceramic in these composites, implants made of these composites can form a strong bond with the host tissue resulting in their integration into the biological system.

6.2.3 Bioactive glass (e.g. 45S5 Bioglass)

History of Bioglass began in 1967 when Professor Larry Hench notice the wounds sustained during the Vietnam War in terms of amputations[21]. The need for the development of materials that would help in the repair of tissues by forming a direct bond with them, rather than the interfacial scar tissue that occurred around metallic and polymeric implants of that time. In the early 1970s, Hench et al.[22–24] reported that particular compositions with the Na_2O–CaO–P_2O_5–SiO_2 system with B_2O_3 and CaF_2 additions formed a strong, adherent bond with bone. In vitro tests showed that the 45S5 Bioglass composition undergoes a surface reaction which occurs very rapidly. The surface reaction is a complex, multi-stage process which results in the formation of a biologically active hydroxy-carbonate apatite (HCA) layer. This HCA phase is chemically and structurally similar to the mineral phase in bone and thus it provides a direct bonding by bridging host tissue with implants[25,26].The rate of bone bonding and the strength and stability of the bond vary with the composition and microstructure of the bioactive materials. Hench et al.[27] reported that for their particular formulation of bioactive glass, bone formed a rapid bond when the silica levels were in the range 42–53%; glasses with 54–60% silica required 2–4 weeks for bone to bond; and bone did not form a direct bond with glasses containing more than 60% silica. Hench defined two classes of bioactive materials (A and B) characterised by the rate of bone regeneration and repair. **Class A** materials are those that lead to both osteoconduction (the growth of bone along the bone–implant interface) and osteoproduction as a result of the rapid reactions on the implant surface[28,29]. **Class B** bioactivity occurs when only osteoconduction occurs[30,31].

6.2.4 Apatite-wollastonite (A-W) glass-ceramic

Kokubo et al.first reported the production and behavior of A-W glass-ceramic in 1982[32]. Apatite-wollastonite (A-W) glass-ceramic became one of the most extensively studied glass ceramics for use as a bone substitute. A dense and homogeneous composite was obtained after heat treatment of the parent glass, which comprised 38 wt% oxyfluorapatite ($Ca_{10}(PO_4)_6(O,F)_2$) and 34 wt% _-wollastonite ($CaO \cdot SiO_2$) crystals, 50–100 nm in size in a MgO-CaO-SiO_2 glassy matrix. Apatite-wollastonite glass-ceramic is an assembly of small apatite particles effectively reinforced by wollastonite.The bending strength, fracture toughness and Young's modulus of A-W glass-ceramic are the highest among bioactive glass and glass ceramics, enabling it to be used in some major compression load bearing applications, such as vertebral prostheses and iliac crest replacement. It combines high bioactivity with suitable mechanical properties [33-38]. The final product contains:

- Wollastonite 28 wt% ($CaO \cdot SiO_2$)
- Oxyfluoroapatite 34 wt% ($Ca_{10} (PO_4)_6 (O,F)_2$
- Glass 28 wt% (MgO 17 wt%, CaO, 24 wt%, SiO_2, 59 wt%)

6.3   Bioresorbable

Bioresorbable refers to a material that upon placement within the human body starts to dissolve (resorbed) and slowly replaced by advancing tissue (such as bone).Common examples of bioresorbable materials are tricalcium phosphate [$Ca_3(PO_4)_2$] and polylactic–polyglycolic acid copolymers.Resorption or biodegradation of calcium phosphate ceramics is caused by (1) physiochemical dissolution, which depends on the solubility product of the material and local pH of its environment (New surface phases may be formed, e.g., amorphous calcium phosphate, dicalcium phosphate dihydrate, octacalcium phosphate ($Ca_4(P04)_3H \cdot_3H_20$), and anionic substituted HA); (2) physical disintegration into small particles due to preferential chemical attack of grain boundaries; and (3) biological factors, such as phagocytosis, which causes a decrease in local pH[39].All calcium phosphate ceramics biodegrade to varying degrees in the following order: αTCP>β TCP>>HA. The rate of biodegradation increases as (1) surface area increases (powders> porous solid>dense solid), (2) crystallinity decreases, (3) crystal perfection decreases, (4) crystal and grain size decrease, and (5) ionic substitutions of CO_3^{2-}, Mg^{2+}, Sr^{2+} in HA take place. Factors for decreasing rate of biodegradation includes (1) F-substitution in HA (2) Mg^{2+} substitution in β-TCP, and (3) decreasing β-TCP/HA ratios in biphasic calcium phosphates. Because of these variables it is necessary to control the microstructure and phase state of a resorbable calcium phosphate bioceramic in addition to achieving precise compositional control to produce a given rate of resorption in the body. As yet, there are few data on the kinetics of these reactions and the variables influencing the kinetics. Most commonly used bioresorbable materials in orthopedics include polylactic acid (PLA); polyglycolic acid (PGA); the L-isotope form of PLA; the copolymer of PLA and PGA: PLGA; and the DL-isotope form of PLA: PDLLA. As mentioned above the degradation process of these materials in the body involves two steps: hydrolysis followed by metabolizations. The process of hydrolysis involves the breaking of the polymer chains within the implant material to produce by-products that include lactic acid and glycolic acid single molecules. These degradation products are metabolized in the liver and produce carbon dioxide as a by-product, which body eliminate. In the area of high strength fracture fixation, PLLA is favoured by product because of its slow rate of complete resorption in to the body. The racemic mixture (mixture of isomers) consisting of PDLLA also shows characteristic that could utilize in the high strength situations. However PLLA,s semicrystalline structure provides much higher strength in the material than PDLLA,s amorphous structure, thus PLLA is the preferred material for use in fracture fixation devices(FFDs).

Table 4: Characteristic Features (physical and mechanical) of Ceramic Biomaterials are summarized as below.

Material	Young's Modulus [GPa]	Compressiv eStrength [MPa]	Bond strength [GPa]	Hardness	Density [g/cm3]
Bio Inert					
Al_2O_3	380	4000	300-400	2000-3000 (HV)	>3.9
ZrO_2 (PS)	150-200	2000	200-500	1000-3000 (HV)	≈6.0
Graphite (LTI)	20-25	138	NA	NA	1.5-1.9
Pyrolitic Carbon	17-28	900	270-500	NA	1.7-2.2
Vitreous Carbon	24-31	172	70-207	150-200 (DPH)	1.4-1.6
Bioactive					
HAP	73-117	600	120	350	3.1
Bioglass	≈75	1000	50	NA	2.5
AW Glass Ceramic	118	1080	215	680	2.8
Bone	3-30	130-180	60-160	NA	NA

PS - Partially Stabilized; HA - Hydroxyapatite; NA - Not Available; AW - Apatite-Wallastonite; HV - Vickers Hardness; DPH - Diamond Pyramid Hardness

Carbonaceous Composite Materials   Materials Research Forum LLC
Materials Research Foundations **42** (2018) 143-177   doi: http://dx.doi.org/10.21741/9781945291975-6

6.4   Porous ceramics

Porous ceramics are for tissue ingrowths, examples are hydroxyapatite-coated metals and Alumina. The potential advantage offered by a porous ceramic implant is its inertness combined with the mechanical stability of the highly convoluted interface developed when bone grows into the pores of the ceramic. When pore sizes exceed 100 pm, bone will grow within the interconnecting pore channels near the surface and maintain its vascularity and long-term viability. In this manner the implant serves as a structural bridge and model or scaffold for bone formation.

7.   Ceramic Materials for Artificial Joints

Ceramic materials have been used for artificial joints since the 1970s when the first generation of alumina products demonstrated superior resistance to wear, compared to the traditional metal and polyethylene materials. Advances in material quality and processing techniques and a better understanding of ceramic design led to the introduction of second generation alumina components in the 1980s that offered even better wear performance.

Advantages of Using Ceramics over Traditional Materials for artificial joints: Traditional metal–polyethylene hip system wear generates polyethylene particulate debris, inducing osteolysis, weakening of surrounding bone and results in loosening of the implant. Ceramic materials generate significantly less polyethylene debris when used in conjunction with polyethylene acetabular components in bearing couples. Ceramic-on-ceramic hip joints received FDA approval in 2003. Where an alumina femoral head is mated with an alumina acetabular cup, totally eliminates polyethylene debris and reduces wear significantly. In addition the use of ceramic-on -ceramic hip systems also alleviates metal ion release into the body if a metal on metal hip system were used. This superior wear performance extends the life of artificial joints, giving ceramic-on-ceramic joints a predicted life of well over 20 years. Serving the needs of the increasing numbers of younger patients for whom such surgery is now a viable operation, these ceramic-on-ceramic joints allow them to leads active lifestyles.

8.   Coatings for medical implants

8.1   Carbon coating

Bokros in 1967, describe the medical use of pyrolytic carbon coatings on metal substrates[40]. Soon after that these coatings were used for heart surgery implants. He also emphasized that the good compatibility of carbonaceous materials with bone and other tissue and the similarity of the mechanical properties of carbon to those of bone indicate that carbonaceus composite is also used for orthopedic implants. Unlike metals, polymers and other ceramics, these carbonaceous materials do not suffer from fatigue. However, their intrinsic brittleness and low tensile strength limits their use in major load bearing applications. In 1969 DeBakey[41] used low-temperature isotropic (LTI) carbon coatings for prosthetic heart valve for the first time in humans. Nowadays all prosthetic heart valves have LTI carbon coatings because of their excellent resistance to blood clot formation and long fatigue life [42].Three types of carbon are used in biomedical devices: the LTI variety of pyrolytic carbon, glassy (vitreous) carbon, and the ultralow-temperature isotropic (ULTI) form of vapor deposited carbon [43-44].These carbon materials are used either as integral and monolithic materials (glassy carbon and LTI carbon) or impermeable thin coatings (ULTI carbon). Above forms do not suffer from the integrity problems typical of other available carbon materials. With the exception of the LTI carbons co-deposited with silicon, all the carbon materials used clinically are pure elemental carbon. Up to 20 wt% silicon has been added to LTI carbon without significantly affecting the biocompatibility. Carbon surfaces are thromboresistant, compatible with the cellular elements of blood and also they do not influence plasma proteins or do not alter the activity of plasma enzymes.

8.2   Hydroxyapatite coating

Is the second most common bioceramic coating material on porous metal surfaces for fixation of orthopedic prostheses? Ducheyne and colleague in 1980 [45] suggested that HA powder in the pores of a porous, coated-metal implant would significantly affect the rate and vitality of bone in growth into the pores. Among various means of applying the HA coating, plasma spray coating generally is being preferred. Ceramic-based coatings, such as diamond-like carbon (DLC), provide a biocompatible, sterilization-compatible, non-leaching, and wear resistant surface for key pivot points and wear surfaces. Such coatings are used to reduce friction, increase surface hardness and prevent ion release from metal implant components.

9.   Failure of metals used for biomedical devices

Failure of metals at implantation site occurs as covered below.

9.1   Corrosion

Metal implants are prone to corrosion due to corrosive medium at implantation site and cyclic loading. Corrosion of metallic implants that occur within the human body constitutes an ion source that may potentially affect the local and systemic host environment. Therefore, corrosion resistance is the important property of the metallic

Carbonaceous Composite Materials   Materials Research Forum LLC
Materials Research Foundations **42** (2018) 143-177   doi: http://dx.doi.org/10.21741/9781945291975-6

implants. Types of corrosion in implant applications are fretting, pitting and fatigue. Fretting corrosion behavior of metallic implant is determined by many factors like Corrosive medium, chemical composition of alloy and level of stress at the contact surfaces. Titanium nitride coating on the metallic implant has been a popular method to improve corrosion resistance of metallic implant such as Ti alloy and Co based alloy by physical vapor deposition, plasma spray process, etc. Other methods to improve the corrosion resistance of the implant are like modification of metallic implant surface by electropolishing, sand blasting or shot peening method (Aparicioa, 2003)[46]. Evaluation of corrosion behavior using methods which resemble the services condition of the metal implants helps in considering the condition for prevention of corrosion. The fatigue corrosion of metallic implant in corrosive medium can be evaluated by ultrasonic frequency which enables the application of very-high stress cycle within reasonable testing period (Papakyriacou, 2000)[47].The fretting corrosion behavior of metallic implant can be evaluated by a typical pin-on disc method in an artificial physiological medium (Tritschler, 1999, Kumar, 2010)[48,49]. Parameters that are monitored includes concentration of corrosive medium, load or friction forces, frequency and number of fretting cycles. Pitting corrosion can be evaluated with the absences of applied forces. It was reported that a good example of pitting corrosion evaluation was obtained in a buffered saline solution using anodic polarisation and electrochemical impedance measurements (Aziz-Kerrzo, 2001)[50].

9.2   Fatigue and fracture

In human body cyclic loading of medical implants is the possible cause for fatigue fracture. Microstructure of the implant materials is one of the factors which determine the fatigue behavior of implant materials; example is Ti6Al4V with equiaxed structure has a better fatigue strength property than the elongated structure (Akahori, 1998)[51]. Other factors are the frequency of the cyclic loading or the cycling rate (Karla 2009, Lee 2009),[52-53] design of the implants and the type of fluid medium of the implant.

9.3   Wear

Implant wear and aseptic loosening are two important failure problems that should be taken into consideration when dealing with long-term prosthetic devices. Corrosion process and wear are the surface degradation properties that limits the use of metallic implant such as Ti alloy (Dearnley, 2004) [54]. An improvement on the wear properties of Ti alloy was achieved due to titanium nitride coating on hip implant (Harman, 1997)[55].Wear failure can be avoided by proper material selection. It was also reported that in knee replacement changing implant material from UHMWPE (ultra-high molecular weight polyethylene) to CoCrMo implant alloys significantly reduces the wear debris process (Harman, 1997)[55]. Similarly metal-on-metal ortho prostheses show better wear performance than metal-on-UHMPWE (Spriano, 2005)[56].

9.4 Metal ions release

There are several factors which play important role on the metal ion release. First, the existence of passive oxide films, once it is broken, metal ions release will be easier to occur (Hanawa, 2004) [57]. Second, pH factor where ion release in both stainless steel and Co are affected by pH of the body fluid at a degree that higher for stainless steel (Okazaki, 2008) [58]. Coating is used to reduce the metal ion release from metallic implant. Titanium nitride layer have an excellent biocompatibility and the formation of hard nitride layer causes a lower ion release on the metallic implant (Ferrari, 1993) [59]. Therefore titanium nitride coating has been implemented on the Ti alloy and Co based alloy using plasma spraying method (Ferrari, 1993)[59]. Hydroxyapatite coating was also reported to decrease the metal ion release (Browne, 2000)[60].

Acknowledgements

The authors are thankful to the Department of Applied Chemistry, Anatomy & orthopedics JNMC, Aligarh, for generous support.

References

[1] B.D. Ratner , A.S. Hoffman , F.J. Schoen Biomaterials science. An introduction to materials in medicine. In Elsevier/Academic Press 2nd edn. 2004 Amsterdam, the Netherlands/New York, NY: Elsevier/Academic Press.

[2] J. Black. Biological performance of tantalum, Clin Materials, 16 (1994) 167-173. https://doi.org/10.1016/0267-6605(94)90113-9

[3] Steven M. Kurtz. The UHMWPE handbook: ultra-high molecular weight polyethylene in total joint replacement (2004), Academic Press. ISBN 978-0-12-429851-4. Retrieved 19 September 2011.

[4] J.D. Bobyn, G. Stackpool, K.K Toh, et. al. Bone in growth characteristics and interface mechanics of a new porous tantalum biomaterial, Journal of Bone Joint Surgery, 81-B (1999) 907-914. https://doi.org/10.1302/0301-620X.81B5.0810907

[5] J.D. Bobyn, S.A. Hacking, S.P Chan, et. al. Characterization of a new porous tantalum biomaterial for reconstructive orthopaedics. Scientific Exhibit, Proc of AAOS, Anaheim CA, (1999).

[6]   J.J. Krygier, J.D. Bobyn, R.A. Poggie, et. al. Mechanical characterization of a new porous tantalum biomaterial for orthopaedic reconstruction. Proc SIROT. Sydney, Australia, (1999).

[7]   Hench, L. Larry. "Bioceramics: From Concept to Clinic". Journal of the American Ceramic Society 74 (1991) 1487. https://doi.org/10.1111/j.1151-2916.1991.tb07132.x

[8]   T. Yamamuro, L.L. Hench, J. Wilson. CRC Handbook of bioactive ceramics vol ii (1990).

[9]   Kassinger, Ruth. Ceramics: From Magic Pots to Man-Made Bones. Brookfield, CT: Twenty- First Century Books, (2003).

[10]   J.F. Shackelford (editor) MSF bioceramics applications of ceramic and glass materials in medicine (1999).

[11]   H. Oonishi, H. Aoki, K. Sawai (editors) Bioceramics vol. 1(1988).

[12]   P. Ducheyne, G.W. Hastings (editors) CRC metal and ceramic biomaterials vol1(1984).

[13]   De Jong, W. F., Le, substance minerale dans le os, Recueil des Travaux Chimiques des pays, 45 (1926) 445-450.

[14]   A. S. Posner, Crystal chemistry of bone mineral. Physiology Review, 49 (1969) 760–792. https://doi.org/10.1152/physrev.1969.49.4.760

[15]   I.M. Kay, R.A. Young, A.S. Posner, Crystal structure of hydroxyapatite, Nature, 204 (1964) 1050–1052. https://doi.org/10.1038/2041050a0

[16]   H.A. Benghurri, P.K. Bajpai, "Sustained Release of Steroid Hormones from Polylactic acid or Polycaprolactone-Impregnated Ceramics", pp 93-110 in Handbook of Bioactive Ceramics. Vol 11, Calcium Phosphate and Hydroxylapatite Ceramics Edited by T. Yarnamuro, L.L. Hench, and J Wilson CRC Press, Boca Raton, FL. (1990).

[17]   L.L. Hench, "Bioactive Ceramics", p. 54 in 610- ceramics' Materials Characteristics Versus In Vivo Behavior. Vol. 523 Edited by P Ducheyne and J Lemons Annals of New York Academy of Sciences, New York, 1988.

[18]   U. Gross, R. Kinne. H.J. Schmitr, V. Strunz, "The Response of Bone to Surface Active Glass/Glass-Ceramics," RC Critical Reviews in Biocompatibility, 4 (1988) 155-179.

Materials Research Forum LLC
doi: http://dx.doi.org/10.21741/9781945291975-6

[19]   M. Wang, "Developing Bio-stable and Biodegradable Composites for Tissue Replacement and Tissue Regeneration", Materials Research Society Symposium 724: Biological and Biomimetic Materials – Properties to Function, San Francisco, USA, 2002.

[20]   W. Bonfield, M.D. Grynpas, A.E. Tully, J. Bowman, J. Abram, "Hydroxyapatite Reinforced Polyethylene - A Mechanically Compatible Implant Material for Bone Replacement", Biomaterials, 2 (1981) 185-186. https://doi.org/10.1016/0142-9612(81)90050-8

[21]   L.L. Hench, The story of bioglass, Journal of Materials Science: Materials in Medicine, 17 (2006) 967-978. https://doi.org/10.1007/s10856-006-0432-z

[22]   L.L. Hench, R.J. Splinter, T.K. Greenlee, W. C. Allen, Bonding mechanisms at the interface of ceramic prosthetic materials, Journal of Biomedical Materials Research, 2 (1971) 117-141. https://doi.org/10.1002/jbm.820050611

[23]   Jr. Greenlee, T. K. Beckham, C. A. Crebo, A.R. Jr., J.C.Malmborg, Glass ceramic bone implants, Journal of Biomedical Materials Research, 6 (1972) 235-244.

[24]   L.L. Hench, H.A. Paschall, Direct chemical bonding of bioactive glass-ceramic materials and bone, Journal of Biomedical Materials Research, 4 (1973) 25-42. https://doi.org/10.1002/jbm.820070304

[25]   C.G. Pantano Jr., A.E. Clark Jr., L.L. Hench, Multilayer corrosion films on glass surfaces, Journal of the American Ceramic Society, 57 (1974) 412–413. https://doi.org/10.1111/j.1151-2916.1974.tb11429.x

[26]   M. Ogino, L.L. Hench, Formation of calcium phosphate films on silicate glasses, Journal Non-Crystalline Solids, 38-39 (1980) 673-678. https://doi.org/10.1016/0022-3093(80)90514-1

[27]   L.L. Hench, Bioceramics: from concept to clinic, Journal of the American Ceramic Society, 74 (1991) 1487-1510. https://doi.org/10.1111/j.1151-2916.1991.tb07132.x

[28]   L.L. Hench, J.K. West, Biological applications of bioactive glasses, Life Chemistry Reports, 13 (1996) 187-241.

[29]   J. Wilson, S.B. Low, Bioactive ceramics for periodontal treatment: comparative studies, Journal of Applied Biomaterials, 3 (1992) 123-129. https://doi.org/10.1002/jab.770030208

[30] H. Oonishi, S. Kutrshitani, E. Yasukawa, H. Iwaki, L.L. Hench, J. Wilson, E. Tsuji, T. Sugihara, Particulate bioglass compared with hydroxyapatite as a bone graft substitute, Clinical Orthopaedics and Related Research, 334 (1997) 316–325. https://doi.org/10.1097/00003086-199701000-00041

[31] H. Oonishi, L.L. Hench, J. Wilson, F. Sugihara, E. Tsuji, S. Kushitani and H. Iwaki, Comparative bone growth behaviour in granules of bioceramic materials of various sizes. Journal of Biomedical Materials Research, 44 (1999) 31-43. https://doi.org/10.1002/(SICI)1097-4636(199901)44:1<31::AID-JBM4>3.0.CO;2-9

[32] T. Kokubo, M. Shigematsu, Y. Nagashima, M. Tashiro, T. Nakamura, T. Yamamuro and S. Higashi, Apatite- and Wollastonite-containing glass ceramics for prosthetic applications. Bulletin of the Institute for Chemical Research, 60. Kyoto University, (1982) 260-268.

[33] N. Ikeda, K. Kawanabe, T. Nakamura, Quantitative comparison of osteoconduction of porous, dense A-W glass-ceramic and hydroxyapatite granules (effects of granule and pore sizes). Biomaterials, 20 (1999) 1087–1095. https://doi.org/10.1016/S0142-9612(99)00005-8

[34] T. Kokubo, A/W glass-ceramic: processing and properties. In AnIntroduction to Bioceramics, ed. L. L. Hench and J. Wilson. World ScientificPublishing Co. Pte. Ltd., Singapore, (1993) pp. 75–88.

[35] T. Kokubo, H.M. Kim, M. Kawashita, and T. Nakamura, What kinds of materials exhibit bone-bonding? In Bone Engineering, ed. J. E. Davies. Em Squared Incorporated, Toronto, (2000) pp. 190–194.

[36] T. Kokubo, S. Ito, Z.T. Huang, T. Hayashi, S. Sakka, T. Kitsugi, T. Yamamuro, Ca, P-rich layer formed on high-strength bioactive glassceramic A-W. Journal of Biomedical Materterials Research banner, 24, (1990) 331-343.

[37] T. Kokubo, S. Ito, M. Shigematsu, S. Sakka, T. Yamamuro, Mechanical properties of a newtype of apatite-containing glass-ceramic for prosthetic application, Journal of Materials Science, 20 (1985) 2001-2004. https://doi.org/10.1007/BF01112282

[38] T. Yamamuro, A/W glass ceramic for clinical applications. In An Introduction to Bioceramics, ed. L. L. Hench and J. Wilson. World Scientific Publishing Co. Pte. Ltd., Singapore, (1993) pp. 8--104.

[39] K. de Groot, R. Le Geros, "Significance of Porosity and Physical Chemistry of Calcium Phosphate Ceramics", in Bioceramics Material Characteristics Versus InVivo Behavior, Vol 523. Edited by P Ducheyne and J Lemons Annals of New York Academy of Sciences, New York (1988) pp 268-77.

[40] J. C. Bokros, W.H. Ellis, U.S. Pat No 3526005 Method of preparing an intravascular defect by implanting a pyrolytic carbon coated prosthesis, 1971

[41] J. C. Bokros, Carbon Biomedical Devices, Carbon, 15 (1977) 355-371. https://doi.org/10.1016/0008-6223(77)90324-4

[42] A. Haubold, H.S. Shim, and J. C. Bokros, "Carbon in Medical Devices", in Biocompatibiiity ofClinicai lmplant Materials, Vol /I CRC Press, Boca Raton, FL, (1981) pp 3-42

[43] J. D. Bokros, Carbon Biomedical Devices, Carbon, 18 (1977) 355-71. https://doi.org/10.1016/0008-6223(77)90324-4

[44] A. D. Haubold, R A. Yapp. and J D. Bokros, "Carbons", in Concise Encyclopedia of Medical and Dental Materiais Edited by DWilliams Pergainon Press, New York. 1990 pp 95-101

[45] P. Ducheyne, L. L. Hench. A. Kagan, M. Martens, A. Burssens, J.C. Muller, The Effect of Hydroxyapatite Impregnation of Skeletal Bonding of Porous Coated Implants, Journal of Biomedical Materials Research, 14 (1980) 225-137. https://doi.org/10.1002/jbm.820140305

[46] C. Aparicioa, F.J. Gil, C. Fonseca, M. Barbosa, J.A. Planell, Corrosion behavior of commercially pure Ti shot blasted with different materials and sizes of shot particles for dental implant applications, Biomaterials, 24 (2003) 263-273. https://doi.org/10.1016/S0142-9612(02)00314-9

[47] M. Papakyriacou, H. Mayer, C. Pypen, H. Plenk Jr, S. Stanzl-Tschegg, Effects of surface treatments on high cycle corrosion fatigue of metallic implant materials, International Journal of Fatigue., 22 (2000) 873-886. https://doi.org/10.1016/S0142-1123(00)00057-8

[48] B. Tritschler, B. Forest, J. Rieu, Fretting corrosion of materials for orthopaedic implants: a study of a metal/polymer contact in an artificial physiological medium, Tribology International, 32 (1999) 587-596. https://doi.org/10.1016/S0301-679X(99)00099-7

[49] S. Kumar, T.S.N.S Narayanan, S.G.S Raman, S.K. Seshadri, Evaluation of fretting corrosion behaviour of CP-Ti for orthopaedic implant applications, Tribology International, 43 (2010) 1245- 1252. https://doi.org/10.1016/j.triboint.2009.12.007

[50] M. Aziz-Kerrzo, K.G. Conroy, A.M. Fenelon, S.T. Farrell, C.B. Breslin, Electrochemical studies on the stability and corrosion resistance of Ti-based implant materials, Biomaterials, 22 (2001) 1531-1539. https://doi.org/10.1016/S0142-9612(00)00309-4

[51] T. Akahori, M. Niinomi, Fracture characteristics of fatigued Ti6Al4V ELI as an implant material, Materials science & engineering. A, Structural materials: properties, microstructure and processing, 243 (1998) 237-243.

[52] M. Karla, J.R. Kelly, Influence of loading frequency on implant failure under cyclic fatigue conditions, Dental Materials, 25 (2009) 1426-1432. https://doi.org/10.1016/j.dental.2009.06.015

[53] C.K. Lee, M. Karl, J.R. Kelly, Evaluation of test protocol variables for dental implant fatigue research, Dental Materials, 25 (2009) 1419-1425. https://doi.org/10.1016/j.dental.2009.07.003

[54] P.A. Dearnley, K.L. Dahma, H. Çimenoglu, The corrosion–wear behaviour of thermally oxidised CP-Ti and Ti6Al4V. Wear, 256 (2004) 469. https://doi.org/10.1016/S0043-1648(03)00557-X

[55] M.K. Harman, S.A. Banks, W.A. Hodge, Wear analysis of a retrieved hip implant with titanium nitride coating, The Journal of Arthroplasty,12 (1997) 938. https://doi.org/10.1016/S0883-5403(97)90164-9

[56] S. Spriano, E. Vernè, M.G. Faga, S. Bugliosi, G. Maina, Surface treatment on an implant cobalt alloy for high biocompatibility and wear resistance, Wear, 259 (2005) 919-925. https://doi.org/10.1016/j.wear.2005.02.011

[57] T. Hanawa, Metal ion release from metal implants, Materials Science and Engineering C, 24 (2204) 745-752. https://doi.org/10.1016/j.msec.2004.08.018

[58] Y. Okazaki, E. Gotoh, Metal release from stainless steel, Co-Cr-Mo-Ni-Fe and Ni-Ti alloys in vascular implants, Corrosion Science50 (2008) 3429-3438. https://doi.org/10.1016/j.corsci.2008.09.002

[59]   F. Ferrari, A. Miotello, L. Pavloski, E. Galvanetto, G. Moschini, S. Galassini, P. Passi, S. Bogdanovie, S. Fazini, M. Jaksi, V. Valkovi, Metal-ion release. From Ti and TiN coated implants in rat bone, Nuclear Instruments and Methods in Physics Research Section B: Beam Interactions with Materials and Atoms, 79 (1993) 421-423. https://doi.org/10.1016/0168-583X(93)95378-I

[60]   M. Browne, P.J. Gregson, Effect of mechanical surface pretreatment on metal ion release, Biomaterials, 21 (2000)385-392. https://doi.org/10.1016/S0142-9612(99)00200-8

Carbonaceous Composite Materials
Materials Research Foundations **42** (2018) 177-204

Materials Research Forum LLC
doi: http://dx.doi.org/10.21741/9781945291975-7

Chapter 7

Purification of Industrial Effluent by Ultrafiltration Ceramic Membrane based on Natural Clays and Starch Powder

Sonia Bouzid Rekik[1,2]*, Jamel Bouaziz[1], André Deratani[2], Samia Baklouti[1]

[1]Laboratory of Industrial Chemistry, National School of Engineering, University of Sfax, BP 1173, 3038 Sfax, Tunisia

[2]European Membrane Institute (IEM), University of Montpellier 2, Place E. Bataillon, 34095 Montpellier Cedex 5, France

bouzidsonia@gmail.com*

Abstract

This chapter discusses the development of a low cost ceramic membrane prepared from kaolin clays for ultrafiltration application to produce clear water from high turbidity water. The manufacturing of low cost tubular supports via extrusion, using kaolin and corn starch as a pore forming agent was the first purpose in this study. Our second objective is to deposit ultrafiltration layer on the optimised tubular support. Finally, the prepared membrane has been applied for treating of industrial effluent under various pressure values.

Keywords

Natural Clays, Starch Powder, Ceramic Membrane, Filtration, Industrial Effluent

Contents

1.   Introduction

In the 21st century, the scarcity of global water resources has become one of most severe challenges for human beings. On the other hand, the rapid industrial growth has resulted in large wastewater production. The necessity to treat wastewater is an inevitable challenge before its discharge into the environment [1-5]. Actually, membrane technologies are considered as one of the most promising new separation techniques for wastewater treatment. Membranes are widely used in several applications due to advantages offered by their relatively high stability, efficiency, low energy requirement and facility of process [6-10].

According to the material of construction, the main types of membranes are polymer and ceramic membranes. Due to good chemical stability, good pressure resistance, excellent temperature stability, better mechanical resistance and long-life performance, mineral membranes are preferred in the professional environment over organic membranes [11-13]. Furthermore, ceramic membrane allows to achieve high water flux and to alleviate fouling problem compared with conventional polymeric membrane [14-17]. Classically, commercial ceramic membranes for separation processes are made from expensive powders such as zirconia, alumina, silica and titania [18-20]. Recently, membranes made from renewable and cheaper raw materials such as apatite powder, phosphates, natural clay, zeolite, kaolin and waste materials such as fly ash have received attention from academia and industry [16, 17, 21-33].

Ceramic membranes with high performance such as high permeability, high permselectivity parameters and good mechanical resistance can only be obtained in an asymmetric configuration, which consists of a multilayer system with a macroporous support (with the largest pore size). The last one is the top layer at which the separation takes place. Various synthesis routes for the preparation of top layer are usually used such as slip casting method, sol- gel process and dip coating [17, 23,25, 28, 33]. The properties of the ceramic membranes are mainly determined by their composition, the pore-former content and the sintering temperature. The microstructure of these membranes is controllable, thanks to the use of different types of pore forming agents. Besides, it is well known that porosity can be effectively controlled by varying sintering conditions. Normally, an improvement in porosity by adding a poreforming agent or by decreasing the sintering temperature cause a deterioration in mechanical reliability [17, 26, 28]. The key point is to break the trade-off limit between permeability and mechanical stability that is to increase water flux while maintaining mechanical properties.

Membrane processes like microfiltration, ultrafiltration, nanofiltration and reverse osmosis have proved their efficiency in the field of water treatment in various industrial sectors [28- 33]. Ultrafiltration (UF), as a low pressure driven membrane process with separation properties between microfiltration (MF) and nanofiltration (NF), has attracted growing interest in many applications, such as desalination and wastewater treatment. Over the past decades, attentions to treat industrial wastewater such as cuttlefish effluent and textile wastewater by micro and ultrafiltration ceramic membrane have been increased rapidly to obtain high effluent qualities [8, 17, 26, 32]. In this respect, ceramic membrane usage is a promising channel for the treatment of sewage.

Kaolin is one of the most widely clay materials and has an important role in numerous industrial applications and could be a preferred raw material for the development of porous ceramics membrane. The latter exhibits also hydrophilic behaviour, which is extremely desired to prepare membranes for water filtration [16,17, 26, 27]. Following the same perspective, the main goal of this study was to develop a low cost effective UF membrane for cleaning industrial effluent generated by seawater product industry. The membrane has been prepared by deposition of kaolin's active layer directly on kaolin-porous support without intermediate layer. This allowed to decrease the membrane resistance and thus to enhance the ultrafiltration process performances. The primary purpose of this study is the development of a porous ceramic support prepared by mixing an optimum amount of natural kaolin and starch powder as a pore forming to improve the porosity and the permeability of the support. In the second part, the elaboration of ultrafiltration layer supported by the kaolin support and their main properties will be discussed. To this end, the performance of the optimized membrane was mainly explored in this study by evaluating filtration and treatment performances.

2. Characterization of the starting materials

The raw material used for the preparation of both support and active layer is kaolin- clays provided by BWW Minerals.

2.1 Chemical composition of the powder

The chemical composition given in weight percentages of oxides is listed in Table 1. The samples have silica as a major component, followed by aluminium oxide while minor amounts of calcium oxide and titanium oxide are also present [17, 34].

Table 1. Chemical composition of the used kaolin (wt %).

	LOI*	SiO_2	Al_2O_3	K_2O	Fe_2O_3	TiO_2	CaO	MgO
Kaolin (%)	11.27	47.85	37.60	0.97	0.83	0.74	0.57	0.17

* LOI: Loss on Ignition at 1000.

2.2 Particle size distribution (PSD) of the kaolin powder used in the tubular support elaboration

PSD analysis (Figure 1) reveals unimodal distribution for natural clay powder used in tubular support preparation. The particle diameters range varied from 2 to 6 µm.

Carbonaceous Composite Materials   Materials Research Forum LLC
Materials Research Foundations **42** (2018) 177-204   doi: http://dx.doi.org/10.21741/9781945291975-7

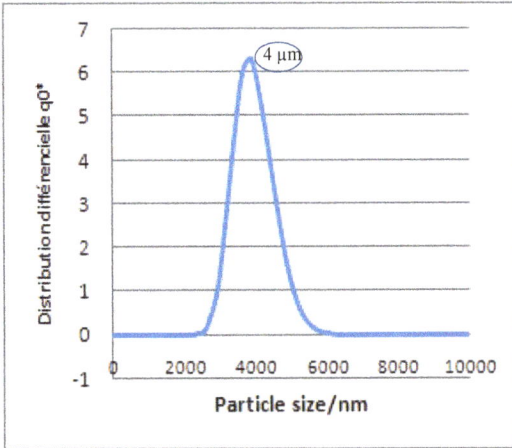

Figure 1. Granulometric repartition of kaolin powder.

2.3   Phase identification

Figure 2 shows the phase composition of used kaolin. It can be seen that the main crystallized phase is kaolinite with minor components of quartz and illite.

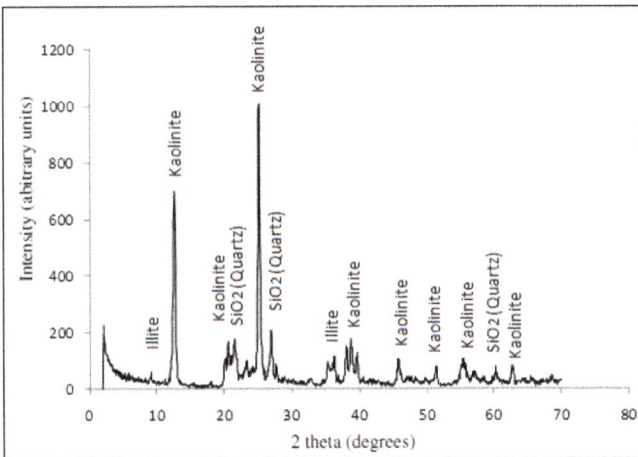

Figure 2. X-ray diffraction of the clay sample, describing the patterns corresponding to the main phases.

Carbonaceous Composite Materials   Materials Research Forum LLC
Materials Research Foundations **42** (2018) 177-204   doi: http://dx.doi.org/10.21741/9781945291975-7

2.4   Thermal analysis

The temperature evolution of the kaolin powder was characterized by TGA as shown in Fig. 3: from room temperature to 150° C for the first one with a loss of 1 % and from 400 °C to 700 °C for the second one, with a total weight loss of 13 %. The first weight loss is related to dehydration whereas the second weight loss is caused by the dehydroxylation of the kaolinite.

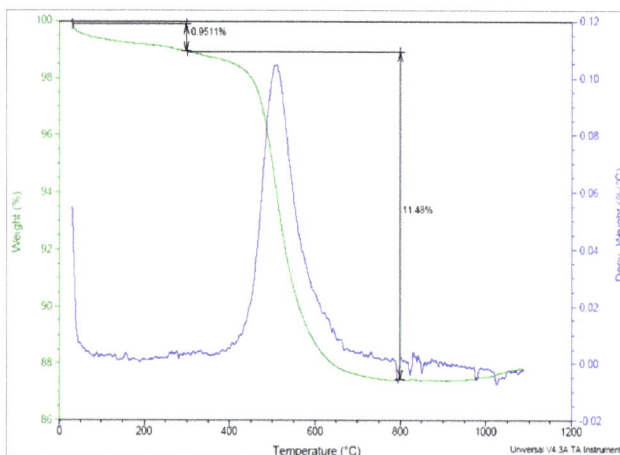

Figure 3. TGA of the kaolin powder [26].

Figure 4 shows the dilatometric measurements of the kaolin powder used in this study. The kaolin began to shrink between 500°C and 650°C. This first shrinkage is related to the hydroxylation of the kaolinite. By studying this figure, the sintering process begins at about 1140°C, and, the shrinkage of 7% was obtained at about 1255°C for the raw kaolin.

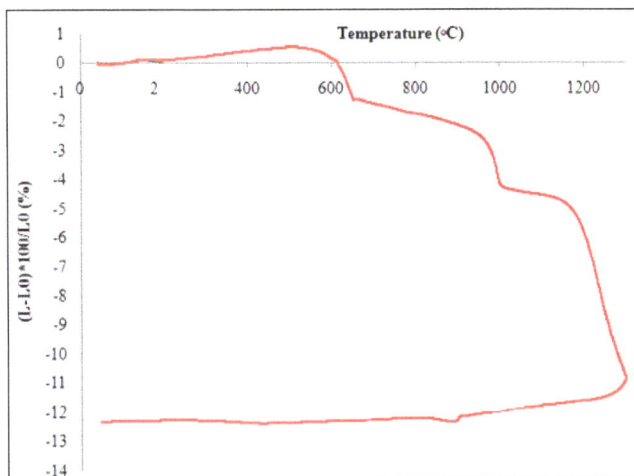

Figure 4. Dilatometric measurements of the kaolin powder used in this study.

3.   Support elaboration and characterization

3.1   Tubular porous support elaboration

Pastes with two different formulations were prepared. The first paste contains only clay powder whereas the solid composition of the second paste was 90 % of kaolin and 10 % of starch powder. Corn starch has been added to clay as a pore former [26]. The support fabrication process is initialized by thorough mixing of dry inorganic raw materials followed by addition of distilled water to prepare a paste (Figure 5). For good dispersion of the water in the paste, this mixture should be covered in a plastic case for at least 24 h. After that, extrusion has been used as the forming process of tubular support. The extruded tube was then dried at room temperature during 24 h on rotating aluminium rolls. The dried tube was then fired at four different temperatures ranging between 1000°C and 1250°C as per the schedule illustrated in Figure 6. Two steps have been determined: the first one for the elimination of organic additives at 500°C. The second heat treatment step involves the heating of the membrane from 500°C to desired sintering temperature at a heating rate of 5°C/min. Then the membrane is kept for 1 hour for sintering. The inner and outer diameters of the tubular support obtained were 11 mm and 16 mm, respectively, while the length was 150 mm.

Carbonaceous Composite Materials   Materials Research Forum LLC
Materials Research Foundations **42** (2018) 177-204   doi: http://dx.doi.org/10.21741/9781945291975-7

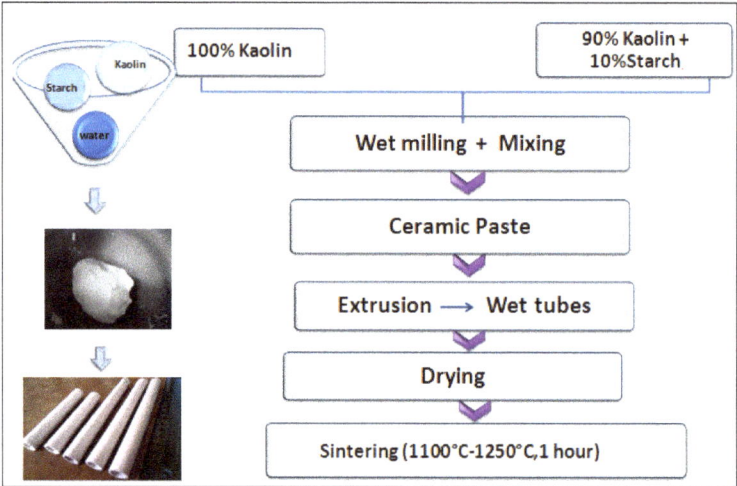

Figure 5. Steps order of membrane supports fabrication.

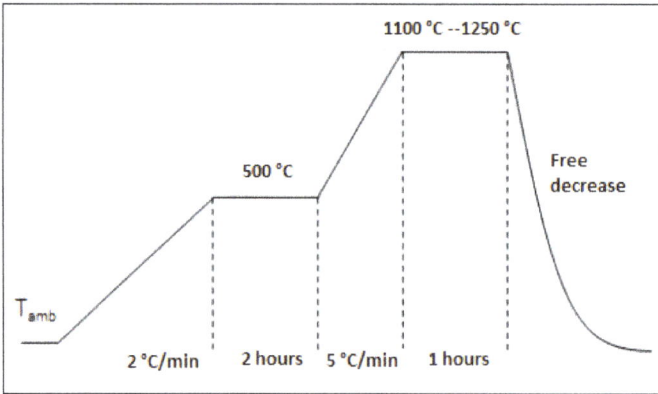

Figure 6. Firing schedule used for the support calcination.

3.2   Support characterization

For the development of high-quality supports, the following properties are of major importance: pore size distribution, porosity, surface texture, mechanical properties and chemical stability.

3.2.1 Optimization of kaolin support composition: Effect of sintering temperature and starch content

To optimize the different processing parameters for making membrane supports, we studied the effect of the content of starch and the sintering temperature on the properties of different prepared supports. The micrographs indicate that the effect of sintering is very marked. A progressive reduction of porosity can be observed when temperature increases. At 1250°C, the phenomenon is important; yielding to an important decrease of the porous volume. In the other hand, we also observed here that the incorporation of organic aditive powder loading in the ceramic has an effect to increase the total porosity of the resulting support. As can be ssen in Figure 7, the samples with starch (K+S) have higher porosities at all temperatures as compared with kaolin alone.

Thus, compared to pristine kaolin support, the presence of organic additive increased the porous nature of the support. It can be described by the fact that porosity was increasing since the number of pores on the surface was increasing by the addition of starch as pore-forming agent. The high quality of the elaborated supports with cracks-free and smooth surface would certainly contribute to an effective deposit of a selective layer.

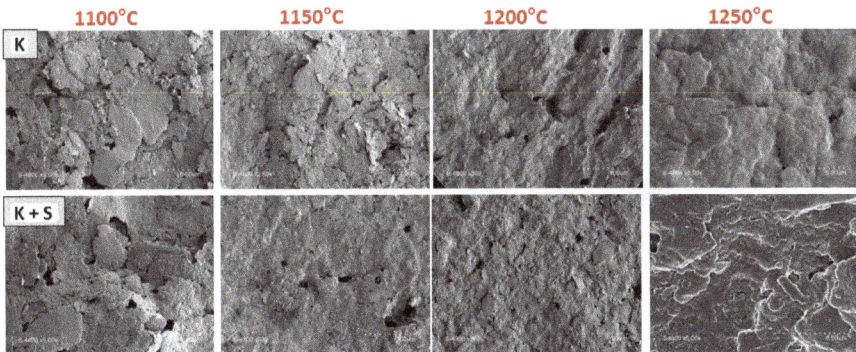

Figure 7. SEM images of tubular supports with and without starch content and sintered at various sintering temperatures [26].

To optimize the temperature required for the sintering of the green supports, a good compromise should be found between porosity and mechanical strength. Results in table 2 indicate that both the pore size and the flexural strength of the supports with and without additive show a gradually-increasing trend with sintering temperature. This gradual increment indicates a process of coarsening of pores. Because high-temperature sintering made the small pores connected with each other and consequently larger pores are formed [16,17,26]. However, at high sintering temperature, the porosity for the specimens decreases. As an example, when the firing temperature reaches to 1250°C, the mean pore size of suupport which contained starch in the starting raw materials mixture was found to be about 1.41 μm while the estimated pore size was decreased to be about 0.83 μm at 1100°C. On the other hand, the porosity increases from 36 to 48 % between 1250°C and 1000°C.

Table 2. Variation of pore size, porosity and the flexural strength of the supports, made from the kaolin and the kaolin+starch system and sintered at different temperatures.

	Temperature (°C)	Pore size (μm)	Porosity (%)	Flexural strength (MPa)
Kaolin	1100 °C	0.41	44.12	10
	1150 °C	0.56	38.23	20.16
	1200 °C	0.67	32.44	25.91
	1250 °C	0.75	26.61	28.41
Kaolin + Starch	1100 °C	0.83	48.89	8.1
	1150 °C	1.02	44	15
	1200 °C	1.21	40	19.2
	1250 °C	1.41	36	21

By examining now the ceramic support containing starch powder, it can be noticed that whatever the temperature, the addition of starch has for effect to increase both the porosity and the average pore diameter of the samples. This phenomen resulted from burning out of starch leading to increasing in the apparent porosity.

It is well known that not only the sample porosity is the factor deciding the suitability of the produced sample for filtration test, but also the mechanical property is a high important factor. Mechanical strength is a critical property in determining the usability of the obtained membranes for filtration. The mechanical resistance test was performed using the three points bending strength to control the resistance of the support (Figure 8).

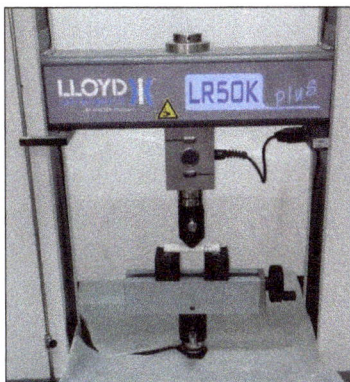

Figure 8. Photo of flexural strength machine.

The influence of both firing temperature and pore forming agent addition on the mechanical stability of supports was illustrated in Figure 9. The greatest flexural strength was found to be associated with the support without corn starch addition and sinteredat 1250°C, which is about 28 MPa. It should be noticed that this higher strength is in a good agreement with the microstructure shown in Figure 7. In fact, the flexural strength was found to be decreased with both decreasing the sintering temperatures and adding starch powder. The decline of the mechanical properties could be explained by the the highly increased in the porosity of the produced sample in both cases. Thus, the mechanical stability increases as porosity decreases.

Carbonaceous Composite Materials   Materials Research Forum LLC
Materials Research Foundations **42** (2018) 177-204   doi: http://dx.doi.org/10.21741/9781945291975-7

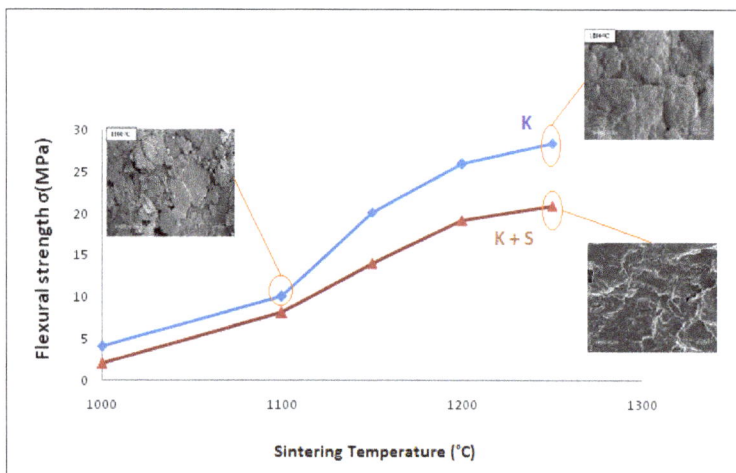

Figure 9. Flexural strength as a function of sintering temperature for tubular ceramic supports, fabricated from kaolin (K) and the mixture (kaolin + 10 wt% starch).

3.2.2 Choice of the support

This study shows that the introduction of starch in the composition of ceramic paste satisfies the compromise between porosity and mechanical properties of sintered support. Hence, the sintering temperature of 1150 °C was considered optimum for the supports made by the optimized compositions 90% kaolin and 10 starch.

The membranes support shows a unimodel distribution (Figure 10). This is necessary for a good integrity of the membrane. In addition, the support is characterized by a flexural strength of 15 MPa, a porosity of 44 % with a mean pore size of 1.02 μm, indicating that it is a good support for ultrafiltration membrane. On the other hand, ceramic supports provide a smoother surface which will allow the deposit of a homogeneous cative layer (Figure 11).

Carbonaceous Composite Materials   Materials Research Forum LLC
Materials Research Foundations **42** (2018) 177-204   doi: http://dx.doi.org/10.21741/9781945291975-7

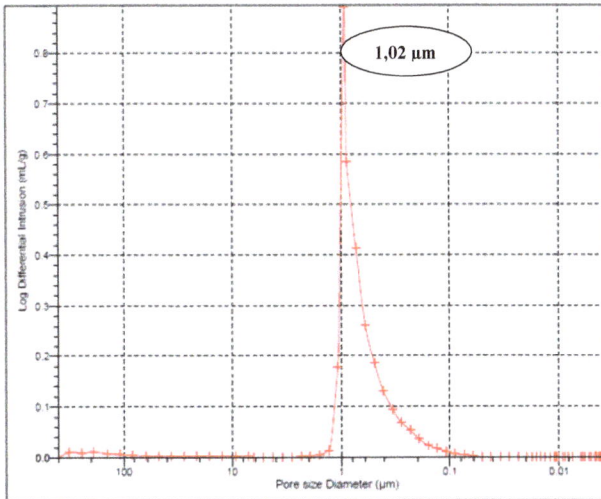

Figure 10. Determination of the pore diameter of the optimised support [17].

Figure 11. Surface and cross sectin photographs of optimised support.

4.   Ultrafiltration layer deposition and characterization

4.1   Ultrafiltration layer deposition

The kaolin powder is crushed for 30 min with a planetary crusher at 250 revolutions/min and calibrated with 50 μm. The particle size distribution of this powder was determined by the Dynamic Laser Scattering (DLS) technique. This method gave an average particle size in the order of 2.8 μm. The obtained particle diameters range from about 1.5 to t 5 μm (Figure 12).

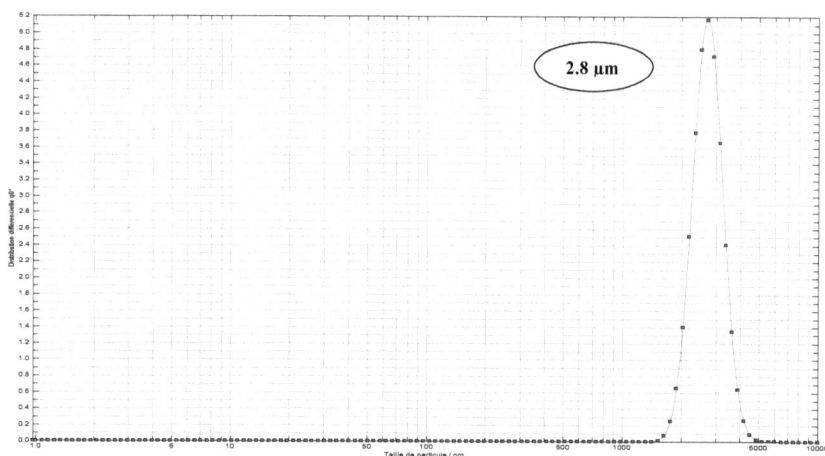

Figure 12. Granulometric repartition of kaolin powder used to prepare the ultrafiltration layer [26].

As depicted in Fig. 13, a defloculated slip was obtained by mixing kaolin powder, PVA (12% w/w aqueous solution) and water. The deposition of the slip on kaolin support was performed by slip casting process. The tubular support was coated with the kaolin dispersion; dried at room temperature and fired at at 650 °C for 3 hours (Fig. 14).

In order to optimise the slip composition suitable for the slip casting, empirical study was performed to select the optimum composition [17]. The slip composition was optimised basing on a rheological study using an Antoon Paar rheometer model MCR301 and the SEM observation of the sintered layer [17]. The optimum suspension composition was done in Table 3.

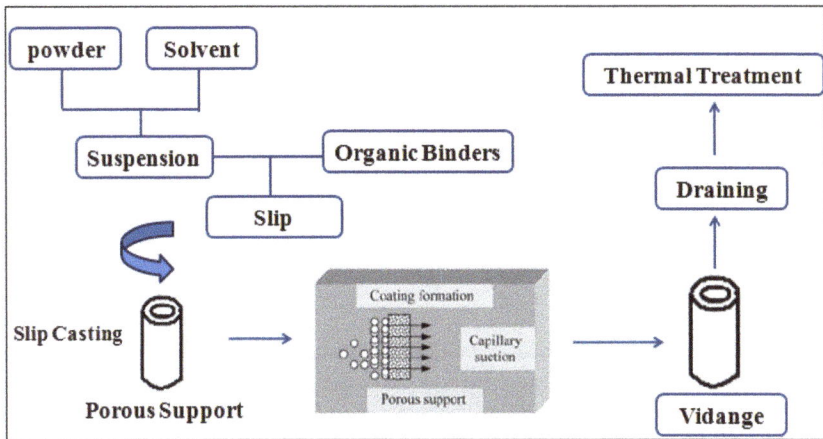

Figure 13. Scheme of slip casting coating process.

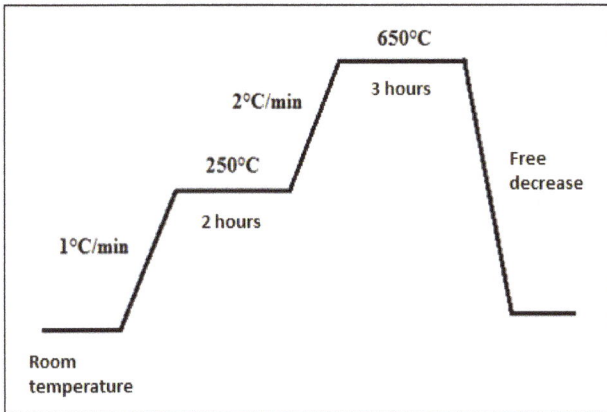

Figure 14. Firing schedule used for the membrane sintering.

Table 3. Composition of UF suspension

Component	conditions	Proportion (Wt %)
Kaolin powder	Particle size ≤ 5 μm	4
PVA	12% aqueous solution	40
Water	Deionised	56

Carbonaceous Composite Materials   Materials Research Forum LLC
Materials Research Foundations **42** (2018) 177-204   doi: http://dx.doi.org/10.21741/9781945291975-7

4.2   Characterization of the slip

The viscosity of the slip elaborated according to the protocol described previously has been studied right before deposition. Figure 15 shows that for the whole range of temperature (25-80°C), as the temperature increases the viscosity decreases. Figure 16 shows the rheogram of the slip used, this figure shows that the optimised slip behave according to a Newtonian type as the shear stress increases with D.

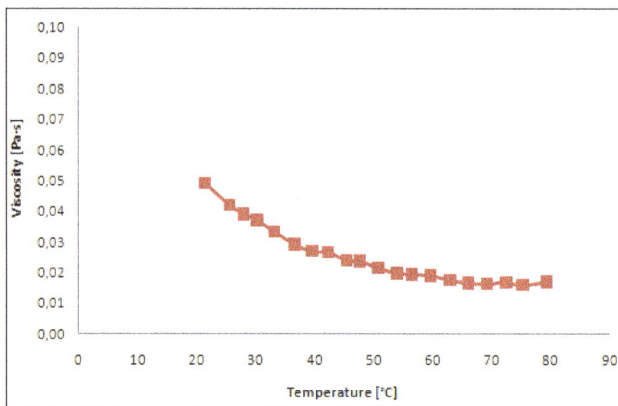

Figure 15. Evolution of viscosity with the temperature.

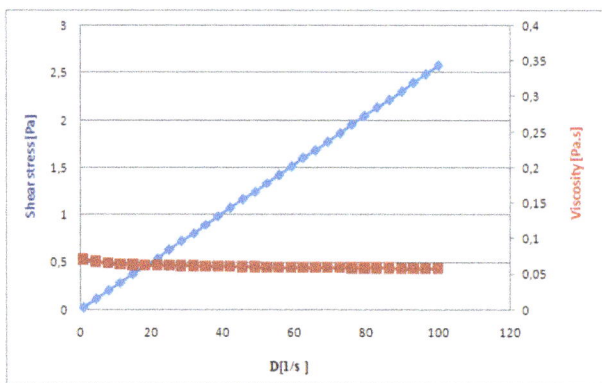

Figure 16. Influence of the shear rate on the rheological curves of the used slip.

4.3 Membrane characterization

4.3.1 SEM analysis and pore size distribution

The morphology, surface quality and the thickness of the active layer were examined by scanning electron microscopy (Figure 17). All samples showed a typical asymmetric structure. The thickness of the ultrafiltration layer may be controlled by the casting time. At 20 min, the membrane thickness reached 30 μm, which seems to be the critical value. The best layer is obtained with 6 min casting time with thickness around 8 μm which is a good thickness for an ultrafiltration layer [17, 35-37]. When focusing on the surface texture of the different samples, it seems that no significant difference arises whatever the enduction time (Figure 18).

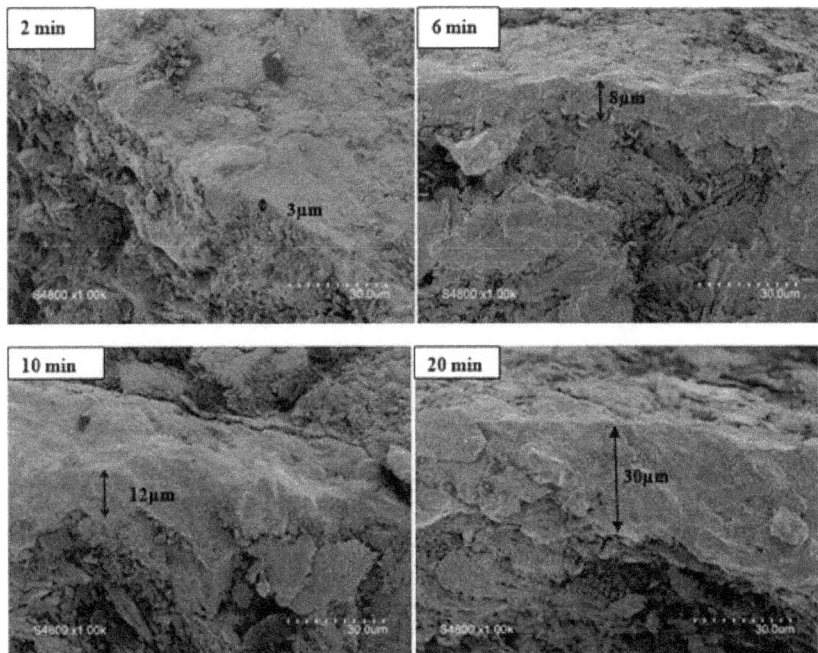

Figure 17. SEM images of membranes cross section obtained with different casting times [17].

The cross-sectional view showed well adhesion between the support and the active layer which is an essential criterion for membrane coating to avoid peeling off of the top layer (Figure 19). UF layer with an average pore diameter of 11 nm (determined by mercury porosimetry) and a thickness of 8 μm was obtained.

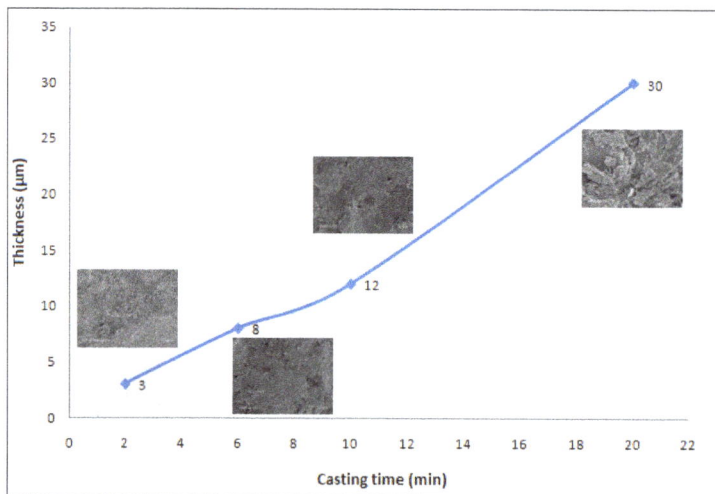

Figure 18. Evolution of the surface morphology and the thickness of the active layer with different casting times.

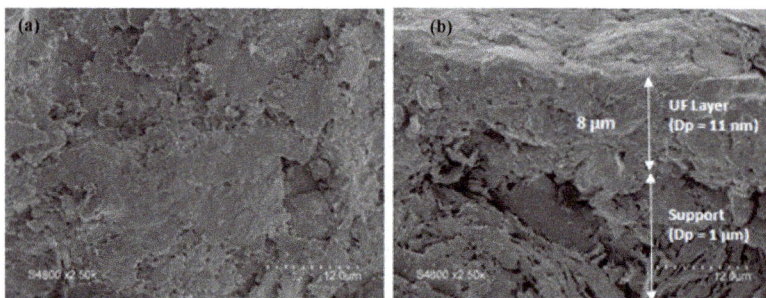

Figure 19. SEM image of the optimised membrane: (a) general view of the surface; (b) cross-section view.

Carbonaceous Composite Materials   Materials Research Forum LLC
Materials Research Foundations **42** (2018) 177-204   doi: http://dx.doi.org/10.21741/9781945291975-7

The mean pore sizes of support and UF layer are 1 µm and 11 nm, respectively (Figure 20). This attribution is based on the comparison of distribution diagrams of both the support sintered at 1150°C/1 h and the membrane sintered at 650°C/3 h. It is also clear that the pore size distribution of the filtration layer is narrower than that of the support.

4.3.2 Water permeability

Figure 21 indicates the effect of the operating pressure on the permeate flux of both coated and uncoated supports. The pure water flux (PWF) increased linearly with the operating pressure. A linear fit was generated. The water flux slope for the optimised membrane was about (79, $R^2 = 0.999$) against 140 $L \cdot h^{-1} \cdot m^{-2} bar^{-1}$ for the simple kaolin support. The deccrease of permeability values in the presence of the active layer was reflected in Figure 21, which can be proves that an UF membrane was actually achieved.

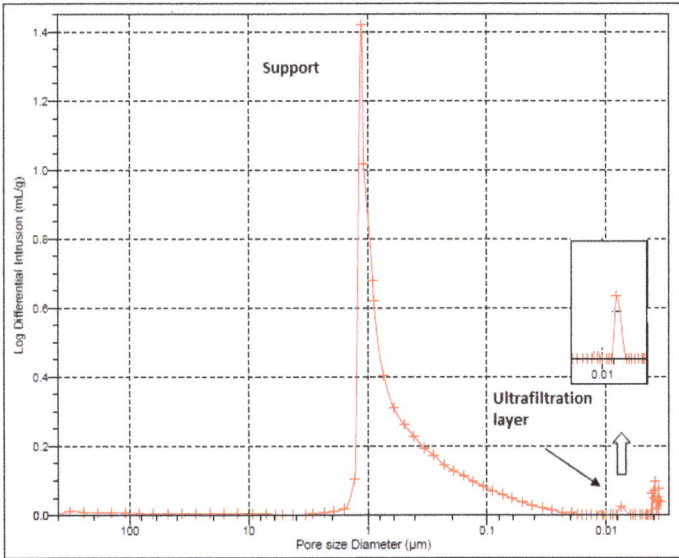

Figure 20. Determination of the pore diameter of UF kaolin membrane [17].

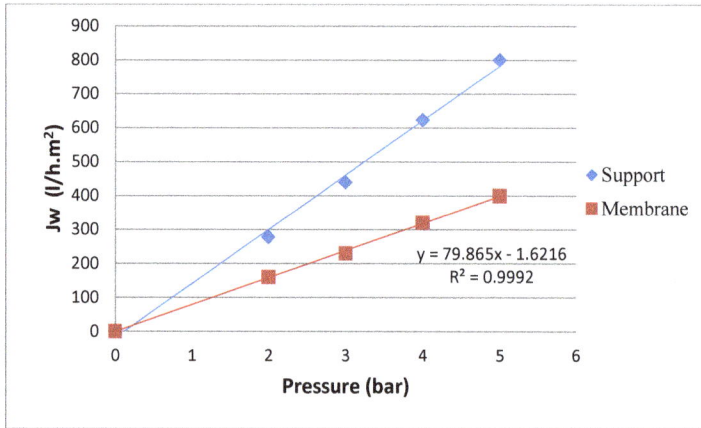

Figure 21. Effect of transmembrane pressure on PWF for both the support and the optimised membrane.

5. Application to the treatment of the industrial wastewater

5.1 Wastewater characteristics

One of the most important industrial applications of the membrane separation process is the clarification and purification of cuttlefish effluent treatment. Samples were collected from Tunisian sea product industry [17].

Table 4. Turbidity, COD and conductuvity of raw and ultrafiltrated effluent by kaolin membrane at different TMP [17].

	Raw effluents	Filtrate at various pressure		
		2 (bar)	4 (bar)	6 (bar)
Turbidity (NTU)	333	0,60	0,93	1,34
COD (mg.L^{-1})	2612	350	458	528
Conductivity (mS.cm^{-1})	202	139	144	146

The tangential ultrafiltration experiments were performed on a laboratory-scale filtration pilot, using recycling configuration at room temperature (Figure 22). It was equipped with a cross-flow filtration system implementing the tubular ceramic tubes.

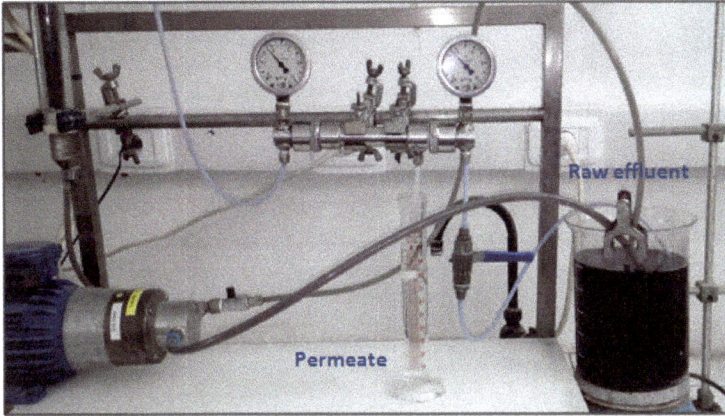

Figure 22. Filtration pilot used for the ultrafiltration experiments.

5.2   Ultrafiltration treatment

The typical measured water flows through the optimised membrane are given as a function of time and of at various working pressure pressure, as illustrated in Figure 23. The flux (PWF) drops fast at the beginning of the filtration until it became stabilized. The fall of the filtration flux observed at the beginning of the operation can be due to a partial fouling of the membrane pores by the melanin particles. The phase of flux stabilisation corresponds to the establishment of the membrane fouling by the formation of a deposit layer [8,17, 36-38]. Furthermore, this figure shows that there is an increase in the stabilized flux values, when the applied pressure is increased.

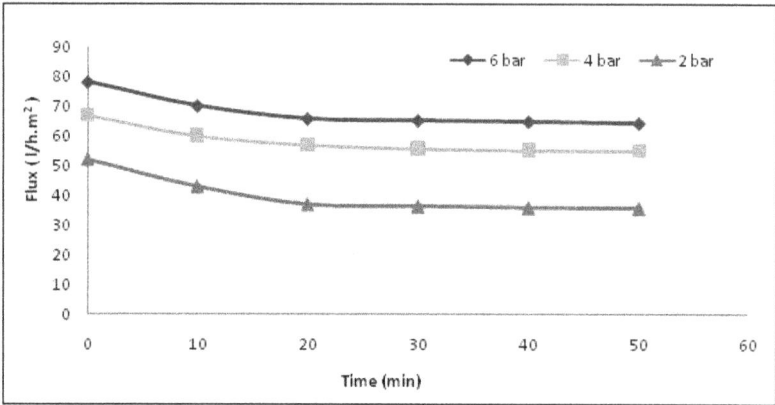

Figure 23. Effect of time of filtration on PWF [17].

5.3   Wastewater characterization

The characteristics of the effluent before and after filtration are presented in Table 4 and Figure 24. The obtained ceramic membrane exhibited superior turbidity removal efficiency for the studied effluent (99%).The result was interesting with a pollutants retention rate of 87% for COD, 28 % for conductivity and almost a total retention of color. As it can be seen in Figure 22, filtration on ceramic membranes improves considerably the appearance and quality of the effluent.

The results prove the industrial feasibility of UF cuttlefish effluent treatment in attendance of the composite Kaolin/Kaolin UF membrane.

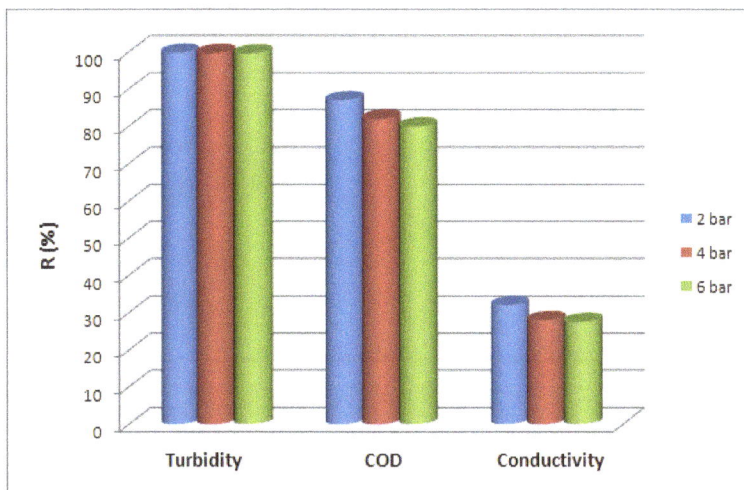

Figure 24. Evolution of retention of the different parameters by kaolin ultrafiltration membrane at different TMP.

5.4   Membrane regeneration

The effectiveness of the regeneration is checked by the determination of the membrane permeability which should be quite similar to that obtained with virgin membrane. Figure 25 shows almost total renovation of the membrane.

Conclusion

In summary, ceramic membranes in a tubular configuration were prepared. The tubular support was prepared by the extrusion procedure using a ceramic paste made with kaolin and corn starch as poreforming agent. A correlation between paste composition, sintering temperatures and physicochemical properties of the obtained supports was discussed. The incorporation of a porogenic reagent (starch powder) in the collodion was the key step to obtain bigger pore sizes and to make the obtained membrane more porous, and thus induce an increase in pure water permeability. Ceramic samples were fired at firing temperatures in range from 1100 to 1250 °C. The optimised supports sintered at 1150°C shows good porosity, good mechanical strength and good chemical stability. These results may allow the deposition of the active layer directly on the porous support by slip-casting process. SEM picturs shows that the obtained membranes have an asymmetric structure and the thickness of the active layer was approximately 9 μm with 6 min casting time. A narrow distribution with an average pore size of 11 nm can be seen for the deposited layer. Finally, this study made it possible to show the effectiveness of the application of UF processes in the clarification of wastewater resulting from the cuttlefish conditioning.

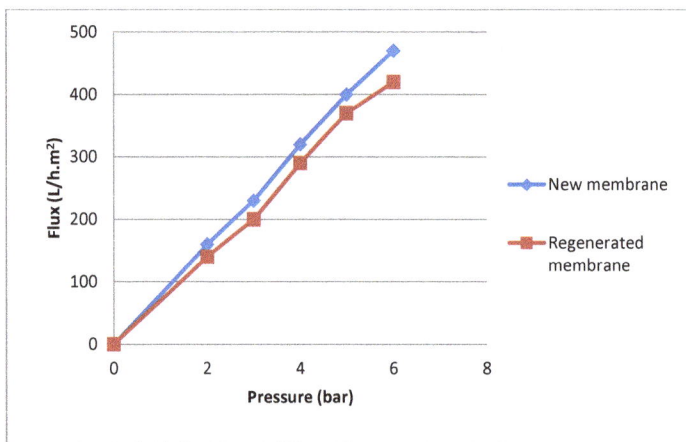

Figure 25. Determination of water permeability for new and regenerated membrane.

References

[1]   G. Sharma, A. Kumar, K. Devi, S. Sharma, M. Naushad, A.A. Ghfar, T. Ahamad, F.J. Stadler, Guar gum-crosslinked-Soya lecithin nanohydrogel sheets as effective adsorbent for the removal of thiophanate methyl fungicide, International journal of biological macromolecules, 114 (2018) 295-305. https://doi.org/10.1016/j.ijbiomac.2018.03.093

[2]   M. Naushad, G. Sharma, A. Kumar, S. Sharma, A.A. Ghfar, A. Bhatnagar, F.J. Stadler, M.R. Khan, Efficient removal of toxic phosphate anions from aqueous environment using pectin based quaternary amino anion exchanger, International journal of biological macromolecules, 106 (2018) 1-10. https://doi.org/10.1016/j.ijbiomac.2017.07.169

[3]   G. Sharma, B. Thakur, M. Naushad, A. Kumar, F.J. Stadler, S.M. Alfadul, G.T. Mola, Applications of nanocomposite hydrogels for biomedical engineering and environmental protection, Environmental Chemistry Letters, 16 (2017) 113-146. https://doi.org/10.1007/s10311-017-0671-x

[4]   G. Sharma, B. Thakur, M. Naushad, A.a.H. Al-Muhtaseb, A. Kumar, M. Sillanpaa, G.T. Mola, Fabrication and characterization of sodium dodecyl sulphate@ironsilicophosphate nanocomposite: Ion exchange properties and selectivity for binary metal ions, Materials Chemistry and Physics, 193 (2017) 129-139. https://doi.org/10.1016/j.matchemphys.2017.02.010

[5]   G. Sharma, M. Naushad, A. Kumar, S. Rana, S. Sharma, A. Bhatnagar, F. J. Stadler, A.A. Ghfar, M.R. Khan, Efficient removal of coomassie brilliant blue R-250 dye using starch/poly(alginic acid-cl-acrylamide) nanohydrogel, Process Safety and Environmental Protection, 109 (2017) 301-310. https://doi.org/10.1016/j.psep.2017.04.011

[6]   L. Xu, W. Li, S. Lu, Z. Wang, Q. Zhu, Y. Ling, Treating dyeing waste water by ceramic membrane in cross flow microfiltration, Desalination, 149 (2002) 199–203. https://doi.org/10.1016/S0011-9164(02)00759-2

[7]   M. Ebrahimi, K. Shams Ashaghi, L. Engel, D. Willershausen, P. Mund, P. Bolduan, P. Czermak, Characterization and application of different ceramic membranes for the oil-field produced water treatment, Desalination, 245 (2009) 533–540. https://doi.org/10.1016/j.desal.2009.02.017

[8]   S. Masmoudi, R. Ben Amar, A. Larbot, H. El Feki, A. Ben Salah, L. Cot, Elaboration of inorganic microfiltration membranes with hydroxyapatite applied to the treatment of waste water from sea product industry, J. Membr.Sci. 247 (2005) 1–9. https://doi.org/10.1016/j.memsci.2004.03.047

[9]   S. Khemakhem, R. Ben Amar, R. Ben Hassen, A. Larbot, M. Medhioub, A. Ben Salah, L. Cot, New ceramic membranes for tangential waste-water filtration, Desalination, 167 (2004) 19–22. https://doi.org/10.1016/j.desal.2004.06.108

[10]   S. Mahesh Kumar, G. M. Madhu, Sukumar Roy, Fouling behaviour, regeneration options and on-line control of biomass-based power plant effluents using micro porous ceramic membranes, Sep. Purif. Technol. 57 (2007) 25–36 https://doi.org/10.1016/j.seppur.2007.03.002

[11] B. K. Nandi, B. Das, R. Uppaluri, M. K. Purkait, Microfiltration of mosambi juice using low cost ceramic membrane, J. Food Eng. 95 (2009) 597–605. https://doi.org/10.1016/j.jfoodeng.2009.06.024

[12] J. Zulewska, M. New bold, D. M. Barbano, Efficiency of serum protein removal from skim milk with ceramic and polymeric membranes at 50°C, J. Dairy Sci. 92 (2009) 1361–1377. https://doi.org/10.3168/jds.2008-1757

[13] V. S. Espina, M. Y. Jaffrin, M. Frappart, L. H. Ding, Separation of case in micelles from whey proteins by high shear microfiltration of skim milk using rotating ceramic membranes and organic membranes in a rotating disk module, J. Membr. Sci. 325 (2008) 872–879. https://doi.org/10.1016/j.memsci.2008.09.013

[14] J.H. Han, E. Oh, B. Bae, I.H. Song, The effect of kaolin addition on the characteristics of a sintered diatomite composite support layer for potential microfiltration applications, Ceram, Int. 39 (2013) 8955–8962. https://doi.org/10.1016/j.ceramint.2013.04.092

[15] J.H. Han, E. Oh, B. Bae, I.H. Song, The fabrication and characterization of sintered diatomite for potential microfiltration for applications, Ceram. Int. 39 (2013) 7641- 7648. https://doi.org/10.1016/j.ceramint.2013.02.102

[16] S. Fakhfakh, S. Baklouti,, J. Bouaziz, Elaboration and characterization of low cost ceramic support membrane, Advances in Applied Ceramics. 108 (2010) 31–38. https://doi.org/10.1179/174367609X422234

[17] S. Rekik., J. Bouaziz., A. Deratani., S. Baklouti, Development of an asymmetric ultrafiltration membrane from naturally-occurring kaolin clays: Application to the cuttlefish effluents treatments, J. Membra. Sci. Technol. 6 (2016) 1–12. https://doi.org/10.4172/2155-9589.1000159

[18] W. Qin, C. Peng, M. Lv, J. Wu, Preparation and properties of high-purity porous alumina support at low sintering temperature, Ceram. Int. 40 (2014) 13741–13746. https://doi.org/10.1016/j.ceramint.2014.05.044

[19] C. Liu, L. Wang, W. Ren, Z. Rong, X. Wang, J. Wang, Synthesis and characterization of a mesoporous silica (MCM-48) membrane on a large-pore α-Al2O3 ceramic tube, Microporous Mesoporous Mater. 106 (2007) 35–39. https://doi.org/10.1016/j.micromeso.2007.02.007

[20] Y.H. Wang, T.F. Tian, X.Q. Liu, G.Y. Meng, Titania membrane preparation with chemical stability for very hash environments applications, J. Membr. Sci. 280 (2006) 261–269. https://doi.org/10.1016/j.memsci.2006.01.027

[21] L. Palacio, Y. Bouzerdi, M. Ouammou, A. Albizane, J. Bennazha, A. Hernández, J.I. Calvo, Ceramic membranes from Moroccan natural clay and phosphate for industrial water treatment, Desalination. 245 (2009) 501-507. https://doi.org/10.1016/j.desal.2009.02.014

[22] S. Ayadi, I. Jedidi, M. Rivallin, F. Gillot, S. Lacour, S. Cerneaux, M. Cretin, R. Ben Amar, Elaboration and characterization of new conductive porous graphite membranes for electrochemical advanced oxidation processes, J. Membr. Sci. 446 (2013) 42–49. https://doi.org/10.1016/j.memsci.2013.06.005

[23] M. Khemakhem, S. Khemakhem, S. Ayedi, M. Cretin, R. Ben Amar, Development of an asymmetric ultrafiltration membrane based on phosphates industry sub-products, Ceram. Int. (2015). https://doi.org/10.1016/j.ceramint.2015.05.101

[24] A. Harabi, F. Zenikheri, B. Boudaira, F. Bouzerara, A. Guechi, L. Foughali, A new and economic approach to fabricate resistant porous membrane supports using kaolin and CaCO3, J. Eur. Ceram. Soc. 34 (2014) 1329–1340. https://doi.org/10.1016/j.jeurceramsoc.2013.11.007

[25] M. Khemakhem, A. Oun, M. Cretin, S. Cerneaux, S. Khemakhem, R. Ben Amar, Decolorization of Dyeing Effl uent by Novel Ultrafi ltration Ceramic Membrane from Low Cost Natural Material, Journal of Membrane Science and Research 4 (2018) 101-107.

[26] S. Rekik, J. Bouaziz., A. Deratani, S. Baklouti, Study of Ceramic Membrane from Naturally Occurring-Kaolin Clays for Microfiltration Applications, Period. Polytech. Chem. Eng. 61 (2017) 206-215. https://doi.org/10.3311/PPch.9679

[27] M. Issaoui, J. Bouaziz, Elaboration of membrane ceramic supports using aluminium powder, Desalin. Water Treat 53 (2015) 1037–1044.

[28] M. Issaoui, L. Limousy, B. Lebeau, J. Bouaziz, M. Fourati, Manufacture and optimization of low-cost tubular ceramic supports for membrane filtration: application to algal solution concentration, Environ Sci Pollut Res. 11 (2017) 9914–9926. https://doi.org/10.1007/s11356-016-8285-6

[29] S. Rekik, S. Gassara, J. Bouaziz, A. Deratani, S. Baklouti, Development and characterization of porous membranes based on kaolin/chitosan composite. Appl. Clay. Sci. 143 (2017) 1–9. https://doi.org/10.1016/j.clay.2017.03.008

[30] W. Aloulou, W. Hamza, H. Aloulou, A. Oun, S. Khemakhem, A. Jada, S. Chakraborty, S. Curcio, R. Ben Amar, Developing of titania-smectite

nanocomposites UF membrane over zeolite based ceramic support, Applied Clay Science 155 (2018) 20–29. https://doi.org/10.1016/j.clay.2017.12.035

[31] P. Rai, C. Rai, G. Majumdara, S. DasGupta, Resistance in series model for ultrafiltration of mosambi (Citrus sinensis (L.) Osbeck) juice in a stirred continuous mode, Journal of Membrane Science 283 (2006) 116–122. https://doi.org/10.1016/j.memsci.2006.06.018

[32] N. Tahri, I. Jedidi, S. Ayadi, S. Cerneaux, M. Cretin, R. Ben Amar, Preparation of an asymmetric microporous carbon membrane for ultrafiltration separation: application to the treatment of industrial dyeing effluent, Desalin. Water Treat (2016) 1–16.

[33] S. Fakhfakh, S. Baklouti, S. Baklouti, J. Bouaziz, Preparation, characterization and application in BSA solution of silica ceramic membranes, Desalination. 262 (2010) 188–195. https://doi.org/10.1016/j.desal.2010.06.009

[34] R. Sahnoun, J. Bouaziz, Sintering characteristics of kaolin in the presence of phosphoric acid binder, Ceramics International. 38 (2012) 1–7. https://doi.org/10.1016/j.ceramint.2011.06.058

[35] P. Belibi, S. e Cerneaux, M. Rivallin, M. Ngassoum, M. Cretin, Elaboration of low-cost ceramic membrane based on local material for microfiltration of particle from drinking water, Journal of Applicable Chemistry 3 (2014) 1991-2003.

[36] M. Khemakhem, S. Khemakhem, S. Ayedi, R. Ben Amar, Study of ceramic ultrafiltration membrane support based on phosphate industry subproduct: application for the cuttlefish conditioning effluents treatment. Ceram Int 37 (2011) 3617-3625. https://doi.org/10.1016/j.ceramint.2011.06.020

[37] M. Khemakhem, S. Khemakhem, S. Ayedi, M. Cretin, R. Ben Amar, Development of an asymmetric ultrafiltration membrane based on phosphates industry sub-products. Ceram. Int. 49 (2015) 10343-10348. https://doi.org/10.1016/j.ceramint.2015.05.101

[38] I. Jedidi, S. Saïdi, S. Khemakhem, A. Larbot, N. Elloumi-Ammar, Elaboration of new ceramic microfiltration membranes from mineral coal fly ash applied to waste water treatment. J. Hazard. Mater. 172 (2009) 152-158. https://doi.org/10.1016/j.jhazmat.2009.06.151

Carbonaceous Composite Materials
Materials Research Foundations **42** (2018) 205-230

Materials Research Forum LLC
doi: http://dx.doi.org/10.21741/9781945291975-8

Chapter 8

Environmental Detoxification Using Carbonaceous Composites

P. Senthil Kumar[1]*, A. Saravanan[2]

[1]Department of Chemical Engineering, SSN College of Engineering, Chennai 603110, India

[2]Department of Biotechnology, Rajalakshmi Engineering College, Chennai 602105, India

senthilkumarp@ssn.edu.in*

Abstract

For decades there has been expanding worldwide worry for the general wellbeing impacted by environmental pollution. The arrival of chemical contaminations into the earth introduces a huge swath of issues related with public health. Carbonaceous composites have been broadly considered for all sorts of contaminants expulsion from wastewater because of its phenomenal and tunable properties. The present and potential utilizations of carbon based materials in wastewater treatment incorporate adsorption, photo catalysis, sanitization and membrane separation. The carbonaceous materials such as activated carbons, carbon nanofibers and carbon nanotubes for the most part managed high extraction effectiveness, great selectivity in complex networks and toughness for continuous adsorption/desorption cycles, basically because of high surface territory, compound security, dispersibility in wastewater and, significantly, multi-sort communication.

Keywords

Environmental Pollution, Carbonaceous Composites, Carbon Nanofibers, Carbon Nanotubes, Water Treatment

Contents

Materials Research Forum LLC
doi: http://dx.doi.org/10.21741/9781945291975-8

1.   Introduction

One of the greatest problems that the world is facing today is that of environmental pollution, increasing with every passing year and causing grave and irreparable damage to the earth. Rising contaminations are a gathering of unregulated mixes introduced in water, which are thought to be extremely hurtful both for the human wellbeing and for the earth [1-4]. The rising toxins have been isolated into five gatherings: pharmaceuticals, steroid hormones, perfluorinated mixes, surfactants and personal care items (Figure 1). In the gathering of pharmaceuticals alone, finished a hundred mixes (barring their metabolites) have been distinguished in effluents and also in surface waters.

Carbonaceous Composite Materials   Materials Research Forum LLC
Materials Research Foundations **42** (2018) 205-230   doi: http://dx.doi.org/10.21741/9781945291975-8

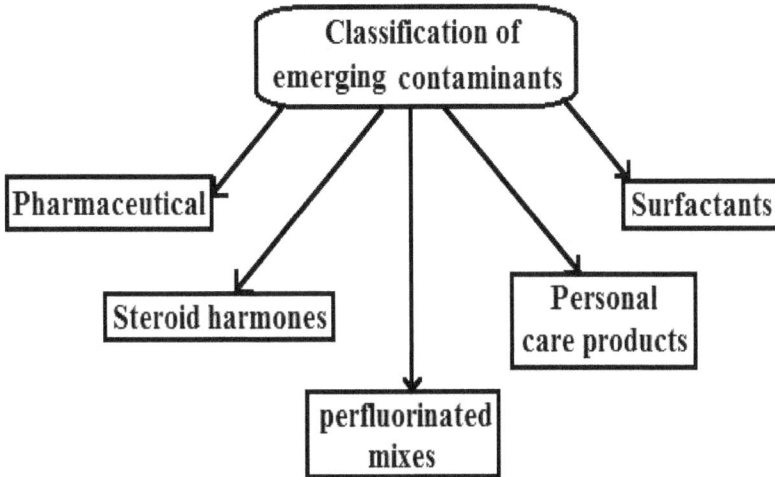

Figure 1: Classification of emerging contaminants.

Regular assets of the world start to get depleted; horticulture needs to fulfil an essential part in a manageable world. Considering the restricted accessibility of water and land assets, accomplishments in farming can be gotten utilizing new advances. Carbon based materials can offer approaches to make critical changes in the farming area in ecological designing and in water assets. The physical, chemical and electrical properties of carbon based materials are considered to be innovative solutions to diligent environmental challenges.

Carbonaceous composites and their practical subsidiaries are used in environmental applications incorporate focused on conveyance of remediation specialists, built evacuation of dangerous contaminants, and novel film structures for water filtration [5-8]. Figure 2 shows applications of carbonaceous composites for wastewater treatment.

Carbon materials is a favoured adsorbent for the expulsion of micro pollutants from the watery stage; be that as it may, its broad utilize is limited because of high related expenses. To diminish treatment costs, endeavours have been made to discover cheap elective carbonaceous material forerunners, for example, squander materials [9]. Carbonaceous composite finds significant applications in agri-sustenance topical zones like: regular assets administration, conveyance components in plants and soils, utilization of rural waste and biomass, in nourishment preparing and sustenance bundling, chance appraisal being likewise assessed. Carbonaceous composite materials are broadly connected as development materials in aerospace and electronics industry and restorative prosthesis, and so on. Of late, they have been utilized as promising impetus backings and special adsorbents [10,11].

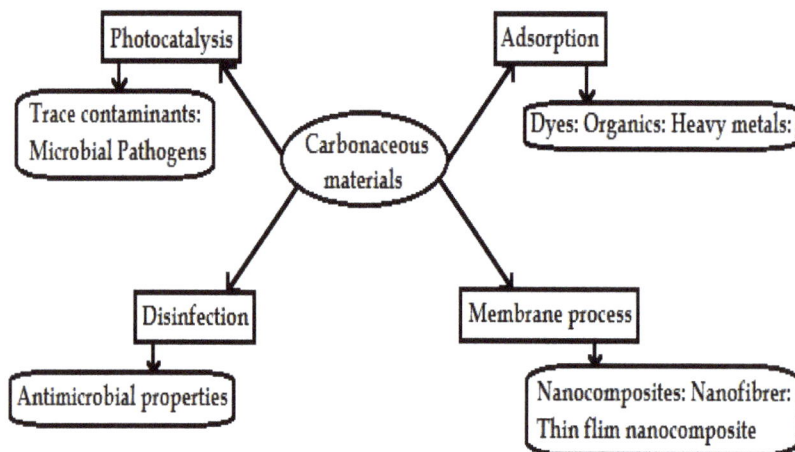

Figure 2: Overall of application of carbonaceous composites in wastewater.

2. Carbon based materials

Carbon-based materials have caught extensive enthusiasm from the materials science group for a considerable length of time. Carbon is a to a great degree light and flexible material that, contingent upon the neighbourhood holding of the constituting carbon molecules, has tremendously changing properties. Surely understood great cases of carbon allotropes are precious stone, undefined carbon and graphite. Carbon-based nanomaterials show phenomenal physical and chemical properties, for example, high quality, fantastic imperviousness to erosion and uncommon electrical and warm conduction and steadiness. As a result of these interesting components, nano-carbon materials are utilized as a part of an extensive variety of fields, including science, vitality stockpiling and prescription. Be that as it may, notwithstanding various advances in the previous decades, specialists still face challenges with respect to the blend, consistency

Carbonaceous Composite Materials   Materials Research Forum LLC
Materials Research Foundations **42** (2018) 205-230   doi: http://dx.doi.org/10.21741/9781945291975-8

and reproducibility in the development of carbon materials. To beat these issues, and to have the capacity to grow new utilizations of this material, particular and thorough portrayal is primordial.

Carbon's exceptional hybridization properties, and the affectability of carbon's structure to irritations in union conditions, take into account custom fitted control to a degree not yet coordinated by inorganic nanostructures. While inorganic nanomaterials are a promising territory of future research. The physical, chemical, and electronic properties of carbonaceous nanomaterials are firmly coupled to carbon's basic compliance and, in this way, its hybridization state [12]. The ground-state orbital design of carbon's six electrons is $1s^2$, $2s^2$, $2p^2$. The tight vitality hole between the 2s and 2p electron shells encourages the advancement of one's orbital electron to the higher vitality p orbital that is void in the ground state. Contingent upon holding connections with the neighbouring particles, this advancement permits carbon to hybridize into a sp, sp^2, or sp^3 design. The vitality picked up from covalent holding with adjoining particles adjusts for the higher vitality condition of the electronic design. This remuneration is almost equivalent for the sp2 what's more, sp3 hybridization states after the out-of-plane holding due to π bonds among unhybridized p orbitals is considered [13].

At high temperatures or weights, carbon accepts the thermodynamically great trigonometric sp^3 design of precious stone. The carbon grasps the planar sp2 consistence and structures monolayer sheets bound by three sigma covalent bonds and a singular π bond. Frail out-of-plane collaborations are an aggregate of van der Waals powers and the connection between covering π orbitals of parallel sheets. Gentle shear powers, physical partition, and compound change disturb these frail interplanar powers and make graphite's planes slip past each other [14]. Along these lines, carbonaceous composites share the same holding setups as plainly visible carbon structures, in any case, their properties and morphology are overwhelmed by the strength of select reverberation structures instead of the mass midpoints of their crystalline structures.

The natural utilizations of carbonaceous nanomaterials laid out in the audit are both proactive (avoiding environmental degradation, enhancing public health, improving vitality productivity) and retroactive (remediation, wastewater reuse, transformation of pollutant). Figure 3 display the environmental application of carbon based materials.

Among carbon-based nanomaterials, contingent upon the hybridization states we can include: fullerenes, carbon nanotubes, carbon black and multi-walled carbon nanotubes and graphene.

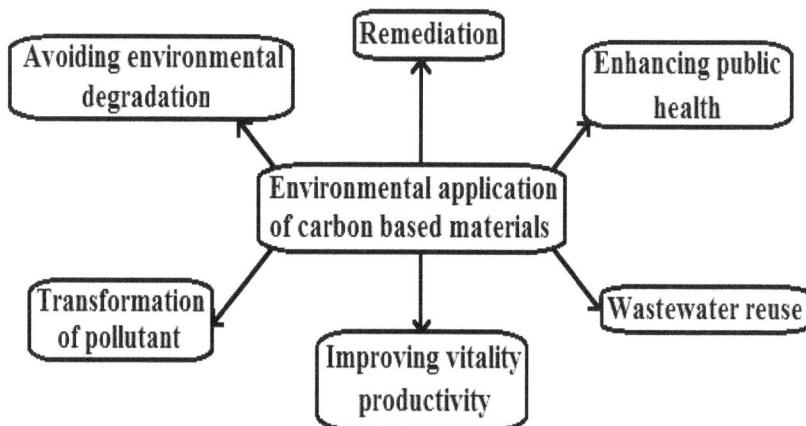

Figure 3: Environmental application of carbon based materials.

2.1   Fullerenes

Fullerenes comprise of 20 hexagonal and 12 pentagonal rings as the premise of an icosohedral symmetry shut enclosure structure. Every carbon particle is attached to three others and is sp2 hybridized. The C60 particle has two bond lengths - the 6:6 ring bonds can be viewed as "twofold bonds" and are shorter than the 6:5 bonds. C60 is not "super aromatic" as it has a tendency to keep away from twofold bonds in the pentagonal rings, bringing about poor electron delocalisation. Accordingly, C60 acts like an electron inadequate alkene, and responds promptly with electron rich species. The geodesic and electronic holding factors in the structure represent the soundness of the particle. In principle, an unbounded number of fullerenes can exist, their structure in light of pentagonal and hexagonal rings, developed by rules for making icosahedra.

The electric and conductive properties of fullerenes and different carbonaceous nanomaterials shape the reason for some of their extraordinary attributes on the nanoscale. Accordingly, adjustment of these electric properties by means of single substitution or, on the other hand endohedral doping has produced critical research consideration.

Application of Fullerenes

Organic Photovoltaics

Fullerene can be utilized as natural photovoltaics. As of now, the record productivity for a mass heterojunction polymer sun based cell is a fullerene/polymer mix. The fullerene goes about as the n-sort semiconductor (electron acceptor). The n-sort is utilized as a part of conjunction with a p-sort polymer (electron contributor), commonly a polythiophene. They are mixed and given a role as the dynamic layer to make what is known as a mass heterojunction [15-16].

Antioxidants and Biopharmaceuticals

Fullerenes are intense cancer prevention agents, responding promptly and at a high rate with free radicals, which are regularly the reason for cell harm or passing. Fullerenes hold awesome guarantee in wellbeing and individual care applications where counteractive action of oxidative cell harm or passing is alluring, and in addition in non-physiological applications where oxidation and radical procedures are damaging (sustenance waste, plastics disintegration, metal erosion). Significant pharmaceutical organizations are investigating the utilization of fullerenes in controlling the neurological harm of such illnesses as Alzheimer's malady and Lou Gehrig's sickness (ALS), which are a consequence of radical harm. Medications for atherosclerosis, photodynamic treatment, and hostile to viral specialists are additionally being developed.

Polymer Additives

Fullerenes and fullerenic dark are synthetically responsive and can be added to polymer structures to make new copolymers with particular physical and mechanical properties. They can likewise be added to make composites. Much work has been done on the utilization of fullerenes as polymer added substances to change physical properties and execution attributes.

2.2 Carbon Nanotubes

Carbon nanotubes (CNTs) have one of a kind properties, yet the external dividers of perfect CNTs are synthetically dormant. In this way, CNTs are required to be altered in order to have biocompatibility and dissolvability [17]. Noncovalent also, covalent surface alterations are two ordinarily utilized systems to alter CNTs. In the primary strategy, long polymer chains (polystyrene sulfonate) are wrapped around the CNTs. CNTs, likewise called buckytubes, are barrel shaped carbon atoms with remarkable properties. They can be partitioned into single-walled CNTs (SWCNTs) and multi-walled CNTs (MWCNTs) on the premise of the tubes number in the CNTs, where SWCNTs are made of single

Carbonaceous Composite Materials   Materials Research Forum LLC
Materials Research Foundations **42** (2018) 205-230   doi: http://dx.doi.org/10.21741/9781945291975-8

layer carbon particles while MWCNTs involve multilateral graphene sheet that moved upon itself with many concentric tubes. As with graphene-based nanomaterials, CNTs display extraordinary properties, for example, huge particular territory, rich hollows and layered structure, which achieve a wide assortment of exceptional applications in wastewater treatment, including the evacuation of natural contaminants and overwhelming metals. Different techniques can be utilized to orchestrate CNTs, for example, synthetic vapour statement (CVD), the bend release strategy, fire union technique and other adjusted strategies. Other nanocomposites joined with polymers and metal oxides could be set up by aqueous technique, microwave illumination and different methodologies. In the later one, coordinate covalent change [18]. A large portion of the present functionalization strategies take after covalent or substance approach and a solid covalent holding framed amongst CNTs and coupling specialist [19]. Comparison of covalent and non-covalent functionalization of CNTS was shown in Table .1

Table 1. Comparison of Covalent and non-Covalent functionalization of CNTs.

S. No	Covalent functionalization	Non-covalent functionalization
1	Arrangement of stable compound bonds	van der Waals cooperation
2	Loss of electronic properties	No loss of electronic properties
3	Done by side divider and end top connection	Wrapping of atoms over the CNT surface
4	Demolition of π-bond change	Basic system is held
5	Done by oxidation, lahogenation, amidation, thiolation, hydrogenation and so on	Done by adsorption of polymer, Surfactants, biomolecules, nanoparticles and so on.

Applications of CNTS

Drug Transporters

CNTs have risen as an effective medication bearer due to their special properties (high medication stacking limit, high mechanical quality) [20]. CNTs have been utilized for conveying different peptides, nucleic acids, and little atomic medications into living cells [21]. Direct conveyance of chemotherapy sedates by physically adsorbed on CNTs, for instance, supramolecular benzene ring structure of CNTs bears shockingly high degree of sweet-smelling particles by π-π stacking.

Another utilization of CNTs for tranquilize conveyance is intravenous infusion. One of the issues with infusing drugs into the body is the danger of veins getting blocked as a result of the extansive size of the medications, which would prompt tissue poisonous quality. It has been recommended that CNTs could be utilized as nano bearers for conveying drugs into the body by means of injectable courses [22]. Medications can either append to the external surface of the CNT by means of practical gatherings or be stacked inside the CNT. Connection of the anticancer medication to the external surface of the CNT can be through either covalent or noncovalent holding [23].

Gene Therapy

In the field of cutting edge quality treatment, promising restorative and preventive impacts were accomplished. Nonetheless, the objective cell specificity and adequacy are basic attributes [24]. Little medication atoms can go into the cells by dissemination yet bio macromolecules, for example, DNA, RNA, and protein can't infiltrate through cell film and along these lines require conveyance vehicles for intracellular conveyance. Both viral and, non-viral vectors can be utilized for treating the hereditary diseases. Viral vectors can accomplish high transfection efficiencies yet have wellbeing concerns, for example, immunogenic and oncogenic. Non-viral vectors are protected yet have low transfection viability. CNTs offer advantages over other non-viral conveyance frameworks. An imperative essential for quality treatment vectors is the capacity to conquer extracellular boundaries, incorporating into vivo leeway instruments and assurance of the nucleic corrosive load from corruption, while accomplishing particular focusing of cells or tissues.

Biocompatibility and biodegradability of CNTs

On the off chance that nano-biomaterial corrupt into the body, the cancer-causing nature and, different types of lethal impacts may incite unfavourable reactions. Furthermore, non-biodegradable nanomaterials can aggregate in tissues and move toward becoming wellbeing dangerous. Poisonous quality related examinations appeared to examine the more beneficial methods for debasement impacts of CNT based medications [25]. Functionalised CNTs appeared to be corrupted inside cells, for example, neutrophils and macrophages by myeloperoxidase (MPO) movement furthermore, in the phagolysosomal reproducing liquid (PSf). Despite this, non-functionalised CNTs were safe (no morphological changes watched) when presented to the same natural oxidative conditions [26]. The show of biodegradability for f-CNTs has critical ramifications for the long haul toxicological profiles of these materials, especially identified with any potential clinical application. A few elements may be dependable, for example, the energy of debasement and the rate of body discharge of these species.

2.3   Graphene based material

Graphene

Graphene and its derivatives are being connected in contamination administration with an importance on gas adsorption and water remediation. Graphene is an unending two-dimensional (2D) monolayer comprising of sp2-hybridized carbon molecules. It could be seen as the central building hinder for other carbon allotropes, for example, fullerenes and carbon nanotubes. It likewise displays inborn layering and a few deformities, for example, topological deformities, opportunities, which would impact the reactivity of graphene with others.

Graphene oxide(GO) and Reduced graphene oxide (RGO)

GO, can be produced using concoction shedding of graphite by responses, is a two-dimensional sheet with plenteous oxygenated practical bunches on their basal plane and edges. The creation procedure of GO breaks the π-conjugated system also, in this manner make it very water-dispersible, in the interim, lessen the conglomeration among individual GO sheets, show a more extensive potential in wastewater treatment. And GO itself has pulled in critical considerations because of numerous fascinating properties and its part as a promising forerunner for mass creation of graphene-based nanomaterials. Amalgamation methodologies of GO contains the Brodie, Staudenmaier, Hummers technique, or altered strategies in light of them. The structure of RGO is amongst graphene and GO, there are just a couple of useful gatherings inserting on the RGO surface, could be lessened from GO by solvothermal and aqueous strategies. The upgraded electronic and conductive properties make it a planned possibility for poisons evacuation in wastewater treatment processes.

Graphene nanocomposites

Lately, plentiful polymers and metal oxides were joined to the graphene-based sheets, creating nanocomposites. These nanocomposites display champion auxiliary execution and multifunctional properties by brushing the two parts attributes. Furthermore, it ought to be noticed that incorporated nanocomposites are not simply the aggregate of diverse segments, however rather another nanomaterial with totally new properties. The functionalization process could encourage the scattering and after that keep the conglomeration of graphene sheet in the fluid stage, and bless them with some prevalent compound and physical properties, which may add to numerous reasonable uses. For instance, attractive nanocomposites of graphene could be effortlessly isolated from treated water and along these lines diminish genuine recontaminations, yet it is troublesome for partition of unique graphene and GO.

Carbonaceous Composite Materials   Materials Research Forum LLC
Materials Research Foundations **42** (2018) 205-230   doi: http://dx.doi.org/10.21741/9781945291975-8

3. Engineered carbon nanomaterials (ECNM)

The natural uses of designed carbonaceous nanomaterials are both proactive (avoiding natural debasement, enhancing general wellbeing, advancing vitality proficiency) and retroactive (remediation, wastewater reuse, toxin change). Carbon's one of a kind hybridization properties and the affectability of carbon's structure to annoyances in combination conditions take into consideration custom-made control to a degree not yet accomplished by inorganic nanostructures. ECNM join the unmistakable properties of sp^2 hybridized carbon bonds with the surprising qualities of material science and science at the nanoscale. From the electrical conductivity of a solitary nanotube to the adsorptive limit of mass nanomaterials, both single particles also, mass properties offer potential advances in ecological frameworks. Carbon nanomaterials are for the most part predictable with conventional physical-compound models also, hypotheses including electrostatics, adsorption, hydrophobicity. Atomic displaying has given an elucidation of physical-compound procedures happening at the nanoscale that are generally blocked off through test methods, however computational requests restrain the scope of length-scale, chirality and layers attainable in the sub-atomic demonstrating of heterogeneous nanotube and graphene tests.

4. Removal of ionic pollutants

Ionic toxins exist in dirtied water in essentially two structures (i) metal particles, for example, arsenic, mercury, cadmium, chromium, cobalt, copper, selenium, lead and (ii) non-metal particles, for example, fluoride, phosphate, nitrate, and sulfide. Nano materials (1– 100 nm) are superb poison adsorbents because of their little size, expansive surface area, high reactant nature and different dynamic locales. Uncovered nanoparticles result in agglomeration because of a high surface free vitality which decreases the dynamic adsorption destinations. Graphene is very steady, expansive in estimate (mm to inch) and has a substantial surface territory. To keep away from total arrangement of exposed nano particles, graphene has been utilized as a supporting material in numerous investigations [27-29].

5. Removal of organic pollutants

For the most part, the ecological uses of graphene based material can be assembled into three angles, to be specific the adsorption, change and identification of natural contaminations. Moreover, graphene based material gadgets, for example, a section stacked with graphene based material, have been essentially connected in smaller than expected scale decontamination and identification. Graphene nanomaterials and their subsidiaries have displayed brilliant execution in ecological poison evacuation, including that of substantial metals, anionic and cationic colours, natural contaminations (phenolic mixes, pesticides, and polycyclic aromatic hydrocarbons (PAHs)), inorganic anionic contaminations, and gas poisons. The expansive applicability of graphene and its high proficiency in the control of different contaminants are principally because of its vast surface range, wealth of surface useful gatherings and quick electron exchange, which make graphene a great adsorbent and reactant medium [30,31].

Graphene-based materials have been effectively connected in the adsorption of natural toxins as colours, polycyclic sweet-smelling hydrocarbons and gas. In these graphene materials, the adsorption is controlled by physisorption between the contaminations and the graphene surface, as such the bigger the surface territory the more noteworthy the adsorption capacity. Because of the benefits of minimal effort, arrangement process capacity and special structure, graphene oxide has been used to manufacture superhydrophobic and superoleophilic materials which can be utilized in the clean-up of oil spillage and concoction spillage. The utilization of graphene-based materials as adsorbents has been stretched out to the evacuation of colours. The favourable position in utilizing this graphene based materials is that they can be isolated from arrangement effortlessly, without attractive or centrifugation partition, along these lines making them effectively regenerable and reusable.

6.   Removal of air pollution

Graphene and its subordinates have been viewed as appealing possibility for air contamination adsorption because of their substantial surface territory. In any case, their adsorption limits are constantly restricted because of the solid stacking of graphene sheets. The unrivalled execution of graphene based materials for vaporous poison adsorption can be credited to two viewpoints. The first is the fundamental components of graphene based materials including its substantial particular surface territory, which uncovered a high amount of locales to adsorption, and the permeable structure giving quick and adaptable transport pathways [32]. The second is its oxygen-containing (and other enhanced polar) utilitarian gatherings, which give concoction adsorption locales and fortify the partiality for polar gas atoms [33].

High adsorption limit is perceived as an essential for compound change, which for the most part happens after adsorption amid contamination debasement. In that route, with its huge surface zone and pore structure graphene based materials perform well during the time spent contamination change. The fundamental explanation behind this (a) Graphene can give inexhaustible synthetic response dynamic locales on its surface utilitarian

Carbonaceous Composite Materials   Materials Research Forum LLC
Materials Research Foundations **42** (2018) 205-230   doi: http://dx.doi.org/10.21741/9781945291975-8

gatherings. Since the dynamic useful gatherings of graphene based material can append and tie with contaminative gasses at high fixations under ordinary conditions, the following stage of gas detoxification or mineralization is encouraged. (b) Graphene can acknowledge electrons and guarantee quick charge transportation in perspective of its high conductivity. The choice and pre-centralization of vaporous toxins by graphene based material is an essential for the change/mineralization and partition of poison blends. The removal mechanism of graphene based material in air pollution purification was shown in Figure 4.

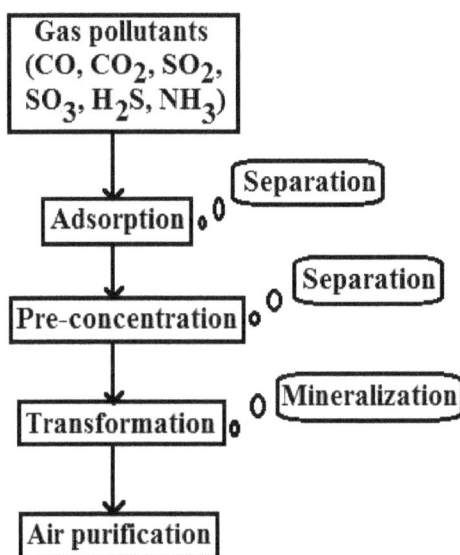

Figure 4: Mechanism of graphene based material in air pollution.

7.   Properties

The remarkable properties of carbonaceous nanomaterials most usually referred to in ecological applications are size, shape, and surface range; sub-atomic collaborations and sorption properties; and electronic, optical, and warm properties. Atomic control infers control over the structure and adaptation of a material. For carbonaceous nanomaterials this incorporates estimate, length, chirality, and the quantity of layers in the fullerene

Carbonaceous Composite Materials   Materials Research Forum LLC
Materials Research Foundations **42** (2018) 205-230   doi: http://dx.doi.org/10.21741/9781945291975-8

confine. Albeit current creation procedures for nanoscale carbon structures need finish accuracy and consistency, interfaces between development conditions and item properties educate the union of tuned nanomaterials. Varieties in blend system, temperature, weight, impetus, electron field, and process gasses upgrade nanomaterial structure, virtue, and physical introduction for particular applications [34].

The physical properties of individual nanomaterials, the size, shape, and surface territory of carbonaceous nanomaterials are profoundly needy upon accumulation (packaging) state and dissolvable science. Scums including vapour, biomolecules, and metals that adsorb to the surface of nanomaterials may in a general sense modify the collection conduct, warm and electric qualities, mechanical quality, and physicochemical properties of the nanomaterials. The physicochemical properties ascribed to optional structures of nanomaterial totals are exceptionally factor and inadequately described. Settling these attributes is basic for across the board use of carbonaceous nanomaterials from both specialized and ecological wellbeing and security viewpoints.

Atomic Interactions and Sorption Properties. Clarifying the atomic cooperation, sorption, and partitioning properties administering fullerenes and nanotubes is a joint exertion amongst scholars and experimentalists. Carbonaceous nanomaterials are for the most part predictable with conventional physical substance models and hypotheses including electrostatics, adsorption, hydrophobicity, and Hansen dissolvability parameters. Sub-atomic demonstrating has given an interpretation of physical-compound procedures at the nanoscale that are generally unavailable through test strategies, however computational difficulties limit the scope of length-scale, chirality, and layers possible in the atomic displaying of heterogonous nanotube tests [35-38].

Electronic, Optical, and Thermal Properties. The holding arrangement of fullerenes and nanotubes presents interesting conductive, optical, and warm properties offering wide guarantee for application in the electronic business. While the importance of exceptional field emanation properties, optical nonlinearity, high warm conductivity, and low temperature quantum wonders may appear to be indirectly identified with customary natural applications, significant roundabout ecological benefits collect from update of vitality and material serious purchaser hardware [39]. Novel electronic properties of carbon-based nanomaterials will likewise contribute to natural detecting gadgets and effective power era in creative sun powered cell designs [40]. At long last, fullerene intervened photo oxidation of determined organics has been investigated as an environmental remediation method

8.   Photocatalysis and sorbents

Photocatalytic system is a propelled oxidation process for expulsion of follow contaminants and pathogens from wastewater, which have been considered as a valuable pretreatment approach for dangerous substance and non-biodegradable contaminations to improve its biodegradability. A photocatalysis upgrade via carbonaceous nanomaterials has been thoroughly analyzed. Carbonaceous nanomaterials could upgrade the photocatalysis execution through three essential systems:

(1) giving fantastic adsorption dynamic destinations;

(2) enhancing the relocation effectiveness of photograph incited electrons and deferring electron-gap recombination;

(3) tuning the band crevice or photosensitization.

A fantastic element of carbonaceous composites as impetus bolsters and adsorbents in respect to customary carbon frameworks is the plausibility of deliberate varieties over wide scopes of fundamental physical and compound properties, for example,

(i) Sizes and volumes of a wide range of pores: small scale (beneath 2 nm in size), meso- (2-50 nm), and macro pores (over 50 nm) morphology of all of small scale, meso-, and full scale components

(ii) Level of crystallinity, thickness of carbon stage; warm and electrical conductivity, and other physico-substance parameters of the surface.

(iii) Mechanical quality: wearing down and affect resistance

Components influencing sorption are the accompanying ones: surface zone, mineral surface properties, natural carbon, dissolvability, temperature, pH, saltiness, co-solvents and broke up natural issue.

Surface region. Adsorption is a surface wonders specifically identified with surface region. Expanding the surface region, the particular adsorption will increment. Sorption is normally announced as a mass property on a for each gram weight premise. Sorption ought to dependably be accounted for on a zone premise, considering microspores and atomic porosity.

Mineral surfaces properties. Surface charge makes surface conditions in which there is an uneven charge appropriation, making a twofold layer of particles, accused natural solutes trading of other counter-particles in the twofold layer, bringing about physisorption.
Natural carbon. It has been discovered that the sorption of hydrophobic natural mixes is emphatically controlled by the nearness of soil natural material. While the response takes after that a sorption and will fit a sorption isotherm, it is parcelling.

Dissolvability. As the solvency of a hydrophobic compound declines, the adsorption coefficient increments from entropy driven collaboration with the surface.

Temperature. Since adsorption is an exothermic procedure, K values for the most part diminish with expanding temperature. All in all, a 10% lessening in K sorp would happen with a temperature ascend from 20 to 30°C.

pH. Just chemicals that have a tendency to ionize are influenced by the pH, on impartial particles the main change will be in the character of the surface, at low pH humic materials being about unbiased, for instance and more hydrophobic. Changes in pH will influence natural acids and bases by evolving dissolvability. Cations coming about because of the protonation of a natural base may more firmly adsorb to soils than nonpartisan species. Sorption of charged species will be influenced by the pH.

Saltiness. An expansion in saltiness can bring down the adsorption coefficient of cations due to the substitution/trade by salt cations. The adsorption of corrosive herbicides increments with saltiness at pH values over the pKa of the corrosive, pH impacting the effects of saltiness. Nonpartisan particles are for the most part less influenced by saltiness yet regularly demonstrate an expanded adsorption with expanding salt fixation, likely because of the expansion in the action coefficient of impartial particles and coming about lessening in watery solvency. Expanding saltiness may likewise change the interlayer dispersing of layer dirts and additionally the morphology of the dirt natural issue.

Co-solvents. Co-solvents are water dissolvable natural solvents, for example, methanol or acetone and they can diminish the sorption steady Ksorp by expanding the clear dissolvability.

9.   Carbonaceous nanomaterials as sorbents

Sorption of natural contaminants to sorbents such as NOM, mud, and initiated carbon represents a noteworthy soak in normal and designed ecological frameworks. Regular drinking water treatment, for instance, depends on physicochemical sorption forms for the evacuation of natural and inorganic contaminants. Many years of research have improved our comprehension of sorption systems also, encouraged streamlining of sorbent properties [41-44].The sorptive limit of customary carbonaceous sorbents is restricted by the thickness of surface dynamic destinations, the actuation vitality of sorptive bonds, the moderate energy and nonequilibrium of sorption in heterogeneous frameworks, and the mass exchange rate to the sorbent surface. The huge measurements of conventional sorbents additionally constrain their vehicle through low porosity situations and confuse endeavours in subsurface remediation. Carbonaceous

Carbonaceous Composite Materials   Materials Research Forum LLC
Materials Research Foundations **42** (2018) 205-230   doi: http://dx.doi.org/10.21741/9781945291975-8

nanosorbents, with their high surface zone to volume proportion, controlled pore estimate circulation, and manipulatable surface science, overcome a considerable lot of these inborn constraints. Sorption examines utilizing carbon-based nanomaterials report fast harmony rates, high adsorption limit, adequacy over a broad pH range, and consistency with BET, Langmuir, or Freundlich isotherms [45-50].

Conventional utilizations of activated carbon in water and wastewater treatment incorporate decrease in natural contaminants, remaining taste, or scent. While carbonaceous nanosorbents are successful in these territories, their cost and conceivable poisonous quality has avoided broad research in coordinate and broad use for water treatment. The incorporation of nanosorbents into customary stuffed bed reactors, however subtle elements on the adequacy of different nanomaterial immobilization procedures have not been exhibited. To date, most research on the ecological utilizations of nanosorbents has focused on the evacuation of particular risky contaminants.

Another favourable position to carbonaceous nanosorbents is the virtual nonattendance of hysteresis amongst adsorption and desorption isotherms for fluids and gasses under environmental weight. Improved air weight pertinent to gas adsorption in hydrogen stockpiling applications may re-establish hysteresis in the framework by lessening the vitality hindrance to fill nonwetting CNT pores and the intraparticle locale of the nanoaggregates. While quick harmony rates and high sorbent limit are intense traits of carbonaceous nanosorbents, their actual progressive potential lies in the different pathways for customized controls of their surface science. Fitting the prevailing physical and synthetic adsorption powers by means of particular functionalization yields carbonaceous nanomaterials that supplement the current suite of moderately unspecific regular sorbents. Functionalized nanosorbents may give a streamlined way to deal with focusing on micro pollutants, evacuating contaminants.

10. Composite filters

While adjusted CNT composite films create profoundly particular, tunable, and quick filtration on the seat scale, they are hard to make are still in the early phases of innovative work. Elective CNT applications in nanocomposite layers use the physical properties of CNTs to enhance the mechanical dependability of the film or as an instrument to disturb polymer pressing of the dynamic layer in customary turn around osmosis layers.

11. Renewable energy

In spite of the fact that nanomaterials and nanocomposites will discover applications crosswise over sustainable power source divisions, the lion's share of utilizations constitute material upgrades to auxiliary segments. CNT anodes, for example, may make strides the affectability and spatial determination of radiation counters utilized as a part of atomic power plants. Symmetric, adjusted nanotube films, may empower the utilization of second rate warm through an osmotic warmth motor. Furthermore, more grounded, lighter materials may yield auxiliary changes in wind gathering gadgets. The best potential for carbonaceous nanomaterials to yield major achievements lies in sun based vitality applications.

In photo electrochemical cells, semiconducting materials create an electron-gap combine and exchange the charge through a circuit to a counter cathode in contact with a redox couple in the electrolyte arrangement. Semiconducting CNTs experience charge division upon introduction to UV light, starting charge exchange to an answer stage reactant. While coordinate utilization of CNTs' photoactive properties in photo electrochemical cells have accomplished unobtrusive photo conversion efficiencies, semiconducting materials, for example, TiO_2 are prevalent reactant specialists. Lamentably, these nanoscale inorganics experience the ill effects of high charge recombination as the free electron moves to the terminal surface. To diminish rates of charge recombination, CNTs have too been proposed as conductive frameworks to pass on charge to the anode. High conductivity and smooth intersections between the anode and SWNTs empower this lessening in control recombination. Once more, union of monochiral nanotubes would significantly enhance the conduction proficiency of CNT platform.

12. Antimicrobials agents

Notwithstanding producing a suite of novel ecological applications, the one of a kind properties and nanoscale measurements of fullerenes and nanotubes have raised concern among toxicologists and natural researchers. While express systems of antimicrobial movement are still under examination, toxicity may depend upon physiochemical and basic qualities, for example, surface science, practical a mass thickness, length, leftover impetus defilement, and width. Various specialists are trusting to misuse these watched antimicrobial properties in ecological and human wellbeing applications. Particular classes of nanomaterials may be appropriate for water sterilization, therapeutic treatment, antimicrobial surface coatings, or research facility procedures in microbiology. Novel antimicrobial surface coatings that adventure the intrinsic defencelessness of microorganisms toward CNTs may give rich building answers for the testing issue of bacterial colonization and biofilm improvement in drinking water frameworks, therapeutic embed gadgets, and other submerged surfaces. Work on CNT harmfulness toward assorted microbial groups is progressing. Various research bunches are examining applications of antimicrobial and antiviral nanoparticles for water treatment also, appropriation frameworks.

Carbonaceous nanomaterials show solid antibacterial action and lower oxidation capacity, could viably inactivate pathogens under obvious light illumination or direct contact, tend to shape cleansing results (DBPs)[51-53]. Graphene-based nanocomposites were broadly explored as disinfectants for a long time. The attractive RGO functionalized with glutaraldehyde nanocomposite (MRGOGA) demonstrated the splendid purification capacity towards *S. aureus* and *E.coli*, up to 99% of gram-positive and gram-negative microbes were killed successfully in 10 minutes upon close infrared laser illumination. The antibacterial movement, substantial particular surface territory and solid conductivity of graphene-based nanomaterials empowered their utilization in sterilization applications. Furthermore, the antibacterial component of these carbonaceous nanomaterials was accounted for to include two viewpoints:

(1) obliterating the honesty of cell films upon coordinate contact;

(2) aggravating the specific microbial advance by means of oxidative anxiety. Moreover, extra inquires about related with augmenting purification execution and natural impacts of nanomaterials ought to be given careful consideration.

13. Sensor based on carbon nanomaterials

CNT based sensors offer various points of interest to existing sensor stages, and peruse are alluded to later basic surveys regarding this matter. Compound, natural, warm, optical, push, strain, weight, and stream sensors draw upon the excellent electrical conductivity, substance soundness, high surface zone, mechanical firmness, and direct functionalization pathways of CNTs to improve conventional carbon cathode sensor stages. In another application, varieties of adjusted MWNTs developed on a SiO_2 substrate fill in as anodes in ionization sensors for gas discovery. The sharp tips of nanotubes encourage the era of high electric fields at low voltages, empowering compact, battery worked, and little scale sensors. Identification happens by means of electric field deterioration of the example took after by cathode enlistment of a special unique mark for each vaporous analyte.

The presentation of CNT nanowire sensors constitutes a real leap forward for the sensor field. Adsorption of charged species to the surface of the CNT changes the nanotube conductance, in this manner setting up a reason for connection between present vacillation and analyte piece or, on the other hand fixation. The energy of the nanowire sensor stage lies in plans for level controlled adjustment to target particular synthetic and organic analytes lacking innate partiality for CNTs. Covalent and supramolecular functionalization enables adjustment with a suite of substance gatherings, metals, compounds, antibodies, DNA particles, and natural receptors. Other natural applications incorporate wellbeing and security observing, framework administration, synthetic and material productivity in assembling, and savvy administrative stages.

Observing ecological microbial biology and identifying microbial pathogens are likewise destinations of biosensor stage examine. A few frameworks use coordinate adsorptive gathering of nucleic corrosive or protein focuses to the surface of the CNT terminal cluster for name free electrical identification of hybridization. Others utilize the one of a kind conductive properties of the CNT to open up flag pathways in both acknowledgment and transduction occasions.

Conclusion

The quick development of this field guarantees us that graphene based material will be another era of materials in toxin administration with extraordinary limits and simple control. With its constant small scale, meso-, and full scale structures, graphene based material shows greatly great potential for the evacuation of natural contaminations by means of adsorption and change forms. Obviously, the natural utilizations of graphene based material ought to be extended from water cleansing, to air cleaning, and on to soil and groundwater remediation. Its high adsorption limit and good selectivity make graphene based material a fantastic bearer for pre-thinking and isolating complex natural poisons. This property can be utilized as a part of the recuperation of contaminations from wastewater and dirtied gas, and to create supportable natural nanotechnology. The particle level instruments of the connection of graphene based material with ecological contaminations should be uncovered. Moreover, graphene based material can be enhanced to meet future natural control needs for all intents and purposes and at the business level by utilizing section operations.

Carbonaceous nanomaterials have high particular surface regions, exceptional electrical, optical, warm and concoction action; have been viewed as a standout amongst the most imminent contender to expel concoction and organic contaminants from wastewater. What's to come advancements and full-scale uses of these nanomaterials still face a wide assortment of difficulties what's more, more careful investigations are positively required.

The present blend techniques for carbonaceous nanomaterials is as yet convoluted and low-proficient, albeit various investigations have been directed to handle it. More basic, strong and proficient manufacture strategies are critically required. In the meantime, business expansive scale generation of carbonaceous nanomaterials is testing and should

Carbonaceous Composite Materials
Materials Research Foundations **42** (2018) 205-230

Materials Research Forum LLC
doi: http://dx.doi.org/10.21741/9781945291975-8

be tended to for expansive range applications. The agglomeration of CNTs and graphene-based nanomaterials in watery stage is another shortcoming in water sterilization. Collected nanomaterials would decrease the surface territory too as dynamic locales and subsequently influence the availability to them, bringing about diminished proficiency on contaminations expulsion. Nanomaterials changed with different useful gatherings and metal oxides have been utilized to beat them, and further looks into should concentrate more on the focused on adjustment furthermore, improve the evacuation proficiency and in addition selectivity and fondness toward particular contaminants. Carbonaceous nanomaterials have many preferences and additionally constraints in wastewater treatment, it is to be sure potential nanomaterials for taking care of assorted natural issues.

References

[1]   J. Virkutyte, R.S. Varma, V. Jegatheesan. Treatment of micropollutants in water and wastewater. IWA Publishing. London. 2010.

[2]   M. Thakur, G. Sharma, T. Ahamad, A.A. Ghfar, D. Pathania, M. Naushad, Efficient photocatalytic degradation of toxic dyes from aqueous environment using gelatin-Zr(IV) phosphate nanocomposite and its antimicrobial activity, Colloids and surfaces. B, Biointerfaces, 157 (2017) 456-463. https://doi.org/10.1016/j.colsurfb.2017.06.018

[3]   S.K. Kahlon, G. Sharma, J.M. Julka, A. Kumar, S. Sharma, F.J. Stadler, Impact of heavy metals and nanoparticles on aquatic biota, Environmental Chemistry Letters, (2018) https://doi.org/10.1007/s10311-018-0737-4.

[4]   G. Sharma, D. Pathania, M. Naushad, N.C. Kothiyal, Fabrication, characterization and antimicrobial activity of polyaniline Th(IV) tungstomolybdophosphate nanocomposite material: Efficient removal of toxic metal ions from water, Chemical Engineering Journal, 251 (2014) 413-421. https://doi.org/10.1016/j.cej.2014.04.074

[5]   N.W.S. Kam, M. O'Connell, J.A. Wisdom, H.J. Dai. Carbon nanotubes as multifunctional biological transporters and near infrared agents for selective cancer cell destruction. Proceedings of the National Academy of Sciences of the United States of America. 102 (2005) 11600-11605. https://doi.org/10.1073/pnas.0502680102

[6]   Y.H. Li, Y.M. Zhao, W.B. Hu, I. Ahmad, Y.Q. Zhu, X.J. Peng, Z.K. Luan. Carbon nanotubes – the promising adsorbent in wastewater treatment. Journal of Physics:

Conference Series. 61 (2007) 698-702. https://doi.org/10.1088/1742-6596/61/1/140

[7]   Y. Patino, E. Diaz, S. Ordonez. Performance of different carbonaceous materials for emerging pollutants adsorption. Chemosphere. 119 (2015) S124-S130. https://doi.org/10.1016/j.chemosphere.2014.05.025

[8]   L. Zhang, F. Pan, X. Liu, L. Yang, X. Jiang, J. Yang, W. Shi. Multi-walled carbon nanotubes as sorbent for recovery of endocrine disrupting compound-bisphenol F from wastewater. Chemical Engineering Journal. 218 (2013) 238-246. https://doi.org/10.1016/j.cej.2012.12.046

[9]   J.M. Dias, M.C.M. Alvim-Ferraz, M.F. Almeida. Waste materials for activated carbon preparation and its use in aqueous-phase treatment: A review. Journal of Environmental Management. 85 (2007) 833-846. https://doi.org/10.1016/j.jenvman.2007.07.031

[10] J.L. Figueiredo, C. Bernardo, R.T.K. Baker, K.J. Huttinger, Carbon fibers filaments and composites. Kluwer academic publishers. Amsterdam. 1990. https://doi.org/10.1007/978-94-015-6847-0

[11] V. A. Likholobov, V.B. Fenelonov, L.G. Okkel, O.V. Goncharova, L.B. Avdeeva, V.I. Zaikovskii, G.G. Kuvshinov, V.A. Semikolenov, V.K. Duplyakin, O.N. Baklanova, G.V. Plaksin. New carbon-carbonaceous composites for catalysis and adsorption. Reaction kinetics and catalysis letters. 54 (1995) 381-411. https://doi.org/10.1007/BF02071033

[12] P.M. Ajayan. Nanotubes from carbon. Chemical Reviews. 99 (1999) 1787-1800. https://doi.org/10.1021/cr970102g

[13] Y. Hu, O. Shenderova, D. Brenner. Carbon nanostructures: morphologies and properties. Journal of Computational and Theoretical Nanoscience. 4 (2007) 199-221. https://doi.org/10.1166/jctn.2007.2307

[14] J.H. Walther, R. Jaffe, T. Halicioglu, P. Koumoutsakos. Carbonnanotubes in water: Structural characteristics and energetics. Journal of Physical Chemistry B. 105 (2001) 9980–9987. https://doi.org/10.1021/jp011344u

[15] P. Tonui, S. Oseni, G. Sharma, Y. Qingfenq, G.T. Mola, Perovskites Photovoltaic Solar Cells: An Overview of Current Status, Renewable & Sustainable Energy Reviews, 91 (2018) 1025-1044. https://doi.org/10.1016/j.rser.2018.04.069

[16] S.O. Oseni, K. Kaviyarasu, M. Maaza, G. Sharma, G. Pellicane, G.T. Mola, ZnO:CNT assisted charge transport in PTB7:PCBM blend organic solar cell,

Journal of Alloys and Compounds, 748 (2018) 216-222. https://doi.org/10.1016/j.jallcom.2018.03.141

[17] J.L. Stevens, A.Y. Huang, H. Peng, I.W. Chiang, V.N. Khabashesku. J.L. Margrave. Sidewall amino-functionalization of single-walled carbon nanotubes through fluorination and subsequent reactions with terminal diamines. Nano Letters. 3 (2003) 331-336. https://doi.org/10.1021/nl025944w

[18] M. Habibizadeh, K. Rostamizadeh, N. Dalali, A. Ramazani. Preparation and characterization of PEGylated multiwall carbon nanotubes as covalently conjugated and non-covalent drug carrier: A comparative study. Materials Science and Engineering: C. 74 (2017) 1-9. https://doi.org/10.1016/j.msec.2016.12.023

[19] Q. Fu, C. Lu, J. Liu. Selective coating of single wall carbon nanotubes with thin SiO_2 layer. Nano Letters. 2 (2002) 329-332. https://doi.org/10.1021/nl025513d

[20] B. Peng, M. Locascio, P. Zapol, S. Li, S.L. Mielke, G.C. Schatz, H.D. Espinosa. Measurements of near-ultimate strength for multiwalled carbon nanotubes and irradiation-induced crosslinking improvements. Nature Nanotechnology. 3 (2008) 626-631. https://doi.org/10.1038/nnano.2008.211

[21] S. Kumar, R. Rani, N. Dilbaghi, K. Tankeshwar, K-H. Kim. Carbon nanotubes: a novel material for multifaceted applications in human healthcare. Chemical Society Reviews. 46 (2017) 158-196. https://doi.org/10.1039/C6CS00517A

[22] S. Beg, M. Rizwan, A.M. Sheikh, M.S. Hasnain, K. Anwer, K. Kohli. Advancement in carbon nanotubes: basics, biomedical applications and toxicity. Journal of Pharmacy and Pharmacology. 63 (2011) 141-163. https://doi.org/10.1111/j.2042-7158.2010.01167.x

[23] Z. Liu, K. Chen, C. Davis, S. Sherlock, Q. Cao, X. Chen. H. Dai. Drug delivery with carbon nanotubes for in vivo cancer treatment. Cancer Research. 68 (2008) 6652-6660. https://doi.org/10.1158/0008-5472.CAN-08-1468

[24] M. Vincent, I. de Lazaro, K. Kostarelos. Graphene materials as 2D non-viral gene transfer vector platforms. Gene Therapy. 24 (2017) 123-132. https://doi.org/10.1038/gt.2016.79

[25] C. Nie, Y. Yang, C. Cheng, L. Ma, J. Deng, L. Wang, C. Zhao. Bioinspired and biocompatible carbon nanotube-Ag nanohybrid coatings for robust antibacterial applications. ActaBiomaterialia. 51 (2017) 479-494. https://doi.org/10.1016/j.actbio.2017.01.027

[26] X. Liu, R.H. Hurt, A.B. Kane. Biodurability of single-walled carbon nanotubes depends on surface functionalization. Carbon. 48 (2010) 1961-1969. https://doi.org/10.1016/j.carbon.2010.02.002

[27] X, Huang, Z. Zeng, Z. Fan, J. Liu, H. Zhang. Graphene-based electrodes. Advanced Materials. 24 (2012) 5979-6004.

[28] Q. Xiang, J. Yu, M. Jaroniec. Graphene-based semiconductor photocatalysts. Chemical Society Reviews. 41 (2012) 782-796. https://doi.org/10.1039/C1CS15172J

[29] W. Choi, I. Lahiri, R. Seelaboyina, Y.S. Kang. Synthesis of graphene and its applications: A review. Critical Reviews in Solid State and Material Sciences. 35 (2010) 52-71. https://doi.org/10.1080/10408430903505036

[30] K.C. Kemp, H. Seema, M. Saleh, N.H. Le, K. Mahesh, V. Chandra, K.S. Kim. Environmental applications using graphene composites: water remediation and gas adsorption. Nanoscale. 5 (2013) 3149-3171. https://doi.org/10.1039/c3nr33708a

[31] G. Zhao, L. Jiang, Y. He, J. Li, H. Dong, X. Wang, W. Hu. Sulfonated graphene for persistent aromatic pollutant management. Advanced Materials. 23 (2011) 3959-3963. https://doi.org/10.1002/adma.201101007

[32] Z.-Y. Sui, Y. Cui, J.-H. Zhu, B.-H. Han. Preparation of three-dimensional graphene oxide-polyethylenimine porous materials as dye and gas adsorbents. ACS Applied Materials and Interfaces. 5 (2013) 9172-9179. https://doi.org/10.1021/am402661t

[33] J. Liang, Z. Cai, L. Li, L. Guo, J. Geng. Scalable and facile preparation of graphene aerogel for air purification. RSC Advances. 4 (2014) 4843-4847. https://doi.org/10.1039/c3ra45147j

[34] O. Jost, A. Gorbunov, X.J. Liu, W. Pompe, J. Fink. Single walled carbon nanotube diameter. Journal of Nanoscience and Nanotechnology. 4 (2004) 61-102. https://doi.org/10.1166/jnn.2004.071

[35] P.Keblinski, S.K. Nayak, P. Zapol, P.M. Ajayan, Chargedistribution and stability of charged carbon nanotubes. Physical Review Letters89 (2002) 255503. https://doi.org/10.1103/PhysRevLett.89.255503

[36] S. Furmaniak, A.P. Terzyk, P.A. Gauden, G. Rychlicki, Simple models of adsorption in nanotubes. Journal of Colloid and Interface Science. 295 (2006) 310–317. https://doi.org/10.1016/j.jcis.2005.12.032

[37] J.H. Walther, R.L. Jaffe, E.M. Kotsalis, T. Werder, T. Halicioglu, P. Koumoutsakos, Hydrophobic hydration of C-60 and carbon nanotubes in water. Carbon. 42 (2004) 1185–1194. https://doi.org/10.1016/j.carbon.2003.12.071

[38] H.T. Ham, Y.S. Choi, I.J. Chung. An explanation of dispersion states of single-walled carbon nanotubes in solvents andaqueous surfactant solutions using solubility parameters. Journal of Colloid and Interface Science.286 (2005) 216–223. https://doi.org/10.1016/j.jcis.2005.01.002

[39] P. Avouris. Carbon nanotube electronics. Chemical Physics. 281 (2002) 429–445. https://doi.org/10.1016/S0301-0104(02)00376-2

[40] P.V. Kamat, M. Haria, S. Hotchandani, C60 Cluster as an electron shuttle in a RuII)-polypyridyl sensitizer-based photochemicalsolar cell. The Journal of Physical Chemistry B.108 (2004) 5166–5170. https://doi.org/10.1021/jp0496699

[41] R.M. Allen-King, P. Grathwohl, W.P. Ball. New modelling paradigms for the sorption of hydrophobic organic chemicalsto heterogeneous carbonaceous matter in soils, sediments,and rocks. Advances in Water Resources.25 (2002) 985–1016. https://doi.org/10.1016/S0309-1708(02)00045-3

[42] M. Naushad, G. Sharma, A. Kumar, S. Sharma, A.A. Ghfar, A. Bhatnagar, F.J. Stadler, M.R. Khan, Efficient removal of toxic phosphate anions from aqueous environment using pectin based quaternary amino anion exchanger, International journal of biological macromolecules, 106 (2018) 1-10. https://doi.org/10.1016/j.ijbiomac.2017.07.169

[43] R. Bushra, M. Naushad, G. Sharma, A. Azam, Z.A. Alothman, Synthesis of polyaniline based composite material and its analytical applications for the removal of highly toxic Hg^{2+} metal ion: Antibacterial activity against E. coli, Korean Journal of Chemical Engineering, 34 (2017) 1970-1979. https://doi.org/10.1007/s11814-017-0076-3

[44] G. Sharma, M. Naushad, A. Kumar, S. Rana, S. Sharma, A. Bhatnagar, F. J. Stadler, A.A. Ghfar, M.R. Khan, Efficient removal of coomassie brilliant blue R-250 dye using starch/poly(alginic acid- cl -acrylamide) nanohydrogel, Process Safety and Environmental Protection, 109 (2017) 301-310. https://doi.org/10.1016/j.psep.2017.04.011

[45] C.S. Lu, Y.L. Chung, K.F. Chang. Adsorption of trihalomethanesfrom water with carbon nanotubes. Water Research.39 (2005) 1183–1189. https://doi.org/10.1016/j.watres.2004.12.033

[46] G. Sharma, M. Naushad, A.a.H. Al-Muhtaseb, A. Kumar, M.R. Khan, S. Kalia, Shweta, M. Bala, A. Sharma, Fabrication and characterization of chitosan-crosslinked-poly(alginic acid) nanohydrogel for adsorptive removal of Cr(VI) metal ion from aqueous medium, International journal of biological macromolecules, 95 (2017) 484-493. https://doi.org/10.1016/j.ijbiomac.2016.11.072

[47] G. Sharma, A. Kumar, C. Chauhan, A. Okram, S. Sharma, D. Pathania, S. Kalia, Pectin-crosslinked -guar gum/SPION nanocomposite hydrogel for adsorption of m-cresol and o-chlorophenol, Sustainable Chemistry and Pharmacy, 6 (2017) 96-106. https://doi.org/10.1016/j.scp.2017.10.003

[48] V.K. Gupta, D. Pathania, N.C. Kothiyal, G. Sharma, Polyaniline zirconium (IV) silicophosphate nanocomposite for remediation of methylene blue dye from waste water, Journal of Molecular Liquids, 190 (2014) 139-145. https://doi.org/10.1016/j.molliq.2013.10.027

[49] M. Naushad, Z.A. Alothman, G. Sharma, Inamuddin, Kinetics, isotherm and thermodynamic investigations for the adsorption of Co(II) ion onto crystal violet modified amberlite IR-120 resin, Ionics, 21 (2014) 1453-1459. https://doi.org/10.1007/s11581-014-1292-z

[50] K. Yang, X.L. Wang, L.Z. Zhu, B.S. Xing. Competitive sorption of pyrene, phenanthrene, and naphthalene on multiwalledcarbon nanotubes. Environmental Science and Technology. 40 (2006) 5804–5810. https://doi.org/10.1021/es061081n

[51] V.K. Gupta, D. Pathania, M. Asif, G. Sharma, Liquid phase synthesis of pectin–cadmium sulfide nanocomposite and its photocatalytic and antibacterial activity, Journal of Molecular Liquids, 196 (2014) 107-112. https://doi.org/10.1016/j.molliq.2014.03.021

[52] M.I. Ahamed, Inamuddin, Lutfullah, G. Sharma, A. Khan, A.M. Asiri, Turmeric/polyvinyl alcohol Th(IV) phosphate electrospun fibers: Synthesis, characterization and antimicrobial studies, Journal of the Taiwan Institute of Chemical Engineers, 68 (2016) 407-414. https://doi.org/10.1016/j.jtice.2016.08.024

[53] G. Sharma, A. Kumar, M. Naushad, D. Pathania, M. Sillanpää, Polyacrylamide@Zr(IV) vanadophosphate nanocomposite: Ion exchange properties, antibacterial activity, and photocatalytic behavior, Journal of Industrial and Engineering Chemistry, 33 (2016) 201-208. https://doi.org/10.1016/j.jiec.2015.10.011

Carbonaceous Composite Materials
Materials Research Foundations **42** (2018) 231-272

Materials Research Forum LLC
doi: http://dx.doi.org/10.21741/9781945291975-9

Chapter 9

Recent Innovation and Advances in Utilization of Graphene Oxide Based Photocatalysis

Ajay Kumar[a], Manisha Chandel[a], Anamika Rana[b], Gaurav Sharma[b], Deepika Jamwal[b,c,] Genene Tessema Mola[d]**, Amit Kumar[b]*

[a]Department of Chemistry, IEC University, Baddi, Himachal Pradesh, India

[b]School of Chemistry, Shoolini University, Himachal Pradesh, India, 173229

[c]Department of Chemistry and Centre of Advanced Studies in Chemistry, Panjab University, Chandigarh 160014, India

[d]School of Chemistry & Physics, University of KwaZulu-Natal, Pietermaritzburg Campus, Private Bag X 01,Scottsville 3209, South Africa

mittuchem83@gmail.com*, genene.mola@gmail.com**

Graphical Abstract

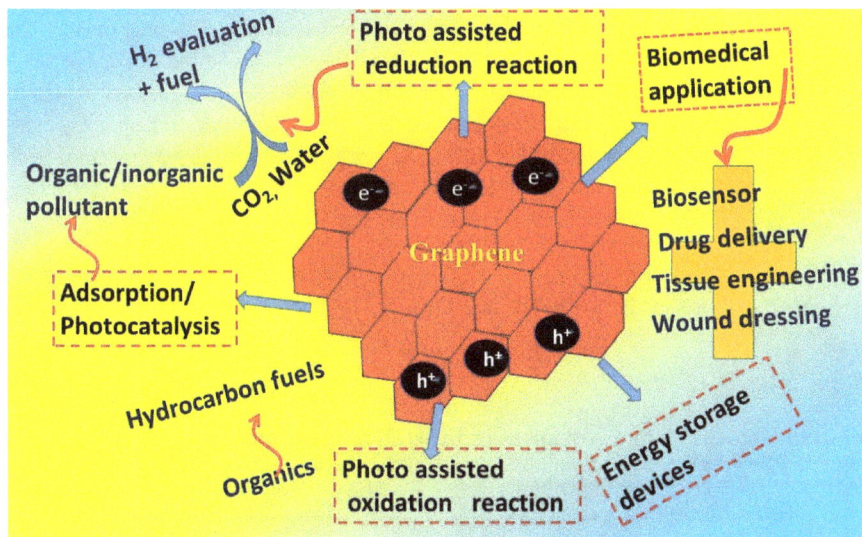

Carbonaceous Composite Materials

Materials Research Foundations **42** (2018) 231-272

Materials Research Forum LLC

doi: http://dx.doi.org/10.21741/9781945291975-9

Abstract

Globally, intensive research and innovation has been carried out for development and designing of photocatalysts with high efficiency and cost-effectiveness for various environmental and energy applications. Photocatalysis has been focused among various advanced oxidation processes for environmental detoxification and energy production via water splitting. Novel photocatalysts have been designed with incorporation of organic and polymeric counterparts for increasing their potential. Fortunately, graphene (carbon allotrope) has found a place and impressive contribution in materials science with outstanding properties for photocatalytic applications. Graphene based nanomaterials, graphene nano-sheets, reduced graphene oxide (RGO) and graphene oxide (GO) have gained immense concern in the perspective of all research fields. These derivatives have unique physiochemical properties such as high surface area, high electron mobility, thermal stability, biocompatibility, 2D scaffold with extended conjugation. All these properties are highly advantageous which trigger the exploration of such materials in the various research applications. From the discovery of graphene to today, it has been engaged in all the research fields such as fabrication of heterogeneous photocatalysts, solar fuel cell, energy storage devices, pollutants removal, hydrogen production and biomedical application. Since the precursor material for the production of graphene derivative is graphite which is a nonprecious and abundant material, thus making the graphene based research approachable for all researcher seekers in the respective field. This review surveys the developments, innovations and challenges involved in use of graphene and its derivatives in photocatalysis. Designing, fabrication of photocatalysts based on graphene and its derivatives and novel strategies have been summarized. Such kind of review helps in promoting awareness and motivation for continuous innovations in utilization of graphene based materials in photocatalytic scene.

Keywords

Graphene, Photocatalysis, Water Treatment, Energy Production, Degradation

Contents

1.   Introduction

Green technologies as semiconductor mediated photocatalysis utilizing the photon energy for various environmental detoxification and clean energy production have been of tremendous interest among scientific and industry [1-3]. The unplanned anthropogenic activities and industrial bloom has led to increase in number of emerging contaminants in water bodies which are often non-regulated. Secondly the dependence on all nations on fossil fuels has given rise to a global energy crisis [4, 5]. There has been an increasing demand for shifting to renewable energy of sun, wind and nuclear power. However clean and cost effective technologies are required to harness energy from renewable sources such as the sun. So keeping this concern in mind various researchers have been continuously looking for alternatives which are ecofriendly as well as cheap. Advance oxidation processes and especially semiconductor photocatalysis has been instrumental in solar energy conversion for water purification, CO_2 reduction and hydrogen production as a clean fuel [5, 6]. With first discovery of photocatalysis TiO_2, ZnO and Iron oxides have been star photocatalysts [7-12]. However their obvious limitations of low efficiencies application [13] led to focus on other metal oxides, sulphides, nitrides, halides, oxyhalides, etc. [14-16].

In recent years, metal free catalysts have gained importance because of their dual characteristics of adsorption and photocatalysis. Carbon based and derived materials such as graphene, carbon nanotubes, activated carbon, graphitic carbon nitride, organic polymers, etc. have been used as catalysts in single and combined forms for various photocatalytic applications [17-19]. Numerous carbon scaffolds have been studied widely owing to their low cost, high strength, ease of preparation, ecofriendly nature, and low toxicity [20]. Graphene oxide, reduced graphene oxide, graphene, activated carbon and biochar are emerging materials which are being utilized as supports for photocatalysts [21]. Despite the supportive properties, the role of these materials are also being explored by some researchers in the field of heterogeneous photocatalysis and they get extraordinary response [22].

Graphene which is a single layer of graphite is a "wonder material" and has a unique structure, high conductivity, high surface area, stability and electron mobility. In addition it can be produced by cost effective approaches. Graphene has been a part of various composite materials, reinforced plastics, nano-composites, heterojunctions, drug delivery vehicles, medical implants, tissue engineering parts, solar cells, energy storage devices and radiation absorbers. Graphene and its derivatives graphene oxide and reduced graphene oxide have been important components for making various functional materials for photocatalytic environmental detoxification and energy conversion.

2. Graphene oxide (GO)

GO is a modernistic 2D motif which is prepared from graphite powder as well as from flakes. GO has been widely explored in the field of photocatalysis due to its admirable optical, mechanical, electrical, and chemical properties [23]. GO is immensely oxygenated and there is huge abundance of epoxy and hydroxyl groups in their basal scaffoled [24]. Despite, the GO is highly dispersible in some solvents and water due to loss in hydrophilicity which is because of the oxygen moieties present on the GO surface which protect it from agglomeration [25]. Peculiarly, -COO are the most abundant groups found on the surface of GO which could provide as nucleation and dock sites for nanocrystal growth which might be beneficial for the synthesis of hybrids photocatalysts [26]. Numerous photocatalysts has been synthesized with the combination of GO and researchers have got satisfying results in the respective fields such as $BiVO_4/TiO_2/GO$[27], ZnO/GO[28], $Ag/AgCl/GO$[29], $Eu/TiO_2/GO$ [30]

2.1   Synthesis of GO from graphite powder/flakes

The instigating work on the synthesis of GO was started in 1859 by Brodie [31]. The method involves, the reaction of Graphite with $KClO_3$ and fuming HNO_3 at 60 °C. However the reaction needs a long time to complete. Further, Hummer developed a method for the preparation of graphene derivatives from graphite powder which has become most important method [32]. This preparation strategies involved harsh treatment of graphite powder with concentrated H_2SO_4 solution containing $KMnO_4$ and $NaNO_3$. However Hummer methods also have some glitch like release of toxic gases such as NO_2 & N_2O_4 during oxidation process and release of Na^+ & NO_3^- ions which are difficult to remove from the reaction mixture [33]. Afterwards, the Hummer method was modified by Tour and co-workers by eliminating the use of $NaNO_3$, and increasing the amount of $KMnO_4$, and operating the reaction with the mixture of H_2SO_4/H_3PO_4 in a 9:1 (v/v). This modification is quite effective which not only increases the yield of the reaction mixture but also diminishes the evolution of toxic gases [34].

3.   Reduced graphene oxide (RGO)

RGO is a 2D material that is stable under ambient conditions; it has a special electronic honeycomb structure with delocalize conjugated π system [19, 35]. RGO seems to be a promising material in the field of heterogeneous photocatalysts owing to its unique structure and properties such as anomalous quantum Hall effect, zero band gap, high charge mobility, etc. [36, 37]. RGO has been utilized as a scaffold for the fabrication of various photocatalysts. It was examined that RGO demonstrate an imperative role in the effective separation of photo generated charge carrier which is beneficial for a photocatalysts to strengthen its optical activity [38, 39]. Thus various heterogeneous photocatalysts has been synthesized in combination with RGO such as $Bi_{25}FeO_{40}/RGO$ [40], CuS/RGO [41], $CdS/CoFe_2O_4/RGO$ [42] TiO_2-RGO [43].

3.1   RGO synthesis

In general, RGO is made from the GO by different methods which includes thermal reduction [44], chemical reduction [45, 46] photocatalysts reduction [47, 48] electro chemical reduction [49, 50], solvothermal reduction [51], etc. Figure 1 shows the synthetic routes to RGO.

Carbonaceous Composite Materials   Materials Research Forum LLC
Materials Research Foundations **42** (2018) 231-272   doi: http://dx.doi.org/10.21741/9781945291975-9

3.2   Thermal reduction

RGO can be prepared from thermal treatment of GO by annealing at 2000 °C. The rapid heat treatment resulted to release CO or CO_2 gases which exert pressure on the stacked layer of graphene sheets and ultimately lead to exfoliation of graphene sheets [52].

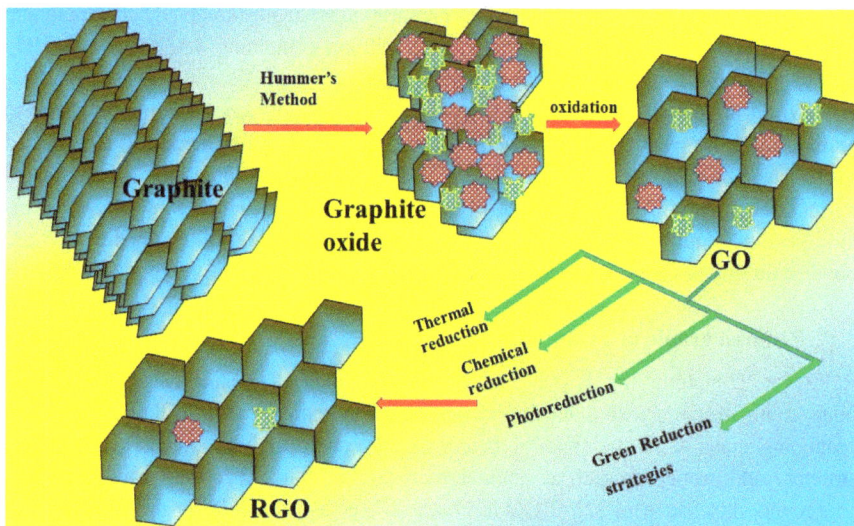

Fig.1: Reduction methods and strategies for RGO synthesis.

The abrupt heating of GO decomposes the oxygen group containing moieties such as – COO, -OH, -OR etc. [53] and removes the carbon atom from the 2D plane which causes the distortion. This ends in formation of small sized and wrinkled graphene sheets [54]. Microwave heating is an alternate method used for the exfoliation of GO in graphene sheets. This method is more popular as compare to the high temperature annealing as it minimize the energy cost. But the productivity of exfoliation is lower as compared to thermal heating [55]. To overcome this complication Voiry et al. annealed GO slightly prior to microwave exposure so that it can absorb more radiation [56].

3.3   Chemical reduction

As compared to thermal annealing chemical reduction method of GO is cheaper and easy to operate and facile at room temperature. Hydrazine and its derivatives such as hydrazine hydrate and dimethylhydrazine have been endorsed as good reducing agents and can be used for the mass production of graphene [57]. The reduction process can be done directly by adding the hydrazine derivative to the aqueous suspension of GO [58].

Metal hydrides are the class of chemicals which have been widely authorized as strong reducing agents. Out of metal hydrides $NaBH_4$ is the most commonly used for the reduction of GO [59]. As NaBH4 is slightly hydrolyzed by the water thus it is recommended that freshly prepared solution of $NaBH_4$ should be used for better reduction. However it is most reactive towards -CO group as compared to –RO, -COOH and -OH, hence after reduction these group remains in the GO [60]. To tackle this hitch, the pre-reduced GO by $NaBH_4$ is treated by conc. H_2SO_4 at 180 °C to achieve complete reduction [61]. As hydrazine and $NaBH_4$ both have some drawbacks such as toxicity of hydrazine and stability of $NaBH_4$ in aqueous medium limiting their use for reduction process. Both these drawbacks have been managed by using ascorbic acid. Which has been recently incorporated for the reduction of GO [62].

The reducing agents used for the reduction of GO to RGO influence its properties such as C/O ratio and conductivity. The influence of different reducing agent on the RGO properties has been summarized in the following table 1.

Table 1: Effect of various reducing agents on the RGO properties.

S.No.	Reducing agent	C/O ratio	Conductivity (S/cm)	Ref.
1	N_2H_4	12.5	99.6	[63]
2	$NaBH_4/H_2SO_4$	8.6	16.6	[61]
3	Ascorbic acid	12.5	77	[63]
4	Hydroiodic acid	15	300	[64]

3.4   Photoreduction of GO

Photoreduction is a 'Green' method used for the synthesis of RGO which not only excludes the use of harmful chemical but also minimizes the cost and time consumption. The method is quite smooth, safe and impregnable as compared to other methods. This method allows reduction process even in liquid phase as a result of which the RGO is also obtained in the solution form which can be directly used for another application [65]. Li et al. reported the photoreduction of GO with $H_3PW_{12}O_{40}$ and isopropanol as sacrificial agent in the UV light (high pressure mercury lamp, WG320 filter, l > 320 nm) for 10 min. The conductivity of RGO reduced by this method is found to be 400 S m^{-1} [66].

Table 2: Green strategies for reduction of GO in various materials.

S.No.	Composite	Strategy for reduction	Application	Ref
1.	rGO-Au$_{nano}$	Rose water	Glucose sensing	[67]
2.	TiO$_2$/RGO	Hydrothermal	Photodegradation of rhodamine B (RhB)	[68]
3.	(rGO)/mono and bimetallic	*Azadirachtaindica* extract	Nonenzymatic hydrogen peroxide sensor	[69]
4.	RGO	Aloe vera	Adsorptive removal of methylene blue	[70]
5.	RGO nano-sheets	Ultrasound & UV-radiation	-	[71]
6.	RGO/Ag/CeO$_2$	Ultrasonication	Reduction of p-nitrophenol	[72]
7.	ZnO-RGO	Hydrothermal reaction	Photo-degradation of RhB	[73]
8.	Ag-ZnFe$_2$O$_4$@rGO	Microwave method	Photodegradation of MB	[74]
9.	Fe3O4/RGO	*Averrhoa carambola* leaf extract	Cr(VI) reduction, Phenol degradation and	[75]
10.	CuO/rGO	Glucose	Antimicrobial activity	[76]
11.	RGO/Fe$_3$O4	*Murrayakoenigii*leaves extract	removal of Pb(II) from aqueous solution	[77]
12.	AgI-RGO	Ultrasonication	degradation of Rhodamine B (RhB)	[78]
13.	Au-Ag-In-rGO	*Piper pedicellatum*	α-glucosidase inhibition and	[79]
14.	RGO/Fe$_3$O$_4$	*Solanum trilobatum*	Degradation of MB	[80]
15.	RGO	Alanine	--	[81]
16.	RGO–AuNP	Glucose	sensing of H$_2$O$_2$	[82]

Generally, during the course of photo-reduction process the energy excitation in the bandgap of GO are liable for reduction. The band gap of GO is about 3.09 eV, during the illumination of aqueous suspension of GO it absorb light energy ≤ the band gap energy

Carbonaceous Composite Materials   Materials Research Forum LLC
Materials Research Foundations **42** (2018) 231-272   doi: http://dx.doi.org/10.21741/9781945291975-9

and lead to the photo generation of electrons and holes [83]. Figure 2 shows photoreduction of GO.

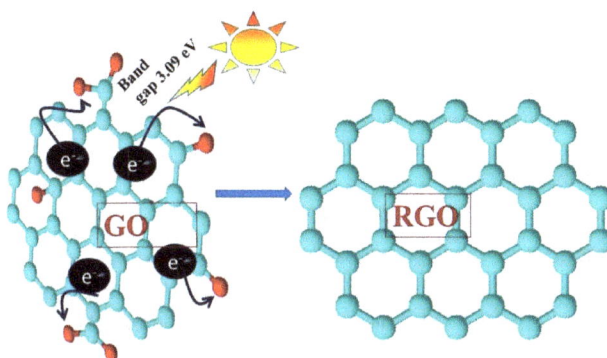

Fig. 2: Mechanism of photoreduction of GO to RGO.

These species are quite reactive towards the reduction of –OH and –RO moieties attached to the basal plane of GO, which ultimately resulted for the reduction of GO. If the reduction is carried in the presence of a semiconductor it depends on the redox potential. Such materials are easily reduced in the presence of light energy and they further react with available oxidant to recover its original state [84]. By this point of view semiconductors such as ZnO, $BiVO_4$, etc. have been utilized for the reduction of GO [85, 86].

3.5   Solvothermal method

Solvothermal method is an environmental benign green approach for the reduction of GO and in the preparation RGO based photocatalysts [87]. Solvothermal reaction has been carried in sealed vessel in the presence of a solvent such as water, ethanol, methanol, etc. The method has many advantages such as it requires less volume solvent and self-generated pressure inside the sealed vessel when heated beyond its boiling point [88]. Various semiconductor/RGO composites have been synthesized by this method such as $RGO/ZnIn_2S_4$ [89], TiO_2/RGO [90] and CoS–RGO [91].

Carbonaceous Composite Materials

Materials Research Forum LLC

Materials Research Foundations **42** (2018) 231-272

doi: http://dx.doi.org/10.21741/9781945291975-9

3.6 Green reduction strategies

Besides all above discussed reduction strategies researcher looking for such alternate methods which are not only cheap but also biocompatible [92]. In this context, the reduction of GO with some plant extract also have been studied along with some greener approaches. The use of plant extract in the reduction process not only eliminates the use of harmful chemicals but also minimizes the cost of the operation. The various plant extracts and greener approaches used for the reduction of GO have been listed in table 2 along with utilization of as prepared RGO.

4. Chemical modification or functionalization

GO and RGO possessed varieties of oxygen containing functional groups thus they can be regarded as excellent applicants for the modification. The modification can be done according to the applicative utilization such as for adsorption [93], drug delivery [94], photocatalysis [95], etc. The modification strategies broaden the area of RGO for diverse application. Table 3 present the different composite fabricated with the modified RGO and application.

5. Utilization and application of GO and RGO

5.1 Role in photocatalysis

Graphene and its derivatives have attracted much attention for the photocatalytic operation owing to its unique properties as discussed earlier. It has a high electron mobility 2×10^5 cm^2 V^{-1}s^{-2} [96] and high surface area $\approx$2965 m^{2}g^{-1} [97]. Additionally, the delocalized π- electrons could not only contribute to admirable conductivity but also act as acceptor of electrons which helps in the suppressing of charge recombination. Thus derivatives of graphene as RGO, GO and graphene itself have been extensively scrutinized for the fabrication of photocatalysts composites (Figure 3 shows charge transfer in RGO during photocatalysis). The use of heterojunction in the photocatalysis is widely considered to separate photoinduced electrons and holes. Figure 4 shows degradation on surface of RGO.

Carbonaceous Composite Materials   Materials Research Forum LLC
Materials Research Foundations **42** (2018) 231-272   doi: http://dx.doi.org/10.21741/9781945291975-9

Table 3: Chemical modification of GO/RGO.

S.No	Composite	Modification	Application	Ref
1.	GO-es-PVA	Esterification	-	[98]
2.	PS-GO	Polysaccharides	Adsorption of MB, Rh_6G, AF and OII dyes	[99]
3.	Cu^{2+}-GO	Cu^{2+}	Mimicking Horseradish Peroxidase and NADH Peroxidase	[100]
4.	NH_2-RGO	Primary amine	-	[101]
5.	GO-DETA	Amine	Catalytical activities for knoevenagel condensation and Michael addition inwater.	[102]
6.	S-rGO	Sulphonation	to improve monovalent anions selectivity andcontrollable resistance of anion exchangemembrane	[103]
7.	Graphene-NH_2	Amine	Thrombo-Protective Safer	[104]
8.	GO	Amide	Hydrogen Sulfide Gas Sensing	[105]
9.	CA/RGO/GCE	Calixarene	Simultaneous determination of Fe(III), Cd(II) and Pb(II) ions	[106]
10.	Fe_3O_4@PEI-RGO	Magnetic polyethyleneimine	Determination of polar acidic herbicides in rice	[107]
11.	pRGO@MS(DOX)-HA	Polydopamine	Targeted Chemo-Photothermal Therapy	[108]
12.	SRGO	Sulfophenyl	-	[109]
13.	ERGO-PA	Polyaniline	Chemiresistive sensor to monitor the pH in real time during microbial fermentations	[110]
14.	RGO-AQ	Anthraquinone	Electrode material for rechargeable batteries	[111]

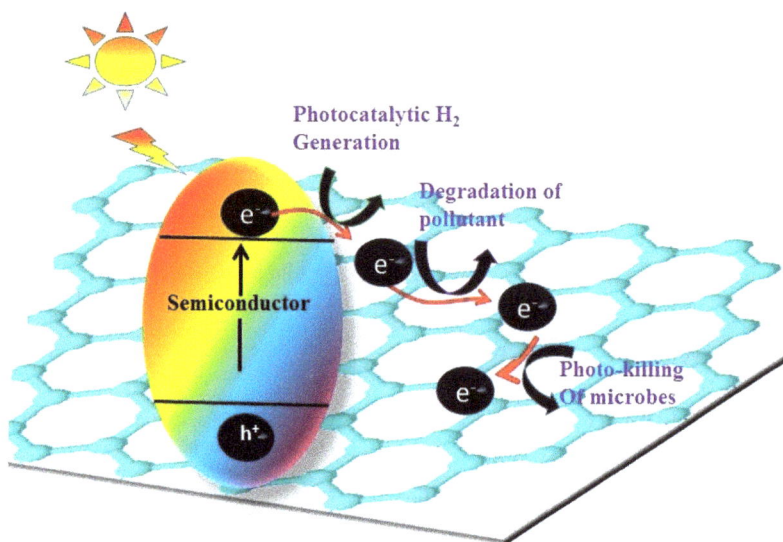

Fig.3: Charge transfer in RGO.

Semiconductor composites with heterojunction may lead to enhanced photocatalytic efficiency [112-115]. Despite, it is reported that graphene oxide could serve as a semiconductor either p-type or n-type owing to its tunable optical properties which depends on the surface composition of the GO. Table 4 & 5 summarizes the research progress in RGO and GO based photocatalysis.

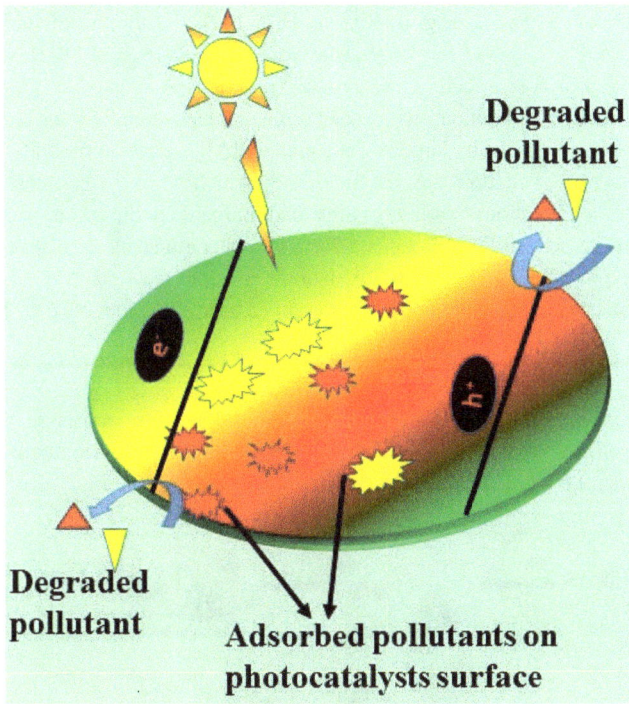

Fig.4: Photodegradation and adsorption on RGO surface.

5.2   Role of GO/RGO in the photocatalytic hydrogen generation and oxidation reduction processes

The present decade is highly dependent on fossil fuel to fulfill their energy need which is very critical by the view point of environmental pollution and other environmental concern. Thus research communities are trying to find such alternate energy sources which are renewable as well come true on the pollution criterion [116]. Thus solar energy has attracted great attention as an alternative to fulfill the energy crises. Among various techniques, photocatalytic hydrogen is a propitious techniques to convert solar illumination into usable chemical energy [117]. Consequently, there is high demand of such photocatalysts which can be capable of harnessing visible spectrum efficiently for the photocatalytic hydrogen production. Various semiconductors based materials have been explored such as Pt, Au, Ag, Ir, Ru, etc. [118-120]. But the use of such expensive material has barred their use for large scale application [116, 121, 122]. Hence, it is influential to find a cost effective alternative. Thus various researchers have explored graphene and its derivative along with other semiconductor materials for photocatalytic hydrogen generation due to its large surface area and high electron mobility. Moreover, photoreduction and oxidation are the processes which utilize solar energy for the reduction/oxidation reactions. Various organic and inorganic pollutants have been treated by this process to reduce their noxious effects. Graphene and its derivative have been extensively explored for such photo oxidation/reduction reaction. Various GO/RGO based photocatalysts along with their utilization have been listed in table 6 & 7.

6.   Mode of biomedical application

The graphene derivative has apprehended the attention of scientists in the area of biomedical application due to its exceptional properties and bio-compatibility [123]. Figure 5 shows such biomedical applications.

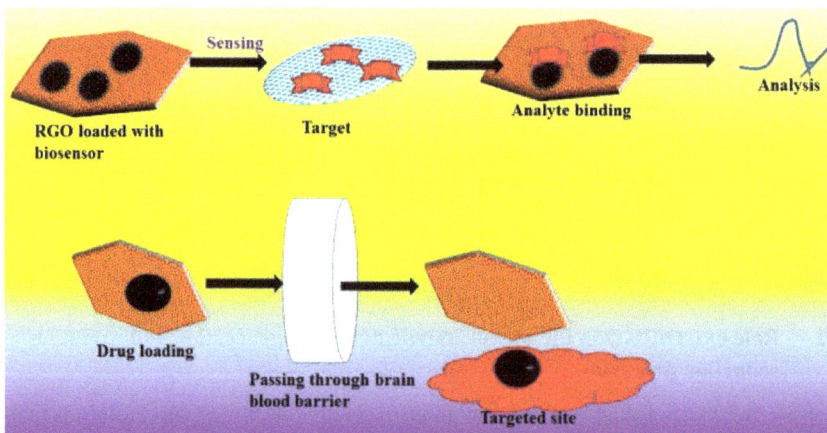

Fig.5: Biomedical applications.

Table 4: Photocatalytic mode of utilization of RGO

S.No.	Composite	Application	Summary Catalyst (mg)	Pollutant (mg L^{-1})	Max. (%)	GO/RGO (wt%)	Ref
1.	g-C$_3$N$_4$/GO/RGO	Phenol decomposition	80	10	28	-	[124]
2.	RGO-C$_3$N$_4$	Degradation of RhB&4-nitrophenol	-	RhB=5.0 4-nitrophenol = 10.0	-	2.5	[125]
3.	Cu$_2$O/RGO	Degradation of RhB	100	10	98.9	-	[126]
4.	g-C$_3$N$_4$/CdS/RGO	Degradation of RhB& CR	100	RhB = 4 CR = 20	-	-	[127]
5.	Ag/Ag$_2$CO$_3$-rGO	methyl orange (MO) and phenol degradation	50	MO= 10 Phenol=10	-	1.0	[128]
6.	Ag/RGO/Fe$_3$O$_4$	reduction of 4-nitrophenol, CR &RhB	5	4-nitrophenol= 350 RhB=10 CR=10	-	-	[129]
7.	Mn$_3$O$_4$–rGO	Orange-II	-	-	100	-	[130]
8.	MIL-101ZnFe$_2$O$_4$–rGO	MO, MV, MB, RhB	20	-	100	-	[131]

245

Table 5: Photocatalytic mode of utilization of GO

S.No.	Composite	Application	Summary Dose (mg)	Pollutant (mg L^{-1})	Degradation (%)	GO/R GO (wt%)	Ref
1.	CuO/Ag$_3$AsO$_4$/GO	Degradation of phenol	200	20	85	5.0	[132]
2.	Fe$_2$O$_3$–TiO$_2$/GO	MB degradation	30	10	-	1.0	[133]
3.	GO/Ag/Ag$_2$S–TiO$_2$	CV Degradation	-	-	80.01	-	[134]
4.	Ag$_3$PO$_4$/GO	Methyl orange	80	20	86.7	15	[135]
5.	MoS$_2$–GO	Degradation of methylene blue	10	10	99	-	[136]
6.	Co$_3$O$_4$/TiO$_2$/GO	Photodegradation of oxytetracycline and Congo Red	50	OTC=10 CR=10	-	-	[137]

No.	Material	Application					Ref.
7.	TiO$_2$-NS/Pt/GO	Degradation of chlortetracycline	20	50	45	-	[138]
8.	g-C$_3$N$_4$/GO/AgBr	RhB degradation	30	10	97	-	[139]
9.	GO/ZnO	MB,MV, MO BB removal	20	-	MB=98.5, 94, 85, 97.5	-	[140]
10.	GO/BiOI	Degradation of RhB	20	10	98	-	[141]
11.	GO/Ag$_3$PO$_4$/g-C$_3$N$_4$	Decomposition of Rhodamine B.	20	10	-	-	[142]
12.	GO/Ag$_3$PO$_4$/AgBr	Photocatalytic degradation of RhB	25	10	90	0.625	[143]
13.	BiVO$_4$/TiO$_2$/GO	Photodegradat-ion of Reactive Blue 19 (RB-19)	30	50	95.87	-	[144]

247

Table 6: Photocatalytic mode of H_2 generation by GO/RGO based photocatalysts

S.No.	Composite	Summary Conc. of photocatalysts Used (mg)	Sacrificial substance	Rate of H_2 production	Ref
1.	Cu_2O/rGO	10	Methanol		[145]
2.	RGO–$ZnIn_2S_4$	50	S^{2-}/SO_3^{2-}	81.6 μmolh^{-1}	[146]
3.	TiO_2/RGO/Cu_2O	-	Methanol	631.6 μmol/hm^2	[147]
4.	AuPd–MnO x /MOF–Graphene	-	-	382.1 molcatalysts^{-1}h^{-1}	[148]
5.	Ni(OH)$_2$-CdS/rGO	10	Na_2S and 0.25 M Na_2SO_3	4731 μmol h^{-1}g^{-1}	[149]
6.	g-C_3N_4-TiO_2/rGO	05	-	23,143 μmol g^{-1}h^{-1}	[150]
7.	Sn_3O_4/rGO	-	-	20 μmol/g/h	[151]
8.	RGO/TiO_2	-	Methanol	149 mol h^{-1} g^{-1}	[152]
9.	Au-CdS/ZnS-RGO	--	-	9.96 mmol h^{-1}g^{-1}	[153]
10.	2D/1D/2D g-C_3N_4/CdS/rGO	50	Triethanolamine	~4800 μmol h^{-1} g^{-1}	[154]
11.	WSe_2/rGO	05	-	-	[155]

Table 7: Reduction/oxidation mode of utilization

S.No.	Composite	Application	Ref
1.	Cu/rGO	CO_2 reduction	[156]
2.	$Pd_{0.50}Au_{0.50}$/PDA-rGO	Interconversion between CO_2 and HCOOH	[157]
3.	2D/2D ZnV_2O_6/RGO	Photo-induced CO_2 reduction to solar fuels	[158]
4.	Cu-Zn/graphene	Direct CO_2 hydrogenation to methanol	[159]
5.	Cu–Zn/N-rGO	Methanol synthesis via CO_2 hydrogenation	[160]
6.	rGO/$SrTi_{0.95}Fe_{0.05}O_{3-\delta}$	Conversion of organic pollutants to methanol and ethanol	[161]
7.	metal/3D-RGO	Electroreduction of CO_2 to Formate	[162]
8.	RGO–TiO_2	Photocatalytic CO_2 reduction	[163]
9.	rGO/Pt-TiO2	Photocatalytic water Splitting	[164]
10.	AgBr@TiO_2/GO	Oxidation of benzyl alcohol	[165]
11.	$CsPbBr_3$ QD/GO	Photocatalytic CO_2 Reduction	[166]
12.	MoS_2/g-C_3N_4/GO	Water splitting	[167]
13.	PtAu/PDA-RGO	Methanol oxidation	[168]
14.	a-Au/CeOx–RGO	N_2 reduction	[169]
15.	TiO_2-RGO	Photo-reduction and removal of Cr(VI)	[170]
16.			

Especially RGO has been explored for drug delivery [171], bio-sensing and tissue engineering [172], bio functionalization of protein [173], antimicrobial agent [174], carcinoma diagnostic and treatment [175], etc. The research in respective fields has created graphene as next generation medical tool. Some modes of utilization of RGO in medicinal fields have been listed in table 8.

Carbonaceous Composite Materials   Materials Research Forum LLC

Materials Research Foundations **42** (2018) 231-272   doi: http://dx.doi.org/10.21741/9781945291975-9

Table 8: Biomedical mode of utilization.

S.No.	Composite	Medical Application	Ref
1.	rGO-PDMS	Wound dressing	[176]
2.	ssDNA-FG	Exhaled Breath Analysis	[177]
3.	rGO-TPA/FeHCF	Immuno-sensor for the determination of SKBR-3 breast cancer	[178]
4.	Graphene/Gold	Applications in cancer diseases	[179]
5.	PDA-Ag@Fe_3O_4/rGO	SPR biosensor	[180]
6.	LZFO-RGO	Applications in hyperthermia	[181]
7.	Anti-IL8/AuNPs–rGO/ITO	Detection of oral cancer	[182]
8.	RGO-Ag	Application as active SERS substrate	[183]
9.	CeO_2/rGO	Oxidase-like Activity and Colorimetric Sensing of Ascorbic Acid	[184]
10.	Ag@rGO	Antibacterial properties	[185]

7. Future perspective and exploration

RGO and GO demonstrated surplus area of applicability such as in water detoxification, hydrogen production, CO_2 reduction, photocatalytic water splitting. Hence, the graphene material can be considered as the future material. Its advantageous properties can be explored in various research fields such as medicinal, sensing, coating of metal surface, energy storage devices, etc. Besides, there are many complications which still have to be dealt with such as development of more efficient methods for the large scale production of graphene derivative, to develop the modification strategies to meet the requirements of the objectives, to alter the morphological characteristics as the performance of a catalyst is highly dependent on morphology. It is also very important to deeply investigate the mechanism of graphene derivatives in the respective field to understand and to tackle the associated hitches.

8.   Conclusion

This chapter explains photocatalytic potential of RGO and GOES including synthesis strategies i.e chemical methods, solvothermal method, thermal annealing, and green reduction method. The various utilization modes of RGO and GO have been presented such as role in photocatalysis, adsorption, medicinal field, hydrogen production, CO_2 reduction, etc. The ease of availability precursor of RGO and GO, a non-precious material graphite make it approachable for all researchers. Graphene oxide and reduced graphene oxide have proven to be highly efficient in various environmental catalytic applications. They are indeed future materials as these studies inspire scientists worldwide to have a multi-pronged approach in designing of such sustainable materials.

References

[1]   L.V. Bora, R.K. Mewada, Visible/solar light active photocatalysts for organic effluent treatment: Fundamentals, mechanisms and parametric review, Renewable and Sustainable Energy Reviews 76 (2017) 1393-1421.
https://doi.org/10.1016/j.rser.2017.01.130

[2]   P. Dhiman, M. Naushad, K.M. Batoo, A. Kumar, G. Sharma, A.A. Ghfar, G. Kumar, M. Singh, Nano $FexZn_{1-x}O$ as a tuneable and efficient photocatalyst for solar powered degradation of bisphenol A from aqueous environment, Journal of Cleaner Production 165 (2017) 1542-1556.
https://doi.org/10.1016/j.jclepro.2017.07.245

[3]   G. Sharma, S. Bhogal, M. Naushad, Inamuddin, A. Kumar, F.J. Stadler, Microwave assisted fabrication of La/Cu/Zr/carbon dots trimetallic nanocomposites with their adsorptional vs photocatalytic efficiency for remediation of persistent organic pollutants, Journal of Photochemistry and Photobiology A: Chemistry 347 (2017) 235-243.
https://doi.org/10.1016/j.jphotochem.2017.07.001

[4]   G. Sharma, A. Kumar, K. Devi, S. Sharma, M. Naushad, A.A. Ghfar, T. Ahamad, F.J. Stadler, Guar gum-crosslinked-Soya lecithin nanohydrogel sheets as effective adsorbent for the removal of thiophanate methyl fungicide, International Journal of Biological Macromolecules 114 (2018) 295-305.
https://doi.org/10.1016/j.ijbiomac.2018.03.093

[5]   G. Sharma, V.K. Gupta, S. Agarwal, A. Kumar, S. Thakur, D. Pathania, Fabrication and characterization of Fe@MoPO nanoparticles: Ion exchange

behavior and photocatalytic activity against malachite green, Journal of Molecular Liquids 219 (2016) 1137-1143. https://doi.org/10.1016/j.molliq.2016.04.046

[6] A. Kumar, A. Rana, G. Sharma, S. Sharma, M. Naushad, G.T. Mola, P. Dhiman, F.J. Stadler, Aerogels and metal–organic frameworks for environmental remediation and energy production, Environmental Chemistry Letters (2018) 1-24. https://doi.org/10.1007/s10311-018-0723-x

[7] S. Kant, D. Pathania, P. Singh, P. Dhiman, A. Kumar, Removal of malachite green and methylene blue by $Fe_{0.01}Ni_{0.01}Zn_{0.98}O$/polyacrylamide nanocomposite using coupled adsorption and photocatalysis, Applied Catalysis B: Environmental 147 (2014) 340-352. https://doi.org/10.1016/j.apcatb.2013.09.001

[8] P. Dhiman, J. Chand, A. Kumar, R.K. Kotnala, K.M. Batoo, M. Singh, Synthesis and characterization of novel Fe@ZnOnanosystem, Journal of Alloys and Compounds 578 (2013) 235-241. https://doi.org/10.1016/j.jallcom.2013.05.015

[9] C. Belver, R. Bellod, A. Fuerte, M. Fernández-García, Nitrogen-containing TiO_2 photocatalysts: Part 1. Synthesis and solid characterization, Applied Catalysis B: Environmental 65 (2006) 301-308. https://doi.org/10.1016/j.apcatb.2006.02.007

[10] A. Kumar, A. Kumari, G. Sharma, M. Naushad, T. Ahamad, F.J. Stadler, Utilizing recycled $LiFePO_4$ from batteries in combination with $B@C_3N_4$ and $CuFe_2O_4$ as sustainable nano-junctions for high performance degradation of atenolol, Chemosphere 209 (2018) 457-469. https://doi.org/10.1016/j.chemosphere.2018.06.117

[11] T. Di, B. Zhu, B. Cheng, J. Yu, J. Xu, A direct Z-scheme $g-C_3N_4/SnS_2$ photocatalyst with superior visible-light CO_2 reduction performance, Journal of Catalysis 352 (2017) 532-541. https://doi.org/10.1016/j.jcat.2017.06.006

[12] A. Kumar, A. Kumar, G. Sharma, A.a.H. Al-Muhtaseb, M. Naushad, A.A. Ghfar, C. Guo, F.J. Stadler, Biochar-templated $g-C_3N_4/Bi_2O_2CO_3/CoFe_2O_4$ nano-assembly for visible and solar assisted photo-degradation of paraquat, nitrophenol reduction and CO_2 conversion, Chemical Engineering Journal 339 (2018) 393-410. https://doi.org/10.1016/j.cej.2018.01.105

[13] G. Sharma, A. Kumar, M. Naushad, A. Kumar, A.a.H. Al-Muhtaseb, P. Dhiman, A.A. Ghfar, F.J. Stadler, M.R. Khan, Photoremediation of toxic dye from aqueous environment using monometallic and bimetallic quantum dots based nanocomposites, Journal of Cleaner Production 172 (2018) 2919-2930. https://doi.org/10.1016/j.jclepro.2017.11.122

[14]   Y. Liu, F. Luo, S. Liu, S. Liu, X. Lai, X. Li, Y. Lu, Y. Li, C. Hu, Z. Shi, Aminated Graphene Oxide Impregnated with Photocatalytic Polyoxometalate for Efficient Adsorption of Dye Pollutants and Its Facile and Complete Photoregeneration, small 13 (2017).

[15]   A. Kumar, M. Naushad, A. Rana, Inamuddin, Preeti, G. Sharma, A.A. Ghfar, F.J. Stadler, M.R. Khan, ZnSe-WO$_3$ nano-hetero-assembly stacked on Gum ghatti for photo-degradative removal of Bisphenol A: Symbiose of adsorption and photocatalysis, International Journal of Biological Macromolecules 104 (2017) 1172-1184. https://doi.org/10.1016/j.ijbiomac.2017.06.116

[16]   A. Kumar, A. Kumar, G. Sharma, A.a.H. Al-Muhtaseb, M. Naushad, A.A. Ghfar, F.J. Stadler, Quaternary magnetic BiOCl/g-C$_3$N$_4$/Cu$_2$O/Fe$_3$O$_4$ nano-junction for visible light and solar powered degradation of sulfamethoxazole from aqueous environment, Chemical Engineering Journal 334 (2018) 462-478. https://doi.org/10.1016/j.cej.2017.10.049

[17]   N. Ding, L. Zhang, M. Hashimoto, K. Iwasaki, N. Chikamori, K. Nakata, Y. Xu, J. Shi, H. Wu, Y. Luo, D. Li, A. Fujishima, Q. Meng, Enhanced photocatalytic activity of mesoporous carbon/C$_3$N$_4$ composite photocatalysts, Journal of Colloid and Interface Science 512 (2018) 474-479. https://doi.org/10.1016/j.jcis.2017.10.081

[18]   J. Liu, H. Wang, M. Antonietti, Graphitic carbon nitride "reloaded": emerging applications beyond (photo)catalysis, Chemical Society Reviews 45 (2016) 2308-2326. https://doi.org/10.1039/C5CS00767D

[19]   A. Kumar, A. Kumar, G. Sharma, M. Naushad, R.C. Veses, A.A. Ghfar, F.J. Stadler, M.R. Khan, Solar-driven photodegradation of 17-β-estradiol and ciprofloxacin from waste water and CO$_2$ conversion using sustainable coal-char/polymeric-g-C$_3$N$_4$/RGO metal-free nano-hybrids, New Journal of Chemistry 41 (2017) 10208-10224. https://doi.org/10.1039/C7NJ01580A

[20]   W. Peng, P. Luo, D. Gui, W. Jiang, H. Wu, J. Zhang, Enhanced anticancer effect of fabricated gallic acid/CdS on the rGO nanosheets on human glomerular mesangial (IP15) and epithelial proximal (HK2) kidney cell lines-Cytotoxicity investigations, Journal of Photochemistry and Photobiology B: Biology 178 (2018) 243-248. https://doi.org/10.1016/j.jphotobiol.2017.11.012

[21]   F. Deng, X. Li, Y. Wang, J. Li, Electropolymerization and Electrochemical Performance of Nickel Schiff Base Complexes on Reduced Graphene Oxide

Carbonaceous Composite Materials   Materials Research Forum LLC
Materials Research Foundations **42** (2018) 231-272   doi: http://dx.doi.org/10.21741/9781945291975-9

(RGO) Support for Supercapacitors, Meeting Abstracts, The Electrochemical Society, 2017, pp. 579-579.

[22] P. Kumar, C. Joshi, A. Barras, B. Sieber, A. Addad, L. Boussekey, S. Szunerits, R. Boukherroub, S.L. Jain, Core–shell structured reduced graphene oxide wrapped magnetically separable $rGO@CuZnO@Fe_3O_4$ microspheres as superior photocatalyst for CO_2 reduction under visible light, Applied Catalysis B: Environmental 205 (2017) 654-665. https://doi.org/10.1016/j.apcatb.2016.11.060

[23] H. Bao, Y. Pan, Y. Ping, N.G. Sahoo, T. Wu, L. Li, J. Li, L.H. Gan, Chitosan-functionalized graphene oxide as a nanocarrier for drug and gene delivery, Small 7 (2011) 1569-1578. https://doi.org/10.1002/smll.201100191

[24] S. Stankovich, D.A. Dikin, G.H. Dommett, K.M. Kohlhaas, E.J. Zimney, E.A. Stach, R.D. Piner, S.T. Nguyen, R.S. Ruoff, Graphene-based composite materials, nature 442 (2006) 282.

[25] A. Erdem, M. Muti, P. Papakonstantinou, E. Canavar, H. Karadeniz, G. Congur, S. Sharma, Graphene oxide integrated sensor for electrochemical monitoring of mitomycin C–DNA interaction, Analyst 137 (2012) 2129-2135. https://doi.org/10.1039/c2an16011k

[26] T. Yu, Z. Xu, S. Liu, H. Liu, X. Yang, Enhanced hydrophilicity and water-permeating of functionalized graphene-oxide nanopores: Molecular dynamics simulations, Journal of Membrane Science (2017).

[27] W. Yu, S. Zhan, Z. Shen, Q. Zhou, A newly synthesized $Au/GO-Co_3O_4$ composite effectively inhibits the replication of tetracycline resistance gene in water, Chemical Engineering Journal 345 (2018) 462-470. https://doi.org/10.1016/j.cej.2018.03.108

[28] S.S.P. Selvin, N. Radhika, O. Borang, I.S. Lydia, J.P. Merlin, Visible light driven photodegradation of Rhodamine B using cysteine capped ZnO/GO nanocomposite as photocatalyst, Journal of Materials Science: Materials in Electronics 28 (2017) 6722-6730. https://doi.org/10.1007/s10854-017-6367-y

[29] L. Liu, J. Deng, T. Niu, G. Zheng, P. Zhang, Y. Jin, Z. Jiao, X. Sun, One-step synthesis of Ag/AgCl/GO composite: A photocatalyst of extraordinary photoactivity and stability, Journal of Colloid and Interface Science 493 (2017) 281-287. https://doi.org/10.1016/j.jcis.2016.11.039

[30] J. Yang, J. Teng, X. Zhao, X. Jiang, F. Jiao, J. Yu, Synthesis, Characterization and Photocatalytic Activities of a Novel $Eu/TiO_2/GO$ Composite, and Its Application

Carbonaceous Composite Materials   Materials Research Forum LLC
Materials Research Foundations **42** (2018) 231-272   doi: http://dx.doi.org/10.21741/9781945291975-9

for Enhanced Photocatalysis of Methylene Blue, Nanoscience and
Nanotechnology Letters 9 (2017) 1622-1631.
https://doi.org/10.1166/nnl.2017.2526

[31]   B.C. Brodie, XIII. On the atomic weight of graphite, Philosophical Transactions of
       the Royal Society of London 149 (1859) 249-259.
       https://doi.org/10.1098/rstl.1859.0013

[32]   W.S. Hummers Jr, R.E. Offeman, Preparation of graphitic oxide, Journal of the
       american chemical society 80 (1958) 1339-1339.
       https://doi.org/10.1021/ja01539a017

[33]   R. Shao, L. Sun, L. Tang, Z. Chen, Preparation and characterization of magnetic
       core–shell $ZnFe_2O_4$@ZnO nanoparticles and their application for the
       photodegradation of methylene blue, Chemical Engineering Journal 217 (2013)
       185-191. https://doi.org/10.1016/j.cej.2012.11.109

[34]   D.C. Marcano, D.V. Kosynkin, J.M. Berlin, A. Sinitskii, Z. Sun, A. Slesarev, L.B.
       Alemany, W. Lu, J.M. Tour, Improved synthesis of graphene oxide, ACS nano 4
       (2010) 4806-4814. https://doi.org/10.1021/nn1006368

[35]   Y.J. Zhang, P.Y. He, Y.X. Zhang, H. Chen, A novel electroconductive
       graphene/fly ash-based geopolymer composite and its photocatalytic performance,
       Chemical Engineering Journal 334 (2018) 2459-2466.
       https://doi.org/10.1016/j.cej.2017.11.171

[36]   X. Yu, J. Zhang, Z. Zhao, W. Guo, J. Qiu, X. Mou, A. Li, J.P. Claverie, H. Liu,
       $NiO–TiO_2$ p–n heterostructurednanocables bridged by zero-bandgap rGO for
       highly efficient photocatalytic water splitting, Nano Energy 16 (2015) 207-217.
       https://doi.org/10.1016/j.nanoen.2015.06.028

[37]   Z. Zhang, B. Chen, M. Baek, K. Yong, Multichannel Charge Transport of a
       $BiVO_4$/(RGO/WO_3)/$W_{18}O_{49}$ Three-Storey Anode for Greatly Enhanced
       Photoelectrochemical Efficiency, ACS applied materials & interfaces 10 (2018)
       6218-6227. https://doi.org/10.1021/acsami.7b15275

[38]   Y.-C. Pu, H.-Y.Chou, W.-S.Kuo, K.-H.Wei, Y.-J. Hsu, Interfacial charge carrier
       dynamics of cuprous oxide-reduced graphene oxide (Cu_2O-rGO)
       nanoheterostructures and their related visible-light-driven photocatalysis, Applied
       Catalysis B: Environmental 204 (2017) 21-32.
       https://doi.org/10.1016/j.apcatb.2016.11.012

Carbonaceous Composite Materials   Materials Research Forum LLC

Materials Research Foundations **42** (2018) 231-272   doi: http://dx.doi.org/10.21741/9781945291975-9

[39]   B. Gomez-Ruiz, P. Ribao, N. Diban, M.J. Rivero, I. Ortiz, A. Urtiaga, Photocatalytic degradation and mineralization of perfluorooctanoic acid (PFOA) using a composite TiO_2- rGO catalyst, Journal of hazardous materials 344 (2018) 950-957. https://doi.org/10.1016/j.jhazmat.2017.11.048

[40]   X. Wang, W. Mao, Q. Wang, Y. Zhu, Y. Min, J. Zhang, T. Yang, J. Yang, X.a. Li, W. Huang, Low-temperature fabrication of $Bi_{25}FeO_{40}$/rGO nanocomposites with efficient photocatalytic performance under visible light irradiation, RSC Advances 7 (2017) 10064-10069. https://doi.org/10.1039/C6RA27025E

[41]   X.-S. Hu, Y. Shen, Y.-T.Zhang, J.-J.Nie, Preparation of flower-like CuS/reduced graphene oxide (RGO) photocatalysts for enhanced photocatalytic activity, Journal of Physics and Chemistry of Solids 103 (2017) 201-208. https://doi.org/10.1016/j.jpcs.2016.12.021

[42]   X. Liu, Y. Qin, Y. Yan, P. Lv, The fabrication of $CdS/CoFe_2O_4$/rGO photocatalysts to improve the photocatalytic degradation performance under visible light, RSC Advances 7 (2017) 40673-40681. https://doi.org/10.1039/C7RA07202C

[43]   A. Morawski, E. Kusiak-Nejman, A. Wanag, J. Kapica-Kozar, R. Wróbel, B. Ohtani, M. Aksienionek, L. Lipińska, Photocatalytic degradation of acetic acid in the presence of visible light-active TiO_2-reduced graphene oxide photocatalysts, Catalysis Today 280 (2017) 108-113. https://doi.org/10.1016/j.cattod.2016.05.055

[44]   S.N. Alam, N. Sharma, L. Kumar, Synthesis of graphene oxide (GO) by modified hummers method and its thermal reduction to obtain reduced graphene oxide (rGO), Graphene 6 (2017) 1-18. https://doi.org/10.4236/graphene.2017.61001

[45]   N.A. Kotov, I. Dekany, J.H. Fendler, Ultrathin graphite oxide–polyelectrolyte composites prepared by self-assembly: Transition between conductive and non-conductive states, Advanced Materials 8 (1996) 637-641. https://doi.org/10.1002/adma.19960080806

[46]   H.J. Shin, K.K. Kim, A. Benayad, S.M. Yoon, H.K. Park, I.S. Jung, M.H. Jin, H.K. Jeong, J.M. Kim, J.Y. Choi, Efficient reduction of graphite oxide by sodium borohydride and its effect on electrical conductance, Advanced Functional Materials 19 (2009) 1987-1992. https://doi.org/10.1002/adfm.200900167

[47]   G. Williams, B. Seger, P.V. Kamat, TiO_2-graphene nanocomposites. UV-assisted photocatalytic reduction of graphene oxide, ACS nano 2 (2008) 1487-1491. https://doi.org/10.1021/nn800251f

[48] P.V. Kamat, Photochemistry on nonreactive and reactive (semiconductor) surfaces, Chemical Reviews 93 (1993) 267-300. https://doi.org/10.1021/cr00017a013

[49] Z. Wang, X. Zhou, J. Zhang, F. Boey, H. Zhang, Direct electrochemical reduction of single-layer graphene oxide and subsequent functionalization with glucose oxidase, The Journal of Physical Chemistry C 113 (2009) 14071-14075. https://doi.org/10.1021/jp906348x

[50] M.N. Chong, B. Jin, C.W.K. Chow, C. Saint, Recent developments in photocatalytic water treatment technology: A review, Water Research 44 (2010) 2997-3027. https://doi.org/10.1016/j.watres.2010.02.039

[51] L. Zhou, H. Deng, J. Wan, J. Shi, T. Su, A solvothermal method to produce RGO-Fe_3O_4 hybrid composite for fast chromium removal from aqueous solution, Applied Surface Science 283 (2013) 1024-1031. https://doi.org/10.1016/j.apsusc.2013.07.063

[52] M.J. McAllister, J.-L. Li, D.H. Adamson, H.C. Schniepp, A.A. Abdala, J. Liu, M. Herrera-Alonso, D.L. Milius, R. Car, R.K. Prud'homme, Single sheet functionalized graphene by oxidation and thermal expansion of graphite, Chemistry of materials 19 (2007) 4396-4404. https://doi.org/10.1021/cm0630800

[53] L. Dong, J. Yang, M. Chhowalla, K.P. Loh, Synthesis and reduction of large sized graphene oxide sheets, Chemical Society Reviews 46 (2017) 7306-7316. https://doi.org/10.1039/C7CS00485K

[54] Y. Dai, X. Qi, W. Fu, C. Huang, S. Wang, J. Zhou, T.H. Zeng, Y. Sun, Graphene sheets manipulated the thermal-stability of ultrasmall Pt nanoparticles supported on porous Fe_2O_3 nanocrystals against sintering, RSC Advances 7 (2017) 16379-16386. https://doi.org/10.1039/C7RA01188A

[55] Y. Zhu, S. Murali, M.D. Stoller, A. Velamakanni, R.D. Piner, R.S. Ruoff, Microwave assisted exfoliation and reduction of graphite oxide for ultracapacitors, Carbon 48 (2010) 2118-2122. https://doi.org/10.1016/j.carbon.2010.02.001

[56] D. Voiry, J. Yang, J. Kupferberg, R. Fullon, C. Lee, H.Y. Jeong, H.S. Shin, M. Chhowalla, High-quality graphene via microwave reduction of solution-exfoliated graphene oxide, Science 353 (2016) 1413-1416. https://doi.org/10.1126/science.aah3398

[57] R. Wang, Y. Wang, C. Xu, J. Sun, L. Gao, Facile one-step hydrazine-assisted solvothermal synthesis of nitrogen-doped reduced graphene oxide: reduction effect

and mechanisms, RSC Advances 3 (2013) 1194-1200.
https://doi.org/10.1039/C2RA21825A

[58] N. Sykam, G.M. Rao, Room temperature synthesis of reduced graphene oxide
      nanosheets as anode material for supercapacitors, Materials Letters 204 (2017)
      169-172. https://doi.org/10.1016/j.matlet.2017.05.114

[59] M. Muda, M.M. Ramli, S.S.M. Isa, M. Jamlos, S. Murad, Z. Norhanisah, M.M.
      Isa, S. Kasjoo, N. Ahmad, N. Nor, Fundamental study of reduction graphene oxide
      by sodium borohydride for gas sensor application, AIP Conference Proceedings,
      AIP Publishing, 2017, pp. 020034. https://doi.org/10.1063/1.4975267

[60] S. Pei, H.-M. Cheng, The reduction of graphene oxide, Carbon 50 (2012) 3210-
      3228. https://doi.org/10.1016/j.carbon.2011.11.010

[61] W. Gao, L.B. Alemany, L. Ci, P.M. Ajayan, New insights into the structure and
      reduction of graphite oxide, Nature chemistry 1 (2009) 403-408.
      https://doi.org/10.1038/nchem.281

[62] S. Abdolhosseinzadeh, H. Asgharzadeh, H.S. Kim, Fast and fully-scalable
      synthesis of reduced graphene oxide, Scientific reports 5 (2015) 10160.
      https://doi.org/10.1038/srep10160

[63] M.J. Fernández-Merino, L. Guardia, J. Paredes, S. Villar-Rodil, P. Solís-
      Fernández, A. Martínez-Alonso, J. Tascon, Vitamin C is an ideal substitute for
      hydrazine in the reduction of graphene oxide suspensions, The Journal of Physical
      Chemistry C 114 (2010) 6426-6432. https://doi.org/10.1021/jp100603h

[64] I.K. Moon, J. Lee, R.S. Ruoff, H. Lee, Reduced graphene oxide by chemical
      graphitization, Nature communications 1 (2010) 73.
      https://doi.org/10.1038/ncomms1067

[65] K. Hatakeyama, K. Awaya, M. Koinuma, Y. Shimizu, Y. Hakuta, Y. Matsumoto,
      Production of water-dispersible reduced graphene oxide without stabilizers using
      liquid-phase photoreduction, Soft matter 13 (2017) 8353-8356.
      https://doi.org/10.1039/C7SM01386H

[66] H. Li, S. Pang, X. Feng, K. Müllen, C. Bubeck, Polyoxometalate assisted
      photoreduction of graphene oxide and its nanocomposite formation, Chemical
      Communications 46 (2010) 6243-6245. https://doi.org/10.1039/c0cc01098g

[67] M.A. Tabrizi, J.N. Varkani, Green synthesis of reduced graphene oxide decorated
      with gold nanoparticles and its glucose sensing application, Sensors and Actuators
      B: Chemical 202 (2014) 475-482. https://doi.org/10.1016/j.snb.2014.05.099

[68] M.S.A. Sher Shah, A.R. Park, K. Zhang, J.H. Park, P.J. Yoo, Green synthesis of biphasic TiO2–reduced graphene oxide nanocomposites with highly enhanced photocatalytic activity, ACS applied materials & interfaces 4 (2012) 3893-3901. https://doi.org/10.1021/am301287m

[69] K.J. Babu, K.S. Nahm, Y.J. Hwang, A facile one-pot green synthesis of reduced graphene oxide and its composites for non-enzymatic hydrogen peroxide sensor applications, RSC Advances 4 (2014) 7944-7951. https://doi.org/10.1039/c3ra45596c

[70] G. Bhattacharya, S. Sas, S. Wadhwa, A. Mathur, J. McLaughlin, S.S. Roy, Aloe vera assisted facile green synthesis of reduced graphene oxide for electrochemical and dye removal applications, RSC Advances 7 (2017) 26680-26688. https://doi.org/10.1039/C7RA02828H

[71] Y. Ding, P. Zhang, Q. Zhuo, H. Ren, Z. Yang, Y. Jiang, A green approach to the synthesis of reduced graphene oxide nanosheets under UV irradiation, Nanotechnology 22 (2011) 215601. https://doi.org/10.1088/0957-4484/22/21/215601

[72] Z. Ji, X. Shen, J. Yang, G. Zhu, K. Chen, A novel reduced graphene oxide/Ag/CeO$_2$ ternary nanocomposite: Green synthesis and catalytic properties, Applied Catalysis B: Environmental 144 (2014) 454-461. https://doi.org/10.1016/j.apcatb.2013.07.052

[73] X. Li, Q. Wang, Y. Zhao, W. Wu, J. Chen, H. Meng, Green synthesis and photo-catalytic performances for ZnO-reduced graphene oxide nanocomposites, Journal of Colloid and Interface Science 411 (2013) 69-75. https://doi.org/10.1016/j.jcis.2013.08.050

[74] A.H. Mady, M.L. Baynosa, D. Tuma, J.-J. Shim, Facile microwave-assisted green synthesis of Ag-ZnFe$_2$O$_4$@rGO nanocomposites for efficient removal of organic dyes under UV-and visible-light irradiation, Applied Catalysis B: Environmental 203 (2017) 416-427. https://doi.org/10.1016/j.apcatb.2016.10.033

[75] D.K. Padhi, T.K. Panigrahi, K. Parida, S.K. Singh, P.M. Mishra, Green Synthesis of Fe$_3$O$_4$/RGO Nanocomposite with Enhanced Photocatalytic Performance for Cr (VI) Reduction, Phenol Degradation, and Antibacterial Activity, ACS Sustainable Chemistry & Engineering 5 (2017) 10551-10562. https://doi.org/10.1021/acssuschemeng.7b02548

[76]   S. Pourbeyram, R. Bayrami, H. Dadkhah, Green synthesis and characterization of ultrafine copper oxide reduced graphene oxide (CuO/rGO) nanocomposite, Colloids and Surfaces A: Physicochemical and Engineering Aspects 529 (2017) 73-79. https://doi.org/10.1016/j.colsurfa.2017.05.077

[77]   C. Prasad, P.K. Murthy, R.H. Krishna, R.S. Rao, V. Suneetha, P. Venkateswarlu, Bio-inspired green synthesis of RGO/Fe$_3$O$_4$ magnetic nanoparticles using Murrayakoenigii leaves extract and its application for removal of Pb (II) from aqueous solution, Journal of environmental chemical engineering 5 (2017) 4374-4380. https://doi.org/10.1016/j.jece.2017.07.026

[78]   D.A. Reddy, S. Lee, J. Choi, S. Park, R. Ma, H. Yang, T.K. Kim, Green synthesis of AgI-reduced graphene oxide nanocomposites: toward enhanced visible-light photocatalytic activity for organic dye removal, Applied Surface Science 341 (2015) 175-184. https://doi.org/10.1016/j.apsusc.2015.03.019

[79]   I. Saikia, M. Hazarika, S. Yunus, M. Pal, M.R. Das, J.C. Borah, C. Tamuly, Green synthesis of Au-Ag-In-rGO nanocomposites and its α-glucosidase inhibition and cytotoxicity effects, Materials Letters 211 (2018) 48-50. https://doi.org/10.1016/j.matlet.2017.09.084

[80]   M. Vinothkannan, C. Karthikeyan, A.R. Kim, D.J. Yoo, One-pot green synthesis of reduced graphene oxide (RGO)/Fe$_3$O$_4$ nanocomposites and its catalytic activity toward methylene blue dye degradation, SpectrochimicaActa Part A: Molecular and Biomolecular Spectroscopy 136 (2015) 256-264. https://doi.org/10.1016/j.saa.2014.09.031

[81]   J. Wang, E.C. Salihi, L. Šiller, Green reduction of graphene oxide using alanine, Materials Science and Engineering: C 72 (2017) 1-6. https://doi.org/10.1016/j.msec.2016.11.017

[82]   P. Zhang, X. Zhang, S. Zhang, X. Lu, Q. Li, Z. Su, G. Wei, One-pot green synthesis, characterizations, and biosensor application of self-assembled reduced graphene oxide–gold nanoparticle hybrid membranes, Journal of Materials Chemistry B 1 (2013) 6525-6531. https://doi.org/10.1039/c3tb21270j

[83]   M. Mohandoss, S.S. Gupta, A. Nelleri, T. Pradeep, S.M. Maliyekkal, Solar mediated reduction of graphene oxide, RSC Advances 7 (2017) 957-963. https://doi.org/10.1039/C6RA24696F

[84]   A. Troupis, A. Hiskia, E. Papaconstantinou, Synthesis of metal nanoparticles by using polyoxometalates as photocatalysts and stabilizers, Angewandte Chemie

International Edition 41 (2002) 1911-1914. https://doi.org/10.1002/1521-3773(20020603)41:11<1911::AID-ANIE1911>3.0.CO;2-0

[85] G. Williams, P.V. Kamat, Graphene– semiconductor nanocomposites: excited-state interactions between ZnO nanoparticles and graphene oxide, Langmuir 25 (2009) 13869-13873. https://doi.org/10.1021/la900905h

[86] Q. Wu, F. Sun, Photochemical synthesis of ZnSnanosheet and its use in photodegradation of organic pollutants, Bioinformatics and Biomedical Engineering (iCBBE), 2010 4th International Conference on, IEEE, 2010, pp. 1-4.

[87] S. Dubin, S. Gilje, K. Wang, V.C. Tung, K. Cha, A.S. Hall, J. Farrar, R. Varshneya, Y. Yang, R.B. Kaner, A one-step, solvothermal reduction method for producing reduced graphene oxide dispersions in organic solvents, ACS nano 4 (2010) 3845-3852. https://doi.org/10.1021/nn100511a

[88] C. Nethravathi, M. Rajamathi, Chemically modified graphene sheets produced by the solvothermal reduction of colloidal dispersions of graphite oxide, Carbon 46 (2008) 1994-1998. https://doi.org/10.1016/j.carbon.2008.08.013

[89] L. Ye, J. Fu, Z. Xu, R. Yuan, Z. Li, Facile one-pot solvothermal method to synthesize sheet-on-sheet reduced graphene oxide (RGO)/ZnIn$_2$S$_4$ nanocomposites with superior photocatalytic performance, ACS applied materials & interfaces 6 (2014) 3483-3490. https://doi.org/10.1021/am5004415

[90] W. Iqbal, B. Tian, M. Anpo, J. Zhang, Single-step solvothermal synthesis of mesoporous anatase TiO$_2$-reduced graphene oxide nanocomposites for the abatement of organic pollutants, Research on Chemical Intermediates 43 (2017) 5187-5201. https://doi.org/10.1007/s11164-017-3049-6

[91] P. Borthakur, G. Darabdhara, M.R. Das, R. Boukherroub, S. Szunerits, Solvothermal synthesis of CoS/reduced porous graphene oxide nanocomposite for selective colorimetric detection of Hg (II) ion in aqueous medium, Sensors and Actuators B: Chemical 244 (2017) 684-692. https://doi.org/10.1016/j.snb.2016.12.148

[92] N. Seyedi, K. Saidi, H. Sheibani, Green Synthesis of Pd Nanoparticles Supported on Magnetic Graphene Oxide by Origanum vulgare Leaf Plant Extract: Catalytic Activity in the Reduction of Organic Dyes and Suzuki–Miyaura Cross-Coupling Reaction, Catalysis Letters 148 (2018) 277-288. https://doi.org/10.1007/s10562-017-2220-4

[93] N.M. Mahmoodi, S.M. Maroofi, M. Mazarji, G. Nabi-Bidhendi, Preparation of Modified Reduced Graphene Oxide nanosheet with Cationic Surfactant and its Dye Adsorption Ability from Colored Wastewater, Journal of Surfactants and Detergents 20 (2017) 1085-1093. https://doi.org/10.1007/s11743-017-1985-1

[94] Y. Oz, A. Barras, R. Sanyal, R. Boukherroub, S. Szunerits, A. Sanyal, Functionalization of Reduced Graphene Oxide via Thiol–Maleimide "Click" Chemistry: Facile Fabrication of Targeted Drug Delivery Vehicles, ACS applied materials & interfaces 9 (2017) 34194-34203. https://doi.org/10.1021/acsami.7b08433

[95] G. Zhao, C. Li, X. Wu, J. Yu, X. Jiang, W. Hu, F. Jiao, Reduced graphene oxide modified NiFe-calcinated layered double hydroxides for enhanced photocatalytic removal of methylene blue, Applied Surface Science 434 (2018) 251-259. https://doi.org/10.1016/j.apsusc.2017.10.181

[96] A.S. Mayorov, R.V. Gorbachev, S.V. Morozov, L. Britnell, R. Jalil, L.A. Ponomarenko, P. Blake, K.S. Novoselov, K. Watanabe, T. Taniguchi, Micrometer-scale ballistic transport in encapsulated graphene at room temperature, Nano letters 11 (2011) 2396-2399. https://doi.org/10.1021/nl200758b

[97] W. Liu, K. Zhang, R.-Z.Wang, J. Zhong, L.-M. Liu, Modulation of the electron transport properties in graphene nanoribbons doped with BN chains, AIP Advances 4 (2014) 067123. https://doi.org/10.1063/1.4883236

[98] H.J. Salavagione, M.A. Gomez, G. Martinez, Polymeric modification of graphene through esterification of graphite oxide and poly (vinyl alcohol), Macromolecules 42 (2009) 6331-6334. https://doi.org/10.1021/ma900845w

[99] Y. Qi, M. Yang, W. Xu, S. He, Y. Men, Natural polysaccharides-modified graphene oxide for adsorption of organic dyes from aqueous solutions, Journal of colloid and interface science 486 (2017) 84-96. https://doi.org/10.1016/j.jcis.2016.09.058

[100] S. Wang, R.m. Cazelles, W.-C. Liao, M. Vázquez-González, A. Zoabi, R. Abu-Reziq, I. Willner, Mimicking horseradish peroxidase and NADH peroxidase by heterogeneous Cu^{2+}-modified graphene oxide nanoparticles, Nano letters 17 (2017) 2043-2048. https://doi.org/10.1021/acs.nanolett.7b00093

[101] F. Zhu, Y. Wang, Y. Zhang, W. Wang, Synthesis of Fe_3O_4 Nanorings/Amine-Functionalized Reduced Graphene Oxide Composites as Supercapacitor Electrode Materials in Neutral Electrolyte, INTERNATIONAL JOURNAL OF

ELECTROCHEMICAL SCIENCE 12 (2017) 7197-7204.
https://doi.org/10.20964/2017.08.53

[102] L. Lai, L. Chen, D. Zhan, L. Sun, J. Liu, S.H. Lim, C.K. Poh, Z. Shen, J. Lin, One-step synthesis of NH_2-graphene from in situ graphene-oxide reduction and its improved electrochemical properties, Carbon 49 (2011) 3250-3257. https://doi.org/10.1016/j.carbon.2011.03.051

[103] Y. Zhao, K. Tang, H. Ruan, L. Xue, B. Van der Bruggen, C. Gao, J. Shen, Sulfonated reduced graphene oxide modification layers to improve monovalent anions selectivity and controllable resistance of anion exchange membrane, Journal of Membrane Science 536 (2017) 167-175. https://doi.org/10.1016/j.memsci.2017.05.002

[104] S.K. Singh, M.K. Singh, P.P. Kulkarni, V.K. Sonkar, J.J. Grácio, D. Dash, Amine-modified graphene: thrombo-protective safer alternative to graphene oxide for biomedical applications, ACS nano 6 (2012) 2731-2740. https://doi.org/10.1021/nn300172t

[105] S. Rani, M. Kumar, R. Garg, S. Sharma, D. Kumar, Amide Functionalized Graphene Oxide Thin Films for Hydrogen Sulfide Gas Sensing Applications, IEEE Sens. J 16 (2016) 2929-2934. https://doi.org/10.1109/JSEN.2016.2524204

[106] C. Göde, M.L. Yola, A. Yılmaz, N. Atar, S. Wang, A novel electrochemical sensor based on calixarene functionalized reduced graphene oxide: Application to simultaneous determination of Fe (III), Cd (II) and Pb (II) ions, Journal of colloid and interface science 508 (2017) 525-531. https://doi.org/10.1016/j.jcis.2017.08.086

[107] N. Li, J. Chen, Y.-P. Shi, Magnetic polyethyleneimine functionalized reduced graphene oxide as a novel magnetic solid-phase extraction adsorbent for the determination of polar acidic herbicides in rice, Analyticachimicaacta 949 (2017) 23-34. https://doi.org/10.1016/j.aca.2016.11.016

[108] L. Shao, R. Zhang, J. Lu, C. Zhao, X. Deng, Y. Wu, Mesoporous silica coated polydopamine functionalized reduced graphene oxide for synergistic targeted chemo-photothermal therapy, ACS applied materials & interfaces 9 (2017) 1226-1236. https://doi.org/10.1021/acsami.6b11209

[109] B.D. Ossonon, D. Bélanger, Synthesis and characterization of sulfophenyl-functionalized reduced graphene oxide sheets, RSC Advances 7 (2017) 27224-27234. https://doi.org/10.1039/C6RA28311J

[110] S. Chinnathambi, G.J.W. Euverink, Polyaniline functionalized electrochemically reduced graphene oxide chemiresistive sensor to monitor the pH in real time during microbial fermentations, Sensors and Actuators B: Chemical 264 (2018) 38-44. https://doi.org/10.1016/j.snb.2018.02.087

[111] S.B. Sertkol, B. Esat, A.A. Momchilov, M.B. Yılmaz, M. Sertkol, An anthraquinone-functionalized reduced graphene oxide as electrode material for rechargeable batteries, Carbon 116 (2017) 154-166. https://doi.org/10.1016/j.carbon.2017.02.005

[112] Y. Chen, J.C. Crittenden, S. Hackney, L. Sutter, D.W. Hand, Preparation of a novel TiO_2-based p–n junction nanotube photocatalyst, Environmental science & technology 39 (2005) 1201-1208. https://doi.org/10.1021/es049252g

[113] J. Yu, W. Wang, B. Cheng, Synthesis and enhanced photocatalytic activity of a hierarchical porous flowerlike p–n junction NiO/TiO_2 photocatalyst, Chemistry–An Asian Journal 5 (2010) 2499-2506. https://doi.org/10.1002/asia.201000550

[114] A. Kumar, C. Guo, G. Sharma, D. Pathania, M. Naushad, S. Kalia, P. Dhiman, Magnetically recoverable ZrO_2/Fe_3O_4/chitosan nanomaterials for enhanced sunlight driven photoreduction of carcinogenic Cr (VI) and dechlorination & mineralization of 4-chlorophenol from simulated waste water, RSC Advances 6 (2016) 13251-13263. https://doi.org/10.1039/C5RA23372K

[115] G. Sharma, V.K. Gupta, S. Agarwal, S. Bhogal, M. Naushad, A. Kumar, F.J. Stadler, Fabrication and characterization of trimetallicnano-photocatalyst for remediation of ampicillin antibiotic, Journal of Molecular Liquids 260 (2018) 342-350. https://doi.org/10.1016/j.molliq.2018.03.059

[116] G. Sharma, Z.A. ALOthman, A. Kumar, S. Sharma, S.K. Ponnusamy, M. Naushad, Fabrication and characterization of a nanocomposite hydrogel for combined photocatalytic degradation of a mixture of malachite green and fast green dye, Nanotechnology for Environmental Engineering 2 (2017) 4. https://doi.org/10.1007/s41204-017-0014-y

[117] G.P. Mane, S.N. Talapaneni, K.S. Lakhi, H. Ilbeygi, U. Ravon, K. Al-Bahily, T. Mori, D.H. Park, A. Vinu, Highly Ordered Nitrogen Rich Mesoporous Carbon Nitrides and Their Superior Performance for Sensing and Photocatalytic Hydrogen Generation, Angewandte Chemie International Edition 56 (2017) 8481-8485. https://doi.org/10.1002/anie.201702386

[118] C.u. Gomes Silva, R. Juárez, T. Marino, R. Molinari, H. García, Influence of excitation wavelength (UV or visible light) on the photocatalytic activity of titania containing gold nanoparticles for the generation of hydrogen or oxygen from water, Journal of the American Chemical Society 133 (2010) 595-602. https://doi.org/10.1021/ja1086358

[119] K. Lalitha, J.K. Reddy, M.V.P. Sharma, V.D. Kumari, M. Subrahmanyam, Continuous hydrogen production activity over finely dispersed Ag_2O/TiO_2 catalysts from methanol: water mixtures under solar irradiation: a structure–activity correlation, International Journal of Hydrogen Energy 35 (2010) 3991-4001. https://doi.org/10.1016/j.ijhydene.2010.01.106

[120] M.S. Lowry, J.I. Goldsmith, J.D. Slinker, R. Rohl, R.A. Pascal, G.G. Malliaras, S. Bernhard, Single-layer electroluminescent devices and photoinduced hydrogen production from an ionic iridium (III) complex, Chemistry of materials 17 (2005) 5712-5719. https://doi.org/10.1021/cm051312+

[121] A. Kumar, A. Kumar, G. Sharma, M. Naushad, F.J. Stadler, A.A. Ghfar, P. Dhiman, R.V. Saini, Sustainable nano-hybrids of magnetic biochar supported g-C_3N_4/$FeVO_4$ for solar powered degradation of noxious pollutants- Synergism of adsorption, photocatalysis& photo-ozonation, Journal of Cleaner Production 165 (2017) 431-451. https://doi.org/10.1016/j.jclepro.2017.07.117

[122] G. Sharma, A. Kumar, S. Sharma, M. Naushad, R. Prakash Dwivedi, Z.A. Alothman, G.T. Mola, Novel development of nanoparticles to bimetallic nanoparticles and their composites: A review, Journal of King Saud University - Science (2017).

[123] H. Dong, S. Qi, Realising the potential of graphene-based materials for biosurfaces–A future perspective, Biosurface and Biotribology 1 (2015) 229-248. https://doi.org/10.1016/j.bsbt.2015.10.004

[124] M. Aleksandrzak, W. Kukulka, E. Mijowska, Graphitic carbon nitride/graphene oxide/reduced graphene oxide nanocomposites for photoluminescence and photocatalysis, Applied Surface Science 398 (2017) 56-62. https://doi.org/10.1016/j.apsusc.2016.12.023

[125] Y. Li, H. Zhang, P. Liu, D. Wang, Y. Li, H. Zhao, Cross-Linked g-C_3N_4/rGO Nanocomposites with Tunable Band Structure and Enhanced Visible Light Photocatalytic Activity, Small 9 (2013) 3336-3344. https://doi.org/10.1002/smll.201203135

[126] A. Wang, X. Li, Y. Zhao, W. Wu, J. Chen, H. Meng, Preparation and characterizations of Cu_2O/reduced graphene oxide nanocomposites with high photo-catalytic performances, Powder Technology 261 (2014) 42-48. https://doi.org/10.1016/j.powtec.2014.04.004

[127] R.C. Pawar, V. Khare, C.S. Lee, Hybrid photocatalysts using graphitic carbon nitride/cadmium sulfide/reduced graphene oxide (g-C_3N_4/CdS/RGO) for superior photodegradation of organic pollutants under UV and visible light, Dalton Transactions 43 (2014) 12514-12527. https://doi.org/10.1039/C4DT01278J

[128] S. Song, B. Cheng, N. Wu, A. Meng, S. Cao, J. Yu, Structure effect of graphene on the photocatalytic performance of plasmonic Ag/Ag_2CO_3-rGO for photocatalytic elimination of pollutants, Applied Catalysis B: Environmental 181 (2016) 71-78. https://doi.org/10.1016/j.apcatb.2015.07.034

[129] M. Maham, M. Nasrollahzadeh, S.M. Sajadi, M. Nekoei, Biosynthesis of Ag/reduced graphene oxide/Fe_3O_4 using Lotus garcinii leaf extract and its application as a recyclable nanocatalyst for the reduction of 4-nitrophenol and organic dyes, Journal of colloid and interface science 497 (2017) 33-42. https://doi.org/10.1016/j.jcis.2017.02.064

[130] Y. Yao, C. Xu, S. Yu, D. Zhang, S. Wang, Facile synthesis of Mn_3O_4–reduced graphene oxide hybrids for catalytic decomposition of aqueous organics, Industrial & Engineering Chemistry Research 52 (2013) 3637-3645. https://doi.org/10.1021/ie303220x

[131] L. Nirumand, S. Farhadi, A. Zabardasti, A. Khataee, Synthesis and sonocatalytic performance of a ternary magnetic MIL-101 (Cr)/RGO/$ZnFe_2O_4$ nanocomposite for degradation of dye pollutants, UltrasonicsSonochemistry 42 (2018) 647-658. https://doi.org/10.1016/j.ultsonch.2017.12.033

[132] M. Rakibuddin, S. Mandal, R. Ananthakrishnan, A novel ternary CuO decorated Ag_3AsO_4/GO hybrid as a Z-scheme photocatalyst for enhanced degradation of phenol under visible light, New Journal of Chemistry 41 (2017) 1380-1389. https://doi.org/10.1039/C6NJ02366E

[133] W.-K. Jo, N.C.S. Selvam, Synthesis of GO supported Fe_2O_3–TiO_2 nanocomposites for enhanced visible-light photocatalytic applications, Dalton Transactions 44 (2015) 16024-16035. https://doi.org/10.1039/C5DT02983J

[134] S. Shuang, R. Lv, X. Cui, Z. Xie, J. Zheng, Z. Zhang, Efficient photocatalysis with graphene oxide/Ag/Ag$_2$S–TiO$_2$ nanocomposites under visible light irradiation, RSC Advances 8 (2018) 5784-5791. https://doi.org/10.1039/C7RA13501G

[135] G. Chen, M. Sun, Q. Wei, Y. Zhang, B. Zhu, B. Du, Ag$_3$PO$_4$/graphene-oxide composite with remarkably enhanced visible-light-driven photocatalytic activity toward dyes in water, Journal of hazardous materials 244 (2013) 86-93. https://doi.org/10.1016/j.jhazmat.2012.11.032

[136] Y. Ding, Y. Zhou, W. Nie, P. Chen, MoS$_2$–GO nanocomposites synthesized via a hydrothermal hydrogel method for solar light photocatalytic degradation of methylene blue, Applied Surface Science 357 (2015) 1606-1612. https://doi.org/10.1016/j.apsusc.2015.10.030

[137] W.-K. Jo, S. Kumar, M.A. Isaacs, A.F. Lee, S. Karthikeyan, Cobalt promoted TiO$_2$/GO for the photocatalytic degradation of oxytetracycline and Congo Red, Applied Catalysis B: Environmental 201 (2017) 159-168. https://doi.org/10.1016/j.apcatb.2016.08.022

[138] S. Liang, Y. Zhou, K. Kang, Y. Zhang, Z. Cai, J. Pan, Synthesis and characterization of porous TiO$_2$-NS/Pt/GO aerogel: A novel three-dimensional composite with enhanced visible-light photoactivity in degradation of chlortetracycline, Journal of Photochemistry and Photobiology A: Chemistry 346 (2017) 1-9. https://doi.org/10.1016/j.jphotochem.2017.05.036

[139] X. Miao, X. Shen, J. Wu, Z. Ji, J. Wang, L. Kong, M. Liu, C. Song, Fabrication of an all solid Z-scheme photocatalyst g-C$_3$N$_4$/GO/AgBr with enhanced visible light photocatalytic activity, Applied Catalysis A: General 539 (2017) 104-113. https://doi.org/10.1016/j.apcata.2017.04.009

[140] S.K. Moorthy, C. Viswanathan, N. Ponpandian, Facile Approach for Synthesis of GO/ZnO Nanocomposite for Highly Efficient Photocatalytic Degradation of Organic Dyes under Visible Light, Nano Hybrids and Composites, Trans Tech Publ, 2017, pp. 121-126.

[141] W. Tong, L. Zhu, J. Xia, J. Di, S. Yin, Q. Zhang, Z. Chen, H. Li, One-pot ionic liquid-assisted strategy for GO/BiOI hybrids with superior visible-driven photocatalysis and mechanism research, Materials Technology 32 (2017) 131-139. https://doi.org/10.1080/10667857.2016.1157914

[142] N. Wang, Y. Zhou, C. Chen, L. Cheng, H. Ding, A g-C$_3$N$_4$ supported graphene oxide/Ag$_3$PO$_4$ composite with remarkably enhanced photocatalytic activity under

visible light, Catalysis Communications 73 (2016) 74-79.
https://doi.org/10.1016/j.catcom.2015.10.015

[143] Z. Dong, J. Pan, B. Wang, Z. Jiang, C. Zhao, J. Wang, C. Song, Y. Zheng, C. Cui,
C. Li, The p-n-type Bi$_5$O$_7$I-modified porous C3N4 nano-heterojunction for
enhanced visible light photocatalysis, Journal of Alloys and Compounds 747
(2018) 788-795. https://doi.org/10.1016/j.jallcom.2018.03.112

[144] Z. Zhu, Q. Han, D. Yu, J. Sun, B. Liu, A novel pn heterojunction of
BiVO$_4$/TiO$_2$/GO composite for enhanced visible-light-driven photocatalytic
activity, Materials Letters 209 (2017) 379-383.
https://doi.org/10.1016/j.matlet.2017.08.045

[145] P.D. Tran, S.K. Batabyal, S.S. Pramana, J. Barber, L.H. Wong, S.C.J. Loo, A
cuprous oxide–reduced graphene oxide (Cu$_2$O–rGO) composite photocatalyst for
hydrogen generation: employing rGO as an electron acceptor to enhance the
photocatalytic activity and stability of Cu 2 O, Nanoscale 4 (2012) 3875-3878.
https://doi.org/10.1039/c2nr30881a

[146] J. Zhou, G. Tian, Y. Chen, X. Meng, Y. Shi, X. Cao, K. Pan, H. Fu, In situ
controlled growth of ZnIn$_2$S$_4$ nanosheets on reduced graphene oxide for enhanced
photocatalytic hydrogen production performance, Chemical Communications 49
(2013) 2237-2239. https://doi.org/10.1039/c3cc38999e

[147] W. Fan, X. Yu, H.-C. Lu, H. Bai, C. Zhang, W. Shi, Fabrication of
TiO$_2$/RGO/Cu$_2$O heterostructure for photoelectrochemical hydrogen production,
Applied Catalysis B: Environmental 181 (2016) 7-15.
https://doi.org/10.1016/j.apcatb.2015.07.032

[148] X. Yan, K. Yuan, N. Lu, H. Xu, S. Zhang, N. Takeuchi, H. Kobayashi, R. Li, The
interplay of sulfur doping and surface hydroxyl in band gap engineering:
Mesoporous sulfur-doped TiO$_2$ coupled with magnetite as a recyclable, efficient,
visible light active photocatalyst for water purification, Applied Catalysis B:
Environmental 218 (2017) 20-31. https://doi.org/10.1016/j.apcatb.2017.06.022

[149] Z. Yan, X. Yu, A. Han, P. Xu, P. Du, Noble-metal-free Ni(OH)$_2$-modified
CdS/reduced graphene oxide nanocomposite with enhanced photocatalytic activity
for hydrogen production under visible light irradiation, The Journal of Physical
Chemistry C 118 (2014) 22896-22903. https://doi.org/10.1021/jp5065402

[150] H.Y. Hafeez, S.K. Lakhera, S. Bellamkonda, G.R. Rao, M. Shankar, D.
Bahnemann, B. Neppolian, Construction of ternary hybrid layered reduced

graphene oxide supported g-C$_3$N$_4$-TiO$_2$ nanocomposite and its photocatalytic hydrogen production activity, international journal of hydrogen energy (2017).

[151] X. Yu, Z. Zhao, D. Sun, N. Ren, J. Yu, R. Yang, H. Liu, Microwave-assisted hydrothermal synthesis of Sn$_3$O$_4$ nanosheet/rGO planar heterostructure for efficient photocatalytic hydrogen generation, Applied Catalysis B: Environmental 227 (2018) 470-476. https://doi.org/10.1016/j.apcatb.2018.01.055

[152] D. Xu, L. Li, R.He, L. Qi, L. Zhang, B. Cheng, Noble metal-free RGO/TiO$_2$ composite nanofiber with enhanced photocatalytic H2-production performance, Applied Surface Science 434 (2018) 620-625. https://doi.org/10.1016/j.apsusc.2017.10.192

[153] S. Kai, B. Xi, X. Liu, L. Ju, P. Wang, Z. Feng, X. Ma, S. Xiong, An Innovative Au-CdS/ZnS-RGO Architecture for Efficiently Photocatalytic Hydrogen Evolution, Journal of Materials Chemistry A (2018). https://doi.org/10.1039/C7TA10958J

[154] S. Tonda, S. Kumar, Y. Gawli, M. Bhardwaj, S. Ogale, g-C$_3$N$_4$ (2D)/CdS (1D)/rGO (2D) dual-interface nano-composite for excellent and stable visible light photocatalytic hydrogen generation, International Journal of Hydrogen Energy 42 (2017) 5971-5984. https://doi.org/10.1016/j.ijhydene.2016.11.065

[155] J. Li, P. Liu, Y. Qu, T. Liao, B. Xiang, WSe$_2$/rGO hybrid structure: A stable and efficient catalyst for hydrogen evolution reaction, International Journal of Hydrogen Energy 43 (2018) 2601-2609. https://doi.org/10.1016/j.ijhydene.2017.12.160

[156] M.N. Hossain, J. Wen, S.K. Konda, M. Govindhan, A. Chen, Electrochemical and FTIR spectroscopic study of CO$_2$ reduction at a nanostructured Cu/reduced graphene oxide thin film, Electrochemistry Communications 82 (2017) 16-20. https://doi.org/10.1016/j.elecom.2017.07.006

[157] H. Zhong, M. Iguchi, M. Chatterjee, T. Ishizaka, M. Kitta, Q. Xu, H. Kawanami, Interconversion Between CO$_2$ and HCOOH Under Basic Conditions Catalyzed by PdAu Nanoparticles Supported by Amine Functionalized Reduced Graphene Oxide as a Dual Catalyst, ACS Catalysis (2018). https://doi.org/10.1021/acscatal.8b00294

[158] A. Bafaqeer, M. Tahir, N.A.S. Amin, Synergistic effects of 2D/2D ZnV$_2$O$_6$/RGO nanosheets heterojunction for stable and high performance photo-induced CO$_2$

reduction to solar fuels, Chemical Engineering Journal 334 (2018) 2142-2153. https://doi.org/10.1016/j.cej.2017.11.111

[159] V. Deerattrakul, P. Puengampholsrisook, W. Limphirat, P. Kongkachuichay, Characterization of supported Cu-Zn/graphene aerogel catalyst for direct CO_2 hydrogenation to methanol: Effect of hydrothermal temperature on graphene aerogel synthesis, Catalysis Today (2017).

[160] V. Deerattrakul, W. Limphirat, P. Kongkachuichay, Influence of reduction time of catalyst on methanol synthesis via CO_2 hydrogenation using Cu–Zn/N-rGO investigated by in situ XANES, Journal of the Taiwan Institute of Chemical Engineers 80 (2017) 495-502. https://doi.org/10.1016/j.jtice.2017.08.011

[161] W.-H.Dong, D.-D.Wu, J.-M.Luo, Q.-J.Xing, H. Liu, J.-P.Zou, X.-B.Luo, X.-B.Min, H.-L.Liu, S.-L. Luo, Coupling of photodegradation of RhB with photoreduction of CO_2 over rGO/SrTi$_{0.95}$Fe$_{0.05}$O$_3$− δ catalyst: A strategy for one-pot conversion of organic pollutants to methanol and ethanol, Journal of Catalysis 349 (2017) 218-225. https://doi.org/10.1016/j.jcat.2017.02.004

[162] G. He, H. Tang, H. Wang, Z. Bian, Highly Selective and Active Pd/In/three-dimensional Graphene with Special Structure for Electroreduction CO_2 to Formate, Electroanalysis 30 (2018) 84-93. https://doi.org/10.1002/elan.201700525

[163] T. Takayama, K. Sato, T. Fujimura, Y. Kojima, A. Iwase, A. Kudo, Photocatalytic CO_2 reduction using water as an electron donor by a powdered Z-scheme system consisting of metal sulfide and an RGO–TiO_2 composite, Faraday discussions 198 (2017) 397-407. https://doi.org/10.1039/C6FD00215C

[164] P. Wang, S. Zhan, Y. Xia, S. Ma, Q. Zhou, Y. Li, The fundamental role and mechanism of reduced graphene oxide in rGO/Pt-TiO_2 nanocomposite for high-performance photocatalytic water splitting, Applied Catalysis B: Environmental 207 (2017) 335-346. https://doi.org/10.1016/j.apcatb.2017.02.031

[165] J. Si, Y. Liu, S. Chang, D. Wu, B. Tian, J. Zhang, AgBr@TiO_2/GO ternary composites with enhanced photocatalytic activity for oxidation of benzyl alcohol to benzaldehyde, Research on Chemical Intermediates 43 (2017) 2067-2080. https://doi.org/10.1007/s11164-016-2747-9

[166] Y.-F.Xu, M.-Z.Yang, B.-X.Chen, X.-D.Wang, H.-Y.Chen, D.-B.Kuang, C.-Y.Su, A CsPbBr$_3$ perovskite quantum dot/graphene oxide composite for photocatalytic CO_2 reduction, Journal of the American Chemical Society 139 (2017) 5660-5663. https://doi.org/10.1021/jacs.7b00489

Carbonaceous Composite Materials   Materials Research Forum LLC
Materials Research Foundations **42** (2018) 231-272   doi: http://dx.doi.org/10.21741/9781945291975-9

[167] M. Wang, P. Ju, J. Li, Y. Zhao, X. Han, Z. Hao, Facile Synthesis of MoS_2/g-C_3N_4/GO Ternary Heterojunction with Enhanced Photocatalytic Activity for Water Splitting, ACS Sustainable Chemistry & Engineering 5 (2017) 7878-7886. https://doi.org/10.1021/acssuschemeng.7b01386

[168] F. Ren, C. Wang, C. Zhai, F. Jiang, R. Yue, Y. Du, P. Yang, J. Xu, One-pot synthesis of a RGO-supported ultrafine ternary PtAuRu catalyst with high electrocatalytic activity towards methanol oxidation in alkaline medium, Journal of Materials Chemistry A 1 (2013) 7255-7261. https://doi.org/10.1039/c3ta11291h

[169] S.J. Li, D. Bao, M.M. Shi, B.R. Wulan, J.M. Yan, Q. Jiang, Amorphizing of Au Nanoparticles by CeOx–RGO Hybrid Support towards Highly Efficient Electrocatalyst for N2 Reduction under Ambient Conditions, Advanced Materials 29 (2017).

[170] Y. Zhao, D. Zhao, C. Chen, X. Wang, Enhanced photo-reduction and removal of Cr (VI) on reduced graphene oxide decorated with TiO_2 nanoparticles, Journal of colloid and interface science 405 (2013) 211-217. https://doi.org/10.1016/j.jcis.2013.05.004

[171] L. He, S. Sarkar, A. Barras, R. Boukherroub, S. Szunerits, D. Mandler, Electrochemically stimulated drug release from flexible electrodes coated electrophoretically with doxorubicin loaded reduced graphene oxide, Chemical Communications 53 (2017) 4022-4025. https://doi.org/10.1039/C7CC00381A

[172] S.R. Shin, C. Zihlmann, M. Akbari, P. Assawes, L. Cheung, K. Zhang, V. Manoharan, Y.S. Zhang, M. Yüksekkaya, K.t. Wan, Reduced graphene oxide gel MA hybrid hydrogels as scaffolds for cardiac tissue engineering, Small 12 (2016) 3677-3689. https://doi.org/10.1002/smll.201600178

[173] Y. Li, Z. Zhang, Y. Zhang, D. Deng, L. Luo, B. Han, C. Fan, Nitidine chloride-assisted bio-functionalization of reduced graphene oxide by bovine serum albumin for impedimetric immunosensing, Biosensors and Bioelectronics 79 (2016) 536-542. https://doi.org/10.1016/j.bios.2015.12.076

[174] Y. Li, J. Han, B. Xie, Y. Li, S. Zhan, Y. Tian, Synergistic degradation of antimicrobial agent ciprofloxacin in water by using 3D CeO_2/RGO composite as cathode in electro-Fenton system, Journal of Electroanalytical Chemistry 784 (2017) 6-12. https://doi.org/10.1016/j.jelechem.2016.11.057

[175] G. Eskiizmir, Y. Baskın, K. Yapıcı, Graphene-based nanomaterials in cancer treatment and diagnosis, Fullerens, Graphenes and Nanotubes, Elsevier 2018, pp. 331-374.

[176] W. Qian, X. Hu, W. He, R. Zhan, M. Liu, D. Zhou, Y. Huang, X. Hu, Z. Wang, G. Fei, Polydimethylsiloxane incorporated with reduced graphene oxide (rGO) sheets for wound dressing application: Preparation and characterization, Colloids and Surfaces B: Biointerfaces 166 (2018) 61-71. https://doi.org/10.1016/j.colsurfb.2018.03.008

[177] Y. Jung, H.G. Moon, C. Lim, K. Choi, H.S. Song, S. Bae, S.M. Kim, M. Seo, T. Lee, S. Lee, Humidity Tolerant Single Stranded DNA Functionalized Graphene Probe for Medical Applications of Exhaled Breath Analysis, Advanced Functional Materials 27 (2017). https://doi.org/10.1002/adfm.201700068

[178] M.A. Tabrizi, M. Shamsipur, R. Saber, S. Sarkar, N. Zolfaghari, An ultrasensitive sandwich-type electrochemical immunosensor for the determination of SKBR-3 breast cancer cell using rGO-TPA/FeHCFnano labeled Anti-HCT as a signal tag, Sensors and Actuators B: Chemical 243 (2017) 823-830. https://doi.org/10.1016/j.snb.2016.12.061

[179] L.A. Al-Ani, M.A. AlSaadi, F.A. Kadir, N.M. Hashim, N.M. Julkapli, W.A. Yehye, Graphene–gold based nanocomposites applications in cancer diseases; Efficient detection and therapeutic tools, European journal of medicinal chemistry 139 (2017) 349-366. https://doi.org/10.1016/j.ejmech.2017.07.036

Carbonaceous Composite Materials

Materials Research Forum LLC

Materials Research Foundations **42** (2018) 273-308

doi: http://dx.doi.org/10.21741/9781945291975-10

Chapter 10

A Critical Review on Spectroscopic Characterization of Sustainable Nanocomposites Containing Carbon Nano Fillers

Teklit Gebregiorgis Amabye[1,2,*], Mabrahtu Hagos[3] , Hayelom Dargo Beyene[4]

[1] IHE Delft Institute for Water Education, P.O. Box 3015, 2601 DA Delft, The Netherlands

[2] Mekelle University, Department of chemistry, Mekelle, Ethiopia

[3] Faculty of Natural and Computational Sciences, Woldia University, Woldia-400, Ethiopia

[4] Department of Chemistry, Adigrat University, Ethiopia

teklitgeb@gmail.com*

Abstract

Nanomaterials are a relatively new class of materials that have at least one dimension in a size range below one hundred nanometers (<100nm) resulting in properties that are significantly different from their bulk-sized analogue materials. These opened new application platforms as reinforcing fillers in plastic composites, functional materials in sensors and energy, and in biomedical applications such as medical diagnosis and prevention, and drug delivery. Nanomaterials have various interesting physicochemical properties such as electrical conductivity, antimicrobial properties, reinforcing capability, photoactivity, optical properties, etc. Thus, characterization of nanomaterials is crucial to fully understand their merits in various material systems. In this review, an overview of various nanomaterial characterizations techniques with an emphasis on spectroscopic techniques is presented. The utilization of scanning tunnelling microscopy (STM), X-ray diffraction (XRD), Fourier transform infrared spectroscopy (FTIR), transmission electron microscopy (TEM), X-ray photoelectron spectroscopy (XPS), and ultraviolet-visible diffuse reflection spectroscopy (UV–vis) are highlighted. Furthermore, future trends of these spectroscopic characterizations for nanomaterials and nanocomposites applications are discussed.

Keywords

Nanomaterials, Spectroscopy, Carbon Nanotubes, Graphene, Nanocomposites

Materials Research Forum LLC
doi: http://dx.doi.org/10.21741/9781945291975-10

Contents

1.   Introduction

The concept of using fillers as strengthening material agents is not novel. It started about 4000 BC, with the utilization of straw to strength mud blocks [1]. The word reinforcement is defined as "the action or process of reinforcing or strengthening, by adding another material to it"; and this addition of another material has mesmerized a large section of the scientific community. The first reinforcement grids, plates or fibers have been incorporated in 1849, and the first patent was approved in 1867 [2]. Very recently, fibers made from alumina, boron, carbon, glass, and silicon carbide materials have been used as fillers in composites. The aforementioned fillers have limited capabilities for mechanical reinforcement due to their mesoscale dimensions [3].

In contrast to micro-scale fillers, nanofillers provide small inter-particle distance because of their size; and this impact the polymer matrix residences even at very low filler concentrations. Furthermore, the reduction of fillers dimensions to nano-scale leads to a large interfacial area for polymer – filler interaction. Nanoscale fillers are also free of most defects that are associated with the hierarchical structure of micro-scale fillers, resulting in improved physical properties that provide them with a unique reinforcing capability in polymer matrices. Polymer nanocomposites came into limelight after their discovery in the 1980s. Prior to their recent widespread prominence for various functional applications, there were various application examples by which nanoparticles were used as fillers. Even though, nanoscale dimensioned black carbons were used as fillers in rubbers they were not specifically referred as nanocomposites until recently. The most announced application of polymer nanocomposites next to the ignition of the renewed interest, was a timing belt cover made from polyamide 6 and exfoliated clay by Toyota for an automotive application [4]. The use of nanofillers can result in improvement in tensile and flexural strength, modulus, and heat distortion temperature and gas barrier properties upon addition of a small fraction of nanofillers in plastic matrices.

Carbon, a remarkable element is known for its tremendous capacity of catenation. The property of carbon to combine with itself and different chemical elements in numerous methods forms the premise of organic chemistry and life. The unique structural and transport properties of carbon-based nanostructures, the first of which was fullerene, have captured the interest of researchers since their discovery [5]. Some other nanostructures of carbon known as graphene have been demonstrated to be the strongest material in nature. Carbon nanotubes (CNT) do not trail back much in their strength. It is clear that carbon nanofillers have become promising ultra-strong, and conducting polymer nanofillers [6].

In this book chapter, we have reviewed various techniques of characterizing nanoparticles and polymer nanocomposites. Analysis of the particle size, shape, aggregation and agglomeration, dispersion in matrices, and interactions were reviewed. Various spectroscopic characterization techniques of carbon nanofillers and their nanocomposites are covered.

Carbonaceous Composite Materials   Materials Research Forum LLC
Materials Research Foundations **42** (2018) 273-308   doi: http://dx.doi.org/10.21741/9781945291975-10

1.1   Carbon nano-fillers

Carbon is abundantly available in nature. It can be found in its elemental form as coal, natural graphite and in smaller quantities as diamonds. Due to its specific hybridization properties [5], there also are several engineered types of carbon such as adsorbent carbon, carbon black, glassy carbon, cokes, artificial graphite and diamonds, carbon and graphitic fibers, etc. and, this makes them applicable in various fields. The birth of carbon science dates back to the mid-1980s by Harry Kroto after the discovery of the first all-carbon molecule, fullerene C_{60} or buckyball [7].

In the 1950s, a report on hollow carbon fibers was presented by Radushkevich and Lukyanovich [8]. Similarly, in the 1970s Oberlin *et al.* showed an image of empty tubular-shaped materials formed by benzene decomposition [8]. These and other researchers gave way to the invention of carbon nanotubes (CNT), which was first observed in 1991 in Japan by Sumio Iijima [9]. Arc-discharge evaporation technique was used to develop needle-like tubes as same as fullerene synthesis by Ijima; and these needles grew on the other end of the electrode.

Also, one of the tubular derivatives of fullerenes is carbon nanotubes (CNTs). They include graphene cylinders that are closed at each ends with caps containing pentagonal earrings. They typically have a diameter of few nanometres, and their length can vary up to numerous centimetres. CNTs may be single, double or multi-walled [8]. Figure 1 depicts the structures of graphene, graphite, fullerene, and carbon nanotubes.

Graphite is one of the allotropes of carbon that has a layered planar shape. Each carbon in graphite layer atoms is arranged in a hexagonal lattice with a separation of 0.142 nm, and the distance among planes is 0.335 nm. Graphene is an allotrope of carbon which is a one atom thick planar sheet (carbon atom is sp^2-bonded) and densely packed in a honeycomb crystal lattice. During 1990 and 2004, numerous efforts had been made to make very thin films of graphite through techniques of mechanical exfoliation [10] to prepare graphene. Although the theoretical concept of graphene was explored since 1947 [11], graphene was officially discovered and reported by Andre Geim and Kostya Novoselov in 2004 [12]. They used micromechanical cleavage or the Scotch tape approach to extract a single atom thick crystallite layer from bulk graphite by pulling out graphene layers and transferring them onto skinny SiO_2 on a silicon wafer. Since its discovery, various forms of carbon nanomaterial have been derived from the basic graphene structure (e.g., fullerene, carbon nanotubes).

Carbonaceous Composite Materials   Materials Research Forum LLC
Materials Research Foundations **42** (2018) 273-308   doi: http://dx.doi.org/10.21741/9781945291975-10

2D Graphene

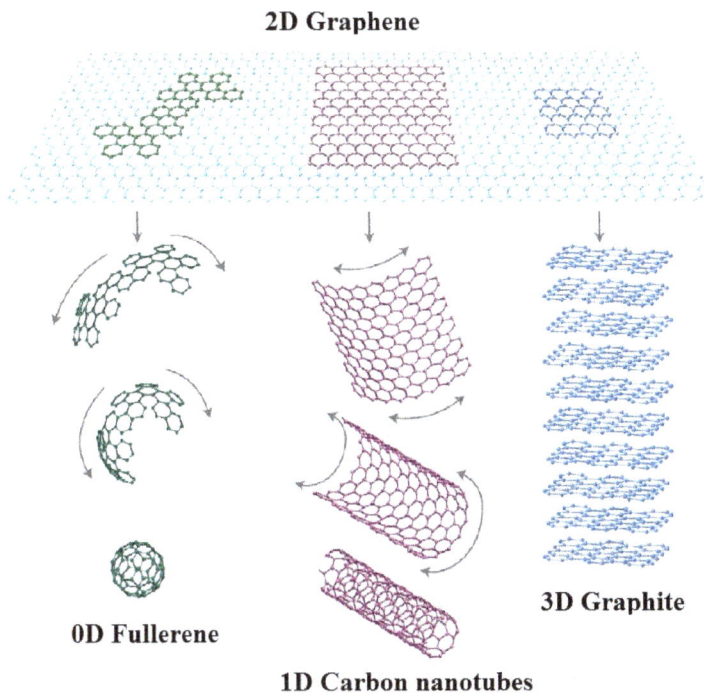

0D Fullerene

3D Graphite

1D Carbon nanotubes

Figure 1 Schematics displaying fullerene and nanotube formation from graphene. As presented in Figure 1, graphene is a 2-D construction block for carbon substances of all different dimensionalities. It could additionally be wrapped into 0-D buckyballs, rolled into 1-D nanotubes or stacked into 3-D graphite Copyright © Geim, A.K.2007 [13].

1.2   Polymer nanocomposites

High performance composites with characteristic reinforcing filler in dimensions of nanoscale (< 100 nm) are known as polymer nanocomposites. Such nanofillers can come in fibrils, whiskers, particulates, platelets, fibroids, and so on. The need for high performance and functional polymeric materials in applications ranging from engineering structures, coatings, biomedical, electronics, to packaging, is the major drive for the rapid increase in the development of nanocomposites. Some of the major assets of nanoscale reinforcement of polymers includes enhancement in stiffness and creep resistance, reduction in the coefficient of thermal expansion, improvement in fracture toughness, improvement in gas and moisture barrier properties, etc. [14, 15]. The use of nanoscale

Carbonaceous Composite Materials   Materials Research Forum LLC
Materials Research Foundations **42** (2018) 273-308   doi: http://dx.doi.org/10.21741/9781945291975-10

fillers may result to unusual polymer properties such as electrical conductivity, electromagnetic wave shielding, fire retardancy, electric, heat insulation, optical transparency, and colour. Moreover, polymer nanocomposites are able to be processed using conventional polymer processing technologies (e.g., extrusion, injection moulding, thermoforming, etc.), thus it avoids the rather costly layup process required for the fabrication of traditional conventional fiber-reinforced composites [16].

As best example, silica or alumina ceramic fillers have been used as cost effective means to enhance the mechanical performance of epoxy resins [17, 18]. The disadvantage is reduction of ductility of the already brittle epoxy while adding rigid particles. On the contrary, the use of elastomeric particles consisting of rubber can enhance the toughness of epoxy matrices, however these can reduce the stiffness [19]. The extraordinary reinforcing capability of nanomaterials by carbon nanotubes, fumed silica, cellulose nanocrystals, graphene, and graphene nanoplatelets (few layers of graphene) provide magnificent mechanical properties as nanofillers of polymer nanocomposites. The production of polymer nanocomposites generally involves the dispersion of the nanofiller phase into the matrix phase [20]. Numerous thermoplastic and thermosetting polymer matrices can be used along with the stated carbon nanofillers to prepare exciting nanocomposites. Characterization of nanofillers, their dispersion in polymer matrices as well as the interfacial interaction between each other is the primary focal point of this chapter.

There are several techniques for manufacturing polymer nanocomposites. The interfacial interaction as well as the dispersion of nanofillers in matrices is critical for the performance of many functional nanocomposites. Nanocomposites have an extensive variety of applications ranging from bicycle frames, wind turbine blades, badminton rackets to micro/nanodevices [21], nanocomposite packaging films [22], smart materials [23], sensors and actuators [24], etc. Bio nanocomposites and load bearing composites are highly significant applications in tissue engineering and bone reconstruction, respectively. Nanocomposites have reduced weight to mechanical performance ratio when compared to metallic material counterparts, and this is one of the most vital reasons driving the use of such materials for structural components in industrial scales.

Nanocomposites also have a vital role in the aerospace industry. For example, the recently launched Boeing 787 Dreamliner is manufactured with 80 % by volume and 50 % by weight polymer-based composite. The result enabled about 20 % fuel saving through magnificent weight reduction. While the majority of the reinforcing agent used in the Boeing 787 Dreamliner is the traditional carbon fiber, nanoparticles can play important roles as a partial substitute or complete replacement of such fibers to enhance

Carbonaceous Composite Materials   Materials Research Forum LLC
Materials Research Foundations **42** (2018) 273-308   doi: http://dx.doi.org/10.21741/9781945291975-10

not only the performance, but also the cost structure.   Other opportunities of nanocomposites in polymer science include the direct integration of functional nanoparticles in electronics, energy generating devices, antimicrobial packaging films, or electromagnetic shielding tools. Figure 2 presents a summary of carbon based nanofillers including conducting films and coatings for devices like displays, touch screens, optical and thermal devices, etc.

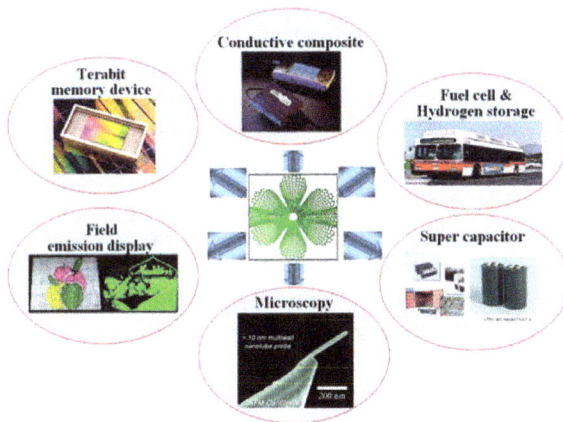

Figure 2 Potential applications of carbon nanofillers as functional components of polymer nanocomposites.

2.   Characterization of nanomaterials

The characterization of nanomaterials is critical for many potential applications. The physicochemical properties of the nanomaterials, which includes size, shape, aspect ratio, crystallinity, hardness, surface properties, purity and composition, molecular weight and weight distribution, rheological properties, stability and solubility in various solvent and temperature environments, colloidal properties, etc. are important for their utilization as a component in nanocomposites. These and other properties dictate their dispersibility, functional properties as well as their reinforcing capability in polymer matrices. Recent progress in nanotechnology has resulted in the development of sophisticated physical methods to characterize nanomaterials. Some of the most important techniques are presented in the following sections.

2.1   X-ray diffraction

X-ray characterization techniques are interesting for material characterization because of the penetration strength and element sensitivity of X-rays. X-ray diffraction (XRD), X-ray photoelectron spectroscopy (XPS), small angle X-ray scattering (SAXS), and wide angle X-ray scattering (WAXS) are extensively used X-ray based techniques for nanomaterial characterizations. XRD, based on wide angle elastic scattering of X-rays, is an indispensable technique for characterizing the size and degree of crystallinity of crystalline materials at the molecular level [25]. According to Bragg's law, the diffraction of X-ray is defined as the reflection of a collimated beam of X-rays incident at the crystalline planes of a specimen [26] in such a way that the diffraction pattern provides structural information of the crystals. Typical XRD device is based on extensive-attitude elastic scattering of X-rays, and it is used to characterize crystalline size, shape and lattice distortion by long-range order, however that is confined to disordered substances [27]. Moreover, it can be applied to identify unknown samples, measure sample purity, textural measurements (e.g., orientation of crystal grains).

In general, XRD is an established non-destructive material structure characterization technique. However, the application of XRD can be limited due to the difficulty in developing crystals and the potential to get results only from single conformation/binding state of the pattern [28]. Furthermore, XRD has a low depth of diffracted X-rays, specifically for low atomic number compounds, compared with electron diffractions [29]. Overall, XRD is effective for characterizing single phase and homogeneous crystalline samples. Contrarily, it has limited use for mixed samples and samples that are not ground into a powder. More recently, femtosecond pulses from a hard-X-ray free-electron laser are getting popularity for structure characterization. This approach can be useful for structural   characterization of macromolecules that do not provide sufficient crystal length, under conventional radiation sources, or other samples that are sensitive to X-ray radiation [30]. The Scherrer formula [29] can be used to determine the crystallite size from the broadening of corresponding X-ray spectral peak. In this formula, the mean size of the crystallite domains (d) is calculated as:

$$d = K\lambda/\beta \cos\theta$$

Where, λ is X-ray radiation wavelength, K is a dimensionless shape factor, is typically taken as 0.89, and β is the line broadening at half the maximum intensity (in radians) after subtraction of the instruments line broadening, and θ is the Bragg angle (in degrees).

For instance, Cheng et al. [31] used XRD to characterize N-doped TiO_2. In this study, the proportion of anatase inside the TiO_2 samples was determined from the integrated XRD peak intensities using a quality factor ratio of anatase to rutile (which is 1.265). Figure 3 displays the XRD patterns of pure TiO_2 and N_3-TiO_2 calcination at a various temperatures.

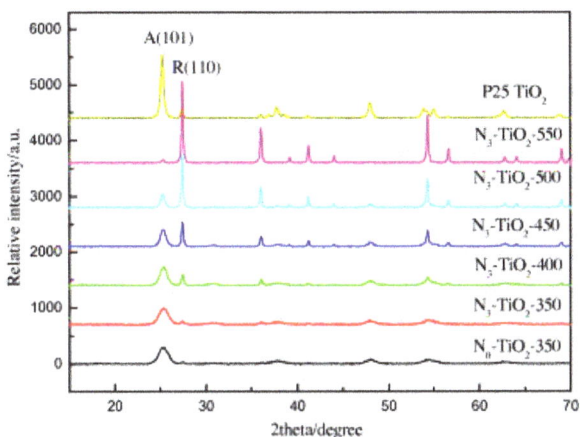

Figure 3 X- ray diffraction patterns of pure TiO₂ and N₃-TiO₂ catalysts calcined at various temperatures. Adapted from [31]. Copyright © Cheng (2016).

Based on the results present in Figure 3, it can be observed that the XRD peak width of the anatase phase at $2\theta = 25.3°$ exhibited gradual sharpness with an increase in the treatment temperature. This is indicative of increase in crystallite size with temperature increase. The N_3-TiO_2 did not display any new XRD peak as compared with the pure TiO_2, demonstrating that no new phase was created. At a calcination temperature of 350 °C, about 5 nm anatase crystallite sized N_3-TiO_2 nanoparticle has a mixed phase of anatase and rutile with compositions of 80% and 20%, respectively. This crystalline phase composition is comparable to commercial TiO_2 (e.g. grade P25 marketed by the chemical company Degussa). High photocatalytic activity was exhibited by the P25, and this is due to its mixed phase composition. Thus, from these results, it can be anticipated that the N_3-TiO_2-350 (as-prepared) possesses superlative photocatalytic activity due to its mixed phase composition. Overall, XRD has a very limited use in disordered or amorphous materials.

2.2   Raman spectroscopy

Raman spectroscopy (RS) is a vibrational spectroscopy technique extensively used for structural analysis (qualitative and quantitative) of molecules, macromolecules, nanomaterials and nanostructures. The approach is preferred for *in-situ* experiments as it offers submicron spatial resolution for transparent materials with limited sample preparation[32]. The working principle of RS involves measuring the inelastic scattering of photons frequency (qualitative analysis) or intensity (for quantitative analysis), from the incident light after interacting with electric powered dipoles of the analyte samples [33] to elucidate structural information.

Frequency variations among the incident photons and the in elastically scattered photons are observed by RS, which are associated with the traits of molecular vibration. In Raman spectroscopy, the in elastically scattered photons emitting frequencies lower than the incident photons are referred to as the Stokes lines; where as anti-Stokes lines emit photon frequencies more than the incident photons [33, 34].

In general, RS is corresponds to infra-red (IR) spectroscopy. While IR transitions result from a net change in the dipole moment of the molecules, Raman transitions result from nuclear motion modulating the polarizability of the molecules [34]. Vibrational modes that are Raman active are IR inactive, and vice versa for most small symmetrical molecules. Hence, RS has many pros for small molecules. For example, since water molecules are weak Raman scatterers, this technique is suitable for studying biological samples in aqueous solution. Furthermore, RS has the potential for detecting tissue abnormalities as the molecular information supplied via RS can be used to investigate conformations and concentrations of tissue components [35].

On the contrary, RS has some limitations. The technique can provide an indirect characterization of nanomaterials; however, it lacks the spatial resolution necessary to delineate domains for nanotechnology [36]. The conventional RS has an extremely small cross-section and is prone to interference from fluorescence. Due to the severe laser excitation in RS, a large sample quantity are usually required to obtain sufficient RS indicators [37].

Surface-enhanced Raman scattering (SERS) are strongly used to enhance the RS signals. This SERS can even increase spatial resolution at the same time as samples adhere to the surface of metal structures, which includes typically gold or silver NP colloid substrates [38-39]. SERS has many other uses. It has been reported in the literature that SERS can be used to (a) study surface functionalization of steel NPs, (b) monitor the conformational alternate in proteins conjugated to the steel NPs, and (c) to track intracellular drug

Carbonaceous Composite Materials   Materials Research Forum LLC
Materials Research Foundations **42** (2018) 273-308   doi: http://dx.doi.org/10.21741/9781945291975-10

movement from the nano platform and size of the pH inside the surrounding medium [40-43].

Another RS method that is getting a lot of popularity is Tip-Enhanced Raman Spectroscopy (TERS). This technique utilizes an aperture less steel tip in place of an optical fiber to obtain surface enhancement of the Raman signals (the SERS effect) [44]. In comparison with standard RS, and SERS strategies, TERS technique provides topological information of the Raman spectroscopy; and it is a very powerful approach for characterization of the microstructural and surface stoichiometric information of inorganic oxide.

Loryuenyong *et al.* [45] (a) used Raman Spectroscopy to study the synthesis of graphene oxide and graphene together with other methods. Raman Spectra results of this study (a) are shown in Figure 4 (a). The Raman spectra of stacks of graphene oxide (XGO) and reduced graphene oxide (RGO) reveal two prominent peaks at G band and D band. The two bands are broader in XGO than those in RGO. The two broader bands in XGO correspond to higher disorder in XGO. The sharpening of the G band and an increase in the peak intensity in RGO are due to the restoring sp^2 domains and the reestablished sp^2 network after the reduction treatment in hot water (b). Thus, the peak width differences were useful in displaying the difference between the XGO and RGO.

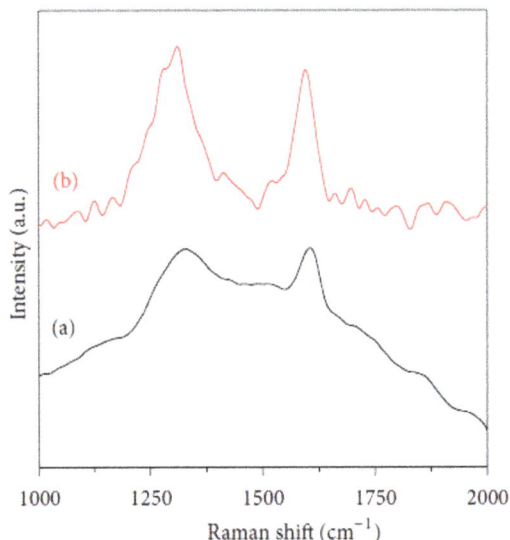

Figure 4 FT-Raman spectra of (a) XGO and (b) RGO[45].

In another example, Raman Spectroscopy was used by Cheng *et al.* [31] to study pure and N_3-TiO_2-350 samples. Spectra's are shown in Figure 5. The peaks positioned at 147, 197, 396, 514, and 636 cm^{-1} in the Raman spectrum of TiO_2 corresponds to $A_{1g} + 2B_{1g} + 3E_g$ modes and these indicated the presence of anatase phase. The other peaks, located at 235, 443 and 608 cm^{-1} are distinctive of the rutile phase [42]. As observed from this figure, the two TiO_2 samples had similar phase composition, each owning a main anatase phase composition with a small quantity of rutile, and this was consistent with XRD observations. Further to this, a great redshift (towards the low wave-variety location) and a touch small full-width at half most (FWHM) at 147 cm^{-1} became observed for the N_3-TiO_2 pattern from the insert sample of Figure 4, predominantly for the main peak around 147 cm^{-1} from 150.6 to 145.3 cm^{-1}. The shift of peak positions and the changes within the width are associated with changes in surface oxygen deficiency [42]. The shift and change in peak spectra are associated to changes in surface oxygen deficiency increased, and this might have resulted from the formation of excessive crystallinity and N doping.

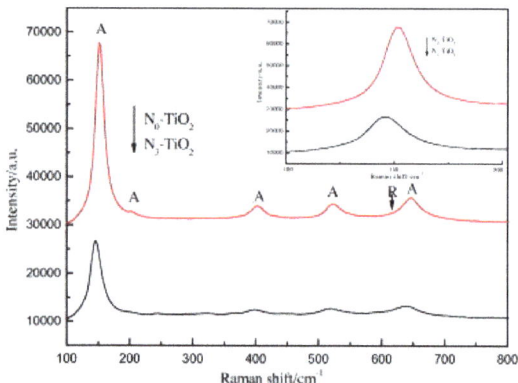

Figure 5 Raman spectra of pure and N doped TiO₂ samples. Adapted from Ref. [38] Copyright © Berkdemir(2013).

2.3   Scanning tunneling microscopy and transmission electron microscopy

2.3.1 Principle of scanning tunneling microscopy

Scanning tunneling microscopy (STM), one of the earliest developed method in the scanning probe microscopy (SPM) family, makes use of quantum tunneling current to generate electron density images for eclectically conductive material surfaces at the atomic scale [46,47]. The working principle for this technique is adapted from the general principle used in other SPM techniques. This involves monitoring the reaction between a property susceptible probe (e.g., current) and the surface of a sample that is placed in close proximity [48]. In general, STM has five essential components. These are listed as follows (i) sharp scanning tip, (ii) XYZ-piezo scanner controlling the lateral and vertical motion of the tip, (iii) coarse manage unit positioning the top close to the sample inside the tunnelling variety, (iv) vibration isolation stage, and (v) feedback regulation electronics [49]. In STM, the end-sample separation is maintained inside a range of 4–7 Å. Tunnelling of electrons occur between the scanning tip and the floor by the small voltage applied. A variation in the response current can then be recorded while the tip moves throughout the sample in the x–y plane to generate a map of charge density [50]. Alternatively, the responding current can be kept unaltered via adjusting the tip height through the use of feedback electronics; and this can produce an image of tip topography throughout the specimen [51].

Samples are often embedded into a matrix to keep their original conformations when biomolecules characterized via STM or EM techniques. The embedded samples will then be covered with a thin steel layer, such as gold, before acquiring pictures [52]. Unlike conventional EM techniques, STM can produce images with an atomic-scale resolution, for instance, the usage of a Pt–Ir tip with a completely sharp quit [52].

Although STM is used to characterize size, shape, morphology, structure, and states of dispersion and aggregation of nanoscale biomaterials due to its high spatial resolution, only few researches had reported the usage of gold or carbon as substrates [53][C]. Most biomaterials are insulating, however, the conductive surface of the pattern and detection of the surface digital shape are the basic requirement to use this technique [53]. Hence, due to the nature of the biomaterials, simple connection to the sample's surface electronic structure may not necessarily exist with its surface topography. Despite this challenge, STM is a favoured technique for investigating conductive atomic structures, for instance, carbon nanotubes, fullerenes and graphene [53]. In addition to the costly STM equipment, the manufacturing of stable probes with atomically sharp tips with well-defined physical and chemical properties is considered to be the major limitation of STM.

2.3.2 Transmission electron microscopy

Transmission electron microscopy (TEM) is one of the most frequently used techniques to characterize nanomaterials in electron microscopy (EM). This technique provides real space images and chemical information of nanomaterials at a spatial resolution all the way down to atomic dimensions (< 1 nm) [54]. The sample should interact with the incident electron beam that is transmitted through a very thin foil specimen during the process. The incident electrons that interacted with the specimen transforms into unscattered electrons, elastically scattered electrons or inelastically scattered electrons. The interactions between the sample and the elections result in an image formation that can be magnified and focused onto a device [55]. The ratio of the gap among the goal lens and the specimen and the distance between the goal lens and its image plane are the determining factors for the magnification of TEM [55]. In TEM, electromagnetic lenses are used to focus the scattered and unscattered electrons. The focused electrons may then be projected on a display to generate an electron diffraction, amplitude-contrast image, a segment-evaluation image or a shadow image of varying darkness in line with the density of unscattered electrons [55]. In addition to the above, it improves the structural and morphological analyses of nanomaterials due to its high spatial resolution. TEM can also be coupled with a wide variety of analytical techniques and be used for different applications. For example, the chemical composition and electronic structure of the nanomaterials can be quantitatively investigated by electron energy loss spectroscopy using chemical analyses and X-ray energy dispersive spectroscopy, respectively [56]. In general, STEM and TEM can provide the size and shape heterogeneity of nanomaterials. They can also reveal the degrees of aggregation and dispersion. Comparing the two techniques, TEM has more advantages in providing better spatial resolution and capability for additional analytical measurements over TEM [57].

TEM has certain drawbacks [55]. For example, in order for the TEM electron –beam penetration and analysis, a high vacuum and thin sample sections need to be prepared [57]. Like many other EM techniques, the destruction of sample and measurement in unnatural/non-physiological conditions are common in TEM. Since TEM usually examines a very small portion of a specimen over a certain period of time, it usually leads to poor statistical sampling results. Furthermore, the TEM image for single sensitivity has no depth due to 3D specimens being probed by the 2D TEM technique in transmission view that results in abundant artifacts by the technique. In general, below 50 nm specimen thicknesses is a requisite for conducting high resolution TEM images. Other limitations of TEM include (ii) long processing time, (iii) intense and high voltage electron beam usage, and (i) deforming sample's structure.

The fact that TEM can be used in aqueous environments to determine the dynamic displacement, particle size, dispersion, and aggregation/agglomeration of nanomaterials [58-60] provided it for various applications. This procedure adapts the function of Environmental Scanning Electron Microscope (ESEM) for analysing samples under partial water vapour strain within the microscope specimen chamber. Currently, wet scanning transmission electron microscopy (STEM) imaging machine, which permits transmission analysis of specimen that are completely submerged in a liquid segment has been developed. STEM also overcomes limitations associated with poor contrast and drifting of objects that could occur during ESEM imaging in liquid [61]. Due to this reason, the wet STEM permits observation in nanoscale resolution and has high contrast despite the presence of several micrometers of water, without the use of contrast agents and stains [62].

2.3.3 Examples of STM and TEM

An STM and TEM images of N_3-TiO_2-350 nanoparticles are shown in Figure 6. As it can be seen from the figure, there were many sphere-like nanoparticles with various caves and heaves. They could have resulted from the aggregation of nanocrystalline TiO_2. N_3-TiO_2 sample that was seen as a sphere with an average particle size of about 5 nm (Fig 6b), and this was in agreement with the XRD results (results not shown here), which shows slight accumulation.

Figure 6 SEM (a) and TEM (b) images of N_3-TiO_2-350. Adapted from ref.[31]. Copyright © Cheng(2016).

2.4 X-ray photoelectron spectroscopy and Fourier transform infrared spectroscopy

2.4.1 X-ray photoelectron spectroscopy

X-ray photoelectron spectroscopy (XPS) is one of the important methods that can be used for studying the surface chemical modification nano materials. This technique is particularly used to study the distribution and bonding of heteroatom dopants in carbon nanomaterials such as graphene, nanocellulose and carbon nanotubes [63]. Sometimes these materials necessarily can be used to surface modify in a controlled way for certain applications, despite their superb intrinsic properties, characterization of the modification is essential. In addition to carbon, there is an interest for the investigation of alternative dopants which are nearest to the carbon, i.e. nitrogen, boron, and phosphorus. XPS has been used by plentiful studies to characterize the concentration and bonding of various modified materials. Among the elements mentioned in the dopants list, it appears that only nitrogen has received a significant research attention, with ongoing effort on the other dopants. The crucial thing with XPS is that, experiments need to be conducted cautiously in order to avoid wrong conclusions that can easily be reached, particularly, during measurement of the binding energies into specific atomic configurations. Hence, care should be taken in the (i) sample preparation, (ii) photoemission response of the material in the intrinsic property in question, and (iii) right spectral analysis [64].

2.4.2 Fourier transform infrared (FTIR) Spectroscopy

A molecule is an IR active, if and only if the molecule in the sample specimen has a dipole moment and its oscillating frequency is analogous to the frequency of incident IR [65]. The absorption of IR radiation is able to transfer energy to the molecule, thereby inducing a corresponding stretching, bending or twisting of the covalent bond in a molecule [66]. On the contrary, molecules deprived of net dipole moments, example diatomic molecules similar to N_2 and O_2, do not absorb IR radiation [67]. As a result of the IR absorption, a spectrum is generated that is a graph of absorption or transmission versus incident IR frequency. The spectrum is the fingerprint of the structure of a molecule of study [67]. The FTIR spectroscopy is often employed for nanomaterial classifications. The spectral bands by FTIR are commonly employed to elucidate characteristics of nanomaterial–polymer conjugation. Moreover, the method can be used to study the nanoscale constituents, such as validation of purposeful molecules covalently attached on the carbon nanotubes [68], [69].

A recently developed FTIR technique, Attenuated total reflection ATR–FTIR spectroscopy, avoids some of the limitations of sample preparation complication as well as spectral irreproducibility that is observed in IR [70]. The method deploys total internal reflection in combination with the IR spectroscopy to probe the structure of adsorbed/deposited species at a solid/air or solid/liquid interface. In this system, the total internal reflectance forms an oscillating wave that extend from the internal reflection element (IRE) crystal–sample interface into the sample with a penetration depth of micrometers (0.5–5 μm) [71]. The changes and modifications in surface properties and identification of chemical properties on the polymer surface can be investigated by ATR–FTIR [70, 72].The method can also be used to study the surface structures of nanomaterials, despite the lack of sensitivity. The penetration depth of ATR–FTIR has the same order of magnitude as the incident IR wavelength and hence, it is not effective to analyze samples at nanometre scale [72].

2.4.3 Examples of measurements of XPS and FTIR

Kharlamova *et al.* [73] used XPS (Figure 7) and other alternative techniques to characterize highly pure semiconducting and metallic single-walled carbon nanotubes (SWCNTs) that have a mean diameter of 1.4 nm. The XPS spectra indicated that the C 1s peak of metallic nanotubes is downshifted from a binding energy of 284.51 to 284.57 eV to make 0.06 eV difference when compared to the semiconducting tube peaks due to different chemical potentials in bulk metallic SWCNTs, as shown in Figure 7(a). On the other hand, The C 1s peak of metallic nanotubes and semiconducting tubes had an asymmetric Doniach–Sunjic profile and Voigtian shape, respectively due to their physical property differences as shown in Figure 7 (b). However, an intense C 1s peak observed in the spectra did not detect oxygen peaks. The overall XPS spectra, Figure 7 (a) confirms the purity of the metallicity sorted samples.

Cheng *et al* [31] reported the XPS spectra comparison of pure TiO_2 and N_3-TiO_2 samples, and results are presented in Figure 8a. XPS was used to examine the chemical composition and state of N_3-TiO_2. The XPS spectra results indicated that the pure TiO_2 sample, contained elemental composition of C, O, and Ti at a concentration of 21.05, 56.35, and 22.60 atomic %, respectively. On the other hand, XPS showed that the N_3-TiO_2 sample does not only contained C, O and Ti elements but also a small concentration of atomic N. The small quantity of N was likely generated from the dopants during the calcination process. The compositions of C, O, Ti and N elements in the N_3-TiO_2 sample were 20.98, 55.47, 22.59 and 0.96 atomic %, respectively.

Two N1s XPS peaks at 401.2 and 399.6 eVs (Figure 8b) were used to investigate the chemical state of N atom in the sample. The former peak, at a binding energy of 401.2 eV was ascribed to the presence of oxidized nitrogen such as Ti–O–N [74], and the latter peak was due to anionic N in the form of N–Ti–O bond [75]. However, there was no typical indication of Ti–N bond formation. The typical binding energy in Ti–N is 396 eV [76]. In comparison with this result, the latter peak was 3.6 eV higher, which was attributed to the lower electron density of TiO_2 when O atom was substituted by N atom. Thus, it is clear that N–Ti–O and Ti–O–N coincide in N_3-TiO_2 nanoparticles.

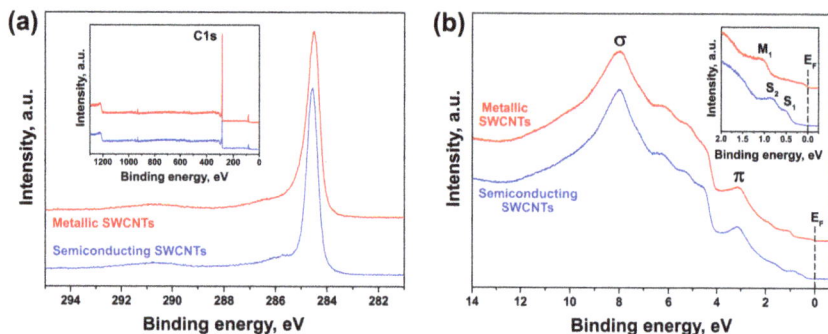

Figure 7 (a) The C 1s XPS spectra of the semiconducting and metallic SWCNTs. The inset shows the overview XPS spectra. The C 1s peak is labeled. (b) The valence band spectra of the separated SWCNTs. π- and σ-resonances are denoted. The inset presents the enlarged.Region of the spectra containing the peaks of van Hove singularities. The labels S_1, S_2, and M_1 mark the first and the second vHs of semiconducting tubes and the first vHs of metallic SWCNTs. E_F is the Fermi level [73].

Carbon atom has a wonderful property in both crystalline and disordered structure forms due to its four valence electrons ($[He]2s^22p^2$) which may hybridize in various ways. The hybridizations forms may be sp^3, sp^2 and sp, allowing carbon to form linear chains, planar sheets and tetrahedral structures. FTIR vibrational spectroscopy is widely used to identify and characterize the chemical structures of the carbon family, such as diamond, amorphous carbon, graphite, graphene, carbon nanotubes (CNTs), fullerene and carbon quantum dots (CQDs). FTIR can be used to confirm the hexagonal structure of the pristine CNT peaks originating from the sp^2-hybridized carbon and the disordered regions of CNTs probably consisting of disordered sp^3-bonded carbon, also serving as nucleation sites for hydrogen, which are assigned to the C=C bond from the CNT skeletal vibration mode. Overall, IR method is advantageous for nanomaterial characterization in that it is non-destructive, real-time measurement and relatively easy to handle.

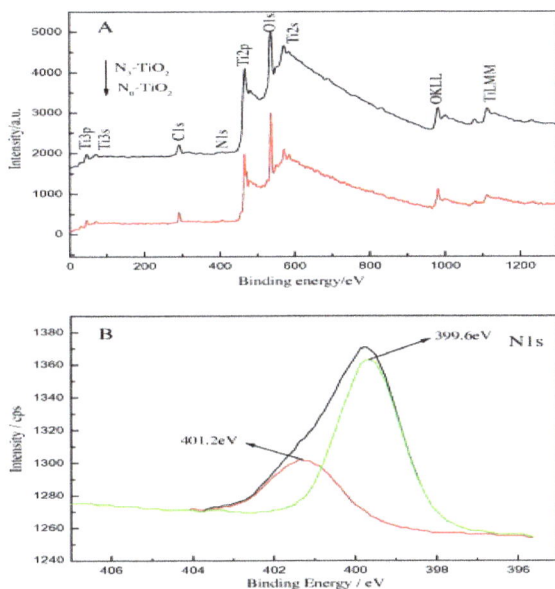

Figure 8 Overall XPS of the as-prepared TiO_2 (A) and N1s spectra for N_3-TiO_2 photocatalyst (B). Adapted from [31]. Copyright © Cheng(2016).

Cheng *et al.* investigated as-synthesized pure and N_3-TiO_2-350 nanoparticles using FTIR, and results are presented in Figure 9. Both samples displayed identical vibrations in the IR region. The intense and broad spectras in between wavenumber 400 and 800 cm^{-1} were attributed to the strong stretching vibrations of Ti–O and Ti–O–Ti bonds [77]. The broad peak at 3400 and a spike at 1620 cm^{-1} were due to O–H stretching on the surface and O–H bending of water molecules, respectively. The intensity of the two absorption bands in the as-synthesized N_3-TiO_2-350 was stronger as compared with that of pure TiO_2 [78]; indicating that the N_3-TiO_2 sample had more OH groups and surface-adsorbed water that contributed greatly in the photocatalytic reaction.

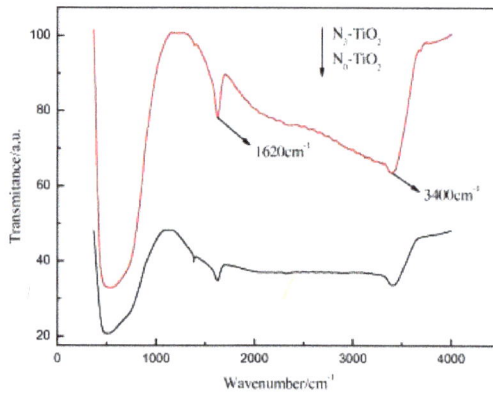

Figure 9 FT-IR spectra for as-prepared pure and N_3-TiO_2 photocatalyst. Adapted from ref [31].

The hydroxyl moities can capture the photo-induced holes (h^+) up on irradiation with light; and then later form hydroxyl radicals ($\cdot OH$) with high oxidation capability. Contrarily, the surface -OH groups can function as absorption centres for oxygen molecules; and form -OH radicals to enhance the photocatalytic activity. In summary, IR spectroscopy is extensively used as a rapid and non-destructive method to characterize materials. It can also be carried out on samples in the solid or liquid state. However, it does not provide information for samples with no dipole moment as these samples will not absorb Infrared rays.

2.5   UV-Visible spectroscopy

Nanoparticles have interesting and novel optical properties, and these properties are dependent on the size, shape, composition, concentration, aggregation and agglomeration state, and refractive index of the nanoparticle. UV-Visible spectroscopy (UV-Vis) that measures the scattering and absorption extinction of light passing through a specimen, is a vital technique for the analysis of nanomaterials and their derivatives [79]. For example, Agilent 8453 single beam diode array spectrometer is one of the modern UV-Vis spectrophotometers, which collects spectra from 200-1100 nm using a slit width of 1 nm [80]. In this, deuterium and tungsten lamps are used to provide illumination across the ultraviolet, visible, and near-infrared electromagnetic spectrum. The instrument is very sensitive, and spectra can be collected from samples as small as 60 µL using a microcell with a path length of 1 cm.

Carbonaceous Composite Materials   Materials Research Forum LLC
Materials Research Foundations **42** (2018) 273-308   doi: http://dx.doi.org/10.21741/9781945291975-10

Figure 10 shows the UV–vis diffuse reflectance spectra of the as-synthesized pure and N_3-TiO_2-350 specimens. The spectrum of commercial TiO_2 was also presented for comparison. The results indicated that the UV–vis adsorption edge of pure TiO_2 was identical to that of the commercial TiO_2, while N_3-TiO_2 shifted to the visible-light region. It was also seen that the optical band edge presented a notable red-shift in comparison with pure TiO_2. In addition to this, two characteristic light absorption edges were shown for the N_3-TiO_2 sample [81]; one direction band and the other originated from the new energy levels in the forbidden band of TiO_2, which was made by N-doping.

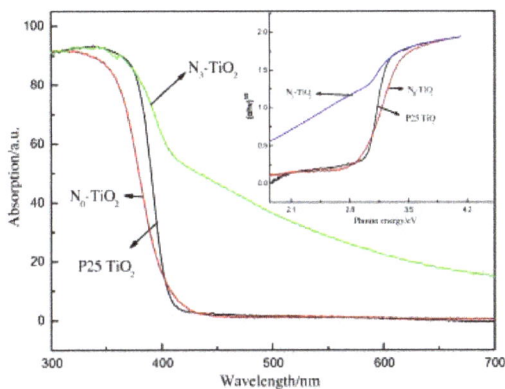

Figure 10 UV–vis reflectance spectra (diffuse) of the as-prepared pure and N_3-TiO_2 samples. Adapted from ref. [31]

The band gap energies of the as-synthesized TiO_2 samples were calculated using the Kubelka–Munk function [82]. $[F(R)h\upsilon]^{1/2}$ versus energy of light was plotted as shown in 10. Bandgap energies of 2.85 and 3.1 eV were obtained for N_3-TiO_2 and TiO_2, respectively, revealing that the band gap of TiO_2 was narrowed by N doping. The band gap narrowing may be attributed to the introduction of nitrogen from ammonium chloride into the lattice of TiO_2. Hence, it can be deduced that the sample of N_3-TiO_2 could display high photocatalytic activity through visible-light irradiation. Overall, UV-vis is effective for characterization of samples that absorb UV or visible light. On the contrary, it is not effective for samples that are insoluble, or samples that shows photo- mediated reactions.

2.6   Other techniques

In addition to the techniques described above, there are other spectroscopic techniques that are fruitful for the study of physiochemical attributes of nanomaterials. For instance, when the absorption profiles of nano-materials are distinct, UV- visible absorbance spectroscopy is useful to investigate the characteristics of nanomaterials (e.g., particle size, concentration of aggregate state as well as bio conjugations) [83]. Fluorescence spectroscopy (FS) is another useful technique. This is a beneficial technique for tracking the ligand binding or conformational variation of macromolecular nanomaterials than light absorption methods. Fluorescence spectroscopy presents detailed information of samples that fluoresce (e.g., amino acids, proteins, organic compounds). FS has high sensitivity, and useful for larger polymer compounds or molecules solution. It is extensively used for amino acid detection and quantification, protein conformation and interactions, enzymatic activities etc. FS finds application in nanomaterials conjugated with biomolecules. The conjugation of an extrinsic fluorophore to the non-intrinsically fluorescent nanomaterials allows this technique to analyze the certain properties, such as concentration, particle size, and spacer composition of the biomolecule on the nanomaterial [84, 85].

There are numerous thermal methods that can assess the thermal stability and the quantity of nanomaterials or their conjugates [26]. For example, thermogravimetric analysis (TGA) can be applied to monitor the temperature dependant weight change of a samples [86]. Also, differential scanning calorimetry (DSC) can be applied to measure thermal transitions, such as melting, crystallization, glass transition temperature of nanostructured materials or their bio conjugates. DSC analysis can also provide other useful structural and stability information of materials [87]. Isothermal titration calorimetry (ITC) is another technique that can be applied to examine the stoichiometry, affinity, as well as enthalpy derivative in nanomaterial and biomolecule conjugates [88]. Thermophoresis can examine the motion of samples to assess the size and apparent potential by locally warming the sample to produce temperature gradient [89]. Nevertheless, the method needs elevated concentrations of the samples than fluorescent spectroscopy to provide strong signals.

A number of other separation methods have been applied routinely as characterization techniques. Amongst these methods, centrifugation is a useful separation technique for mixed colloidal materials in liquid. Analytical ultracentrifugation (AUC), a class of centrifugation methods, can also be used to examine the conformation, structure, and stoichiometry as well as self-aggregation nature of nanomaterials. It can further be used to determine the size and size distribution, shape and the molecular mass of nanomaterials [90].

Chromatography methods, such as hydrodynamic chromatography (HDC) and also high-performance liquid chromatography (HPLC), can be applied for the purification, and identification of nanomaterial bio conjugates. Chromatography techniques can also show the circulation of nanomaterial to biomolecule ratios and the stability as well as the purity of post-products owing to their ability to differentiate the different nanomaterial bio conjugates.

3. Future trend

Nanomaterials can be composed of metallic, ceramic, and organic particles or molecules [91]. In addition to this, they may be conjugated or grafted with other molecules, coupling agents, surfactants etc. to boost their functionality, biocompatibility and tune their application [91]. All nanomaterials have a large surface area to volume ratio that is usually several orders of magnitude larger than their counterpart macroscopic materials. Most nanomaterials are generated using a 'bottoms up' synthesis process from a molecular precursor in order to develop distinct structural and functional characteristics. During the synthesis process, the reactant composition, concentrations, reaction conditions (temperature, pressure, and reaction time), additives (coupling agents, dispersants, surfactants), and solvent medium are very critical for their final features including particle size, morphology, structure, functional properties etc. of common nanomaterials. Prior to developing applications for nanomaterials, developing reproducible characterization methods of both the synthesis process and the final product is crucial.

Nanomaterials are generally categorized based on their types or application areas. As an illustration, based on the applications to nanomedicines, they have various divisions such as drug delivery, drugs, and therapies, *in vivo* imaging, *in vitro* diagnostics, biomaterials, and implants [92]. Regardless of their category, nanomaterials have common physicochemical features related to, for example their small particle size, and high surface area and shares similar characterization techniques in these cases. Application could be communality. For instance, most drug delivery system nanomaterials, including carbon nanotubes, dendrimers, fullerenes, liposomes, Nano colloids, nanoparticles, nanosuspensions, and nanostructured biopolymers are utilized to furnish optimal bioavailability of functional compounds at specific physiological destinations over a period of time. Moreover, they are targeted to cause minimal drug toxicity, enhance drug-therapeutic effect, complement or replace invasive administration routes [92]. The drug

Carbonaceous Composite Materials   Materials Research Forum LLC
Materials Research Foundations **42** (2018) 273-308   doi: http://dx.doi.org/10.21741/9781945291975-10

carriers in nano-drug delivery systems can be devised by regulating the composition, size, shape, and morphology [92]. In such delivery systems, the size of the nanomaterial can affect the bioavailability and circulation time in the bloodstream, partly resulting from the impact of surface area-to-volume ratios on the solubility of the drug delivery systems [93]. The optimal size of nano-drug delivery systems as suggested by many studies is 10–100 nm [94]. Recent studies reported that the shape of the drug carrier nanomaterial is also crucial in the distribution, absorption, avoiding phagocytosis and extended circulation of bioactive in the bloodstream [95]. Moreover, the pharmacokinetics of drugs can be affected by the surface charge of a nano-drug delivery system [96], while drug delivery efficiency can be influenced by the structural characteristics of the delivery systems [97].

Assessments related to the size and size distribution of nanomaterials for drug delivery systems or other applications such as polymer reinforcement applications can be operated by using Atomic Force Microscopy (AFM), Dynamic Light Scattering (DLS), FCS, FS, Magnetic Resonance (NMR), Near-Field Scanning Optical Microscopy (NSOM), Nuclear XRD, RS, SEM, STM, Small Angle X-ray Scattering (SAXS), TEM, and several other separation techniques. The most suitable methods for shape measurement include NSOM, SEM, TEM, STM, AFM, XRD, and SAXS, while the appropriate methods for surface charge measurement involve zeta potential measurement (ELS), ATR–FTIR, and Cation Exchange (CE). The structural properties of the nanomaterials are usually studied using techniques such as AFM, CD, FS, IR, Mass spectroscopy (MS), NMR, SAXS, STM, Thermal, Tip-Enhanced Raman Spectroscopy (TERS), XRD, and other separation techniques. The chemical and structural properties of GO, RGO and their derivatives are characterized by various spectroscopic and non-spectroscopic methods, such as AFM, Elementary Analysis (EA), FTIR, Raman spectroscopy, NMR, SEM, STM, TEM, Thermogravimetric Analysis (TGA), plasmon spectroscopy, quasi-elastic neutron scattering, X-ray photoelectron spectroscopy (XPS), and XRD [F].

Besides, the physicochemical properties descripted above (size, shape, surface charge), the thermal stability of nanomaterials play a vital role in the functionality and efficacy of nanomaterial based drug delivery systems. Thermogravimetric analysis (TGA) is extensively used for the analysis of thermal stability. Furthermore, other characterization techniques such as DSC, ITC, and thermophoresis are receiving an increasing attention.

Molecular diagnostics, detection of specific sequence of DNA and RNA that may not be directly associated with a disease, targets diagnosing a disease at a molecular level prior to the manifestation of symptoms [92]. Compared to traditional molecular imaging agents, utilization of nanomaterial-derived contrast agents could enhance the signal intensity of a single particle [98]. Such strong signal produced by the nanomaterial-derived contrasting agents, could assist in overcoming the critical limitations of low sensitivity in Magnetic Resonance Imaging (MRI) and the depth penetration disadvantages of optical imaging to some extent [98]. Few nanomaterials have specific novel properties that allow them to perform as imaging probes. The list includes quantum dots with distinct electronic and optical properties, up conversion phosphors consisting of phosphor nanocrystals doped with rare earth metals, super-paramagnetic iron oxide nanoparticles containing an iron oxide core of magnetite and/or magnetite enclosed in a polysaccharide, synthetic polymer or monomer coatings etc. [99]. Further to the features of conventional imaging probes, including structure, purity and solubility, some physicochemical characteristics of nanomaterial-derived imaging contrasting agents were studied.

Conclusions

Due to the unique and novel properties of materials at nanoscale, nanomaterials have great potential to provide very useful functional properties in an array of application platforms. The physicochemical properties of nanomaterials such as the structure, morphology, chemical composition, crystallinity, shape, size, aggregation and agglomeration state, surface properties, thermal stability, etc. determines the applicability and unique functional benefit of the nanomaterial. Applications of nanomaterials range from biology to medicine, electronics, optical devices, polymeric materials, or other industrial processes. Although there are several techniques used to investigate the structure of nanomaterials, spectroscopic techniques such as electron microscopy, Raman spectroscopy, and X-ray scattering have proved among the most widely utilized techniques. These techniques cover the analysis of a wide range of length (from sub-nm to μm) scales. These techniques are beneficial in analyzing the dispersion, morphological state, and alignment of nanomaterials in a polymer matrix, and to study the interfacial adhesion between a nanomaterial and polymer matrices. The use of these spectroscopic techniques for the characterization of nanomaterials in medicine (e.g. drug delivery, toxicology assessment, imaging etc.) and other biology applications cannot be overstated. This book chapter provided a description of the most important spectroscopic properties of nanomaterials, and introduces various spectroscopic techniques commonly applied for characterizing nanomaterials. In fact, characterizing nanomaterials intended for different use in both its originally manufactured condition and after introduction into an application is unquestionable. The brief description of each technique, as well as its strengths and drawbacks gives a picture for identifying suitable techniques for spectroscopic characterization of potential nanomaterial and/or nanocomposite materials.

References

[1]   J.N. Coleman, U. Khan, W.J. Blau, Y.K. Gun'ko, Small but strong: a review of the mechanical properties of carbon nanotube–polymer composites, Carbon 44 (2006) 1624-1652. https://doi.org/10.1016/j.carbon.2006.02.038

[2]   L. Hollaway, The evolution of and the way forward for advanced polymer composites in the civil infrastructure, Construction and Building Materials 17 (2003) 365-378. https://doi.org/10.1016/S0950-0618(03)00038-2

[3]   S. Chatterjee, F. Nüesch, B.T. Chu, Comparing carbon nanotubes and graphene nanoplatelets as reinforcements in polyamide12 composites, Nanotechnology 22 (2011) 275714. https://doi.org/10.1088/0957-4484/22/27/275714

[4]   D. Paul, L.M. Robeson, Polymer nanotechnology: nanocomposites, Polymer 49 (2008) 3187-3204. https://doi.org/10.1016/j.polymer.2008.04.017

[5]   S. Chatterjee, F.A. Reifler, B. Chu, R. Hufenus, Investigation of crystalline and tensile properties of carbon nanotube-filled polyamide-12 fibers melt-spun by industry-related processes, Journal of Engineered Fibers and Fabrics 7 (2012).

[6]   S. Chatterjee, J. Wang, W. Kuo, N. Tai, C. Salzmann, W. Li, R. Hollertz, F. Nüesch, B. Chu, Mechanical reinforcement and thermal conductivity in expanded graphene nanoplatelets reinforced epoxy composites, Chemical Physics Letters 531 (2012) 6-10. https://doi.org/10.1016/j.cplett.2012.02.006

[7]   H. Kroto, JR Health, SC O'Brien, RF Curl, and RE Smalley, Nature 318 (1985) I985. https://doi.org/10.1038/318162a0

[8]   L. Radushkevich, V. Lukyanovich, About the structure of carbon formed by thermal decomposition of carbon monoxide on iron substrate, J. Phys. Chem.(Moscow) 26 (1952) 88-95.

[9]   S. Iijima, Helical microtubules of graphitic carbon, nature 354 (1991) 56.

[10]   A.K. Geim, P. Kim, Carbon wonderland, Scientific American 298 (2008) 90-97. https://doi.org/10.1038/scientificamerican0408-90

[11]   P.R. Wallace, The band theory of graphite, Physical Review 71 (1947) 622. https://doi.org/10.1103/PhysRev.71.622

[12]   K.S. Novoselov, A.K. Geim, S.V. Morozov, D. Jiang, Y. Zhang, S.V. Dubonos, I.V. Grigorieva, A.A. Firsov, Electric field effect in atomically thin carbon films, Science 306 (2004) 666-669. https://doi.org/10.1126/science.1102896

[13] A.K. Geim, K.S. Novoselov, The rise of graphene, Nature materials 6 (2007) 183-191. https://doi.org/10.1038/nmat1849

[14] T. Mekonnen, M. Misra, A. Mohanty, Processing, performance, and applications of plant and animal protein-based blends and their biocomposites, Biocomposites, Elsevier2015, pp. 201-235. https://doi.org/10.1016/B978-1-78242-373-7.00017-2

[15] S. Ahmed, F. Jones, A review of particulate reinforcement theories for polymer composites, Journal of Materials Science 25 (1990) 4933-4942. https://doi.org/10.1007/BF00580110

[16] N. Taranu, G. Oprisan, I. Entuc, M. Budescu, V. Munteanu, G. Taranu, COMPOSITE AND HYBRID SOLUTIONS FOR SUSTAINABLE DEVELOPMENT IN CIVIL ENGINEERING, Environmental Engineering & Management Journal (EEMJ) 11 (2012). https://doi.org/10.30638/eemj.2012.101

[17] C. Zhao, G. Hu, R. Justice, D.W. Schaefer, S. Zhang, M. Yang, C.C. Han, Synthesis and characterization of multi-walled carbon nanotubes reinforced polyamide 6 via in situ polymerization, Polymer 46 (2005) 5125-5132. https://doi.org/10.1016/j.polymer.2005.04.065

[18] C.A. Folsom, F.W. Zok, F.F. Lange, D.B. Marshall, Mechanical Behavior of a Laminar Ceramic/Fiber-Reinforced Epoxy Composite, Journal of the American Ceramic Society 75 (1992) 2969-2975. https://doi.org/10.1111/j.1151-2916.1992.tb04373.x

[19] B. Johnsen, A. Kinloch, R. Mohammed, A. Taylor, S. Sprenger, Toughening mechanisms of nanoparticle-modified epoxy polymers, Polymer 48 (2007) 530-541. https://doi.org/10.1016/j.polymer.2006.11.038

[20] P.M. Ajayan, L.S. Schadler, P.V. Braun, Nanocomposite science and technology, John Wiley & Sons2006.

[21] Q. Chen, G. Chai, B. Li, Exploration study of multifunctional metallic nanocomposite utilizing single-walled carbon nanotubes for micro/nano devices, Proceedings of the Institution of Mechanical Engineers, Part N: Journal of Nanomaterials, Nanoengineering and Nanosystems 219 (2005) 67-72.

[22] J.E. Morris, Nanopackaging: nanotechnologies and electronics packaging, High Density Microsystem Design and Packaging and Component Failure Analysis, 2006. HDP'06. Conference on, IEEE, 2006, pp. 199-205.

[23] N. Hu, Y. Karube, C. Yan, Z. Masuda, H. Fukunaga, Tunneling effect in a polymer/carbon nanotube nanocomposite strain sensor, Acta Materialia 56 (2008) 2929-2936. https://doi.org/10.1016/j.actamat.2008.02.030

[24] I. Kang, Y.Y. Heung, J.H. Kim, J.W. Lee, R. Gollapudi, S. Subramaniam, S. Narasimhadevara, D. Hurd, G.R. Kirikera, V. Shanov, Introduction to carbon nanotube and nanofiber smart materials, Composites Part B: Engineering 37 (2006) 382-394. https://doi.org/10.1016/j.compositesb.2006.02.011

[25] P.-C. Lin, S. Lin, P.C. Wang, R. Sridhar, Techniques for physicochemical characterization of nanomaterials, Biotechnology advances 32 (2014) 711-726. https://doi.org/10.1016/j.biotechadv.2013.11.006

[26] K.E. Sapsford, K.M. Tyner, B.J. Dair, J.R. Deschamps, I.L. Medintz, Analyzing nanomaterial bioconjugates: a review of current and emerging purification and characterization techniques, Anal. Chem 83 (2011) 4453-4488. https://doi.org/10.1021/ac200853a

[27] T.L. Jennings, S.G. Becker-Catania, R.C. Triulzi, G. Tao, B. Scott, K.E. Sapsford, S. Spindel, E. Oh, V. Jain, J.B. Delehanty, Reactive semiconductor nanocrystals for chemoselective biolabeling and multiplexed analysis, ACS nano 5 (2011) 5579-5593. https://doi.org/10.1021/nn201050g

[28] E. Oh, J.B. Delehanty, K.E. Sapsford, K. Susumu, R. Goswami, J.B. Blanco-Canosa, P.E. Dawson, J. Granek, M. Shoff, Q. Zhang, Cellular uptake and fate of PEGylated gold nanoparticles is dependent on both cell-penetration peptides and particle size, Acs Nano 5 (2011) 6434-6448. https://doi.org/10.1021/nn201624c

[29] X. Liu, X. Wu, H. Cao, R. Chang, Growth mechanism and properties of ZnO nanorods synthesized by plasma-enhanced chemical vapor deposition, Journal of Applied Physics 95 (2004) 3141-3147. https://doi.org/10.1063/1.1646440

[30] M.M. Seibert, T. Ekeberg, F.R. Maia, M. Svenda, J. Andreasson, O. Jönsson, D. Odić, B. Iwan, A. Rocker, D. Westphal, Single mimivirus particles intercepted and imaged with an X-ray laser, Nature 470 (2011) 78-81. https://doi.org/10.1038/nature09748

[31] X. Cheng, X. Yu, Z. Xing, L. Yang, Synthesis and characterization of N-doped TiO2 and its enhanced visible-light photocatalytic activity, Arabian Journal of Chemistry 9 (2016) S1706-S1711. https://doi.org/10.1016/j.arabjc.2012.04.052

[32] Z. Popović, Z. Dohčević-Mitrović, M. Šćepanović, M. Grujić-Brojčin, S. Aškrabić, Raman scattering on nanomaterials and nanostructures, Annalen der Physik 523 (2011) 62-74. https://doi.org/10.1002/andp.201000094

[33] E. Le Ru, P. Etchegoin, Principles of Surface-Enhanced Raman Spectroscopy: and related plasmonic effects, Elsevier2008.

[34] X. Li, C.W. Magnuson, A. Venugopal, R.M. Tromp, J.B. Hannon, E.M. Vogel, L. Colombo, R.S. Ruoff, Large-area graphene single crystals grown by low-pressure chemical vapor deposition of methane on copper, Journal of the American Chemical Society 133 (2011) 2816-2819. https://doi.org/10.1021/ja109793s

[35] C.S. Kumar, Raman spectroscopy for nanomaterials characterization, Springer Science & Business Media2012. https://doi.org/10.1007/978-3-642-20620-7

[36] X. Gong, Y. Bao, C. Qiu, C. Jiang, Individual nanostructured materials: fabrication and surface-enhanced Raman scattering, Chemical Communications 48 (2012) 7003-7018. https://doi.org/10.1039/c2cc31603j

[37] Q. Tu, C. Chang, Diagnostic applications of Raman spectroscopy, Nanomedicine: Nanotechnology, Biology and Medicine 8 (2012) 545-558. https://doi.org/10.1016/j.nano.2011.09.013

[38] A. Berkdemir, H.R. Gutiérrez, A.R. Botello-Méndez, N. Perea-López, A.L. Elías, C.-I. Chia, B. Wang, V.H. Crespi, F. López-Urías, J.-C. Charlier, Identification of individual and few layers of WS2 using Raman Spectroscopy, Scientific reports 3 (2013). https://doi.org/10.1038/srep01755

[39] A.J. Wilson, K.A. Willets, Surface-enhanced Raman scattering imaging using noble metal nanoparticles, Wiley Interdisciplinary Reviews: Nanomedicine and Nanobiotechnology 5 (2013) 180-189. https://doi.org/10.1002/wnan.1208

[40] A.F. Palonpon, J. Ando, H. Yamakoshi, K. Dodo, M. Sodeoka, S. Kawata, K. Fujita, Raman and SERS microscopy for molecular imaging of live cells, Nature protocols 8 (2013) 677-692. https://doi.org/10.1038/nprot.2013.030

[41] A.C. Ferrari, D.M. Basko, Raman spectroscopy as a versatile tool for studying the properties of graphene, Nature nanotechnology 8 (2013) 235-246. https://doi.org/10.1038/nnano.2013.46

[42] J.F. Li, Y.F. Huang, Y. Ding, Z.L. Yang, S.B. Li, X.S. Zhou, F.R. Fan, W. Zhang, Z.Y. Zhou, B. Ren, Shell-isolated nanoparticle-enhanced Raman spectroscopy, nature 464 (2010) 392-395. https://doi.org/10.1038/nature08907

[43] J. Kumar, K.G. Thomas, Surface-enhanced Raman spectroscopy: investigations at the nanorod edges and dimer junctions, The Journal of Physical Chemistry Letters 2 (2011) 610-615. https://doi.org/10.1021/jz2000613

[44] Y. Wang, J. Irudayaraj, Surface-enhanced Raman spectroscopy at single-molecule scale and its implications in biology, Phil. Trans. R. Soc. B 368 (2013) 20120026. https://doi.org/10.1098/rstb.2012.0026

[45] V. Loryuenyong, K. Totepvimarn, P. Eimburanapravat, W. Boonchompoo, A. Buasri, Preparation and Characterization of Reduced Graphene Oxide Sheets via Water-Based Exfoliation and Reduction Methods, Advances in Materials Science and Engineering 2013 (2013) 1-5. https://doi.org/10.1155/2013/923403

[46] M.J. Miles, H.J. Carr, T.C. McMaster, K.J. I'Anson, P.S. Belton, V.J. Morris, J.M. Field, P.R. Shewry, A.S. Tatham, Scanning tunneling microscopy of a wheat seed storage protein reveals details of an unusual supersecondary structure, Proceedings of the national academy of sciences 88 (1991) 68-71. https://doi.org/10.1073/pnas.88.1.68

[47] M. Hersam, N. Guisinger, J. Lyding, Silicon-based molecular nanotechnology, Nanotechnology 11 (2000) 70. https://doi.org/10.1088/0957-4484/11/2/306

[48] S.S. Elnashaie, F. Danafar, H.H. Rafsanjani, From Nanotechnology to Nanoengineering, Nanotechnology for Chemical Engineers, Springer2015, pp. 79-178. https://doi.org/10.1007/978-981-287-496-2_2

[49] H.-J. Gèuntherodt, R. Wiesendanger, Scanning tunneling microscopy I: general principles and applications to clean and adsorbate-covered surfaces, Springer-Verlag1994. https://doi.org/10.1007/978-3-642-79255-7

[50] S.V. Kalinin, D.A. Bonnell, Temperature dependence of polarization and charge dynamics on the BaTiO 3 (100) surface by scanning probe microscopy, Applied Physics Letters 78 (2001) 1116-1118. https://doi.org/10.1063/1.1348303

[51] J.C. Koepke, J.D. Wood, D. Estrada, Z.-Y. Ong, K.T. He, E. Pop, J.W. Lyding, Atomic-scale evidence for potential barriers and strong carrier scattering at graphene grain boundaries: a scanning tunneling microscopy study, ACS nano 7 (2013) 75-86. https://doi.org/10.1021/nn302064p

[52] C. Koçum, E.K. Çimen, E. Pişkin, Imaging of poly (NIPA-co-MAH)–HIgG conjugate with scanning tunneling microscopy, Journal of Biomaterials Science, Polymer Edition 15 (2004) 1513-1520. https://doi.org/10.1163/1568562042459706

[53] J.W. Wang, Y. He, F. Fan, X.H. Liu, S. Xia, Y. Liu, C.T. Harris, H. Li, J.Y. Huang, S.X. Mao, Two-phase electrochemical lithiation in amorphous silicon, Nano Letters 13 (2013) 709-715. https://doi.org/10.1021/nl304379k

[54] M. O'keefe, C. Hetherington, Y. Wang, E. Nelson, J. Turner, C. Kisielowski, J.-O. Malm, R. Mueller, J. Ringnalda, M. Pan, Sub-Ångstrom high-resolution transmission electron microscopy at 300keV, Ultramicroscopy 89 (2001) 215-241. https://doi.org/10.1016/S0304-3991(01)00094-8

[55] D.B. Williams, C.B. Carter, High Energy-Loss Spectra and Images, Transmission Electron Microscopy (2009) 715-739. https://doi.org/10.1007/978-0-387-76501-3_39

[56] M. Hasellöv, J.W. Readman, J.F. Ranville, K. Tiede, Nanoparticle analysis and characterization methodologies in environmental risk assessment of engineered nanoparticles, Ecotoxicology 17 (2008) 344-361. https://doi.org/10.1007/s10646-008-0225-x

[57] G.R. Chalmers, R.M. Bustin, I.M. Power, Characterization of gas shale pore systems by porosimetry, pycnometry, surface area, and field emission scanning electron microscopy/transmission electron microscopy image analyses: Examples from the Barnett, Woodford, Haynesville, Marcellus, and Doig units, AAPG bulletin 96 (2012) 1099-1119. https://doi.org/10.1306/10171111052

[58] C.E. Carlton, S. Chen, P.J. Ferreira, L.F. Allard, Y. Shao-Horn, Sub-nanometer-resolution elemental mapping of "Pt3Co" nanoparticle catalyst degradation in proton-exchange membrane fuel cells, The Journal of Physical Chemistry Letters 3 (2012) 161-166. https://doi.org/10.1021/jz2016022

[59] X. Chen, K. Noh, J. Wen, S. Dillon, In situ electrochemical wet cell transmission electron microscopy characterization of solid–liquid interactions between Ni and aqueous NiCl 2, Acta Materialia 60 (2012) 192-198. https://doi.org/10.1016/j.actamat.2011.09.047

[60] Y. Tachibana, L. Vayssieres, J.R. Durrant, Artificial photosynthesis for solar water-splitting, Nature Photonics 6 (2012) 511-518. https://doi.org/10.1038/nphoton.2012.175

[61] A. Ponce, S. Mejía-Rosales, M. José-Yacamán, Scanning transmission electron microscopy methods for the analysis of nanoparticles, Nanoparticles in Biology and Medicine: Methods and Protocols (2012) 453-471.

[62]　N. De Jonge, F.M. Ross, Electron microscopy of specimens in liquid, Nature nanotechnology 6 (2011) 695-704. https://doi.org/10.1038/nnano.2011.161

[63]　D. Yang, A. Velamakanni, G. Bozoklu, S. Park, M. Stoller, R.D. Piner, S. Stankovich, I. Jung, D.A. Field, C.A. Ventrice, Chemical analysis of graphene oxide films after heat and chemical treatments by X-ray photoelectron and Micro-Raman spectroscopy, Carbon 47 (2009) 145-152. https://doi.org/10.1016/j.carbon.2008.09.045

[64]　P. Xiao, M.A. Sk, L. Thia, X. Ge, R.J. Lim, J.-Y. Wang, K.H. Lim, X. Wang, Molybdenum phosphide as an efficient electrocatalyst for the hydrogen evolution reaction, Energy & Environmental Science 7 (2014) 2624-2629. https://doi.org/10.1039/C4EE00957F

[65]　K. Edmonds, G. van der Laan, N. Farley, R. Campion, B. Gallagher, C. Foxon, B. Cowie, S. Warren, T. Johal, Magnetic linear dichroism in the angular dependence of core-level photoemission from (Ga, Mn) As using hard X rays, Physical review letters 107 (2011) 197601. https://doi.org/10.1103/PhysRevLett.107.197601

[66]　B. Das, K.E. Prasad, U. Ramamurty, C. Rao, Nano-indentation studies on polymer matrix composites reinforced by few-layer graphene, Nanotechnology 20 (2009) 125705. https://doi.org/10.1088/0957-4484/20/12/125705

[67]　D.M. Marquis, E. Guillaume, C. Chivas-Joly, Properties of nanofillers in polymer, Nanocomposites and Polymers with Analytical Methods, Intech2011.

[68]　E. Perevedentseva, C.-Y. Cheng, P.-H. Chung, J.-S. Tu, Y.-H. Hsieh, C.-L. Cheng, The interaction of the protein lysozyme with bacteria E. coli observed using nanodiamond labelling, Nanotechnology 18 (2007) 315102. https://doi.org/10.1088/0957-4484/18/31/315102

[69]　R. Othman, A.W. Mohammad, M. Ismail, J. Salimon, Application of polymeric solvent resistant nanofiltration membranes for biodiesel production, Journal of Membrane Science 348 (2010) 287-297. https://doi.org/10.1016/j.memsci.2009.11.012

[70]　Q. Husain, S.A. Ansari, F. Alam, A. Azam, Immobilization of Aspergillus oryzae β galactosidase on zinc oxide nanoparticles via simple adsorption mechanism, International journal of biological macromolecules 49 (2011) 37-43. https://doi.org/10.1016/j.ijbiomac.2011.03.011

[71]　M.S. Johal, Understanding nanomaterials, CRC Press2011.

[72] H. Liu, T.J. Webster, Nanomedicine for implants: a review of studies and
      necessary experimental tools, Biomaterials 28 (2007) 354-369.
      https://doi.org/10.1016/j.biomaterials.2006.08.049

[73] M.V. Kharlamova, C. Kramberger, M. Sauer, K. Yanagi, T. Pichler,
      Comprehensive spectroscopic characterization of high purity metallicity-sorted
      single-walled carbon nanotubes, physica status solidi (b) 252 (2015) 2512-2518.
      https://doi.org/10.1002/pssb.201552251

[74] N. Wang, X. Zhang, N. Han, S. Bai, Effect of citric acid and processing on the
      performance of thermoplastic starch/montmorillonite nanocomposites,
      Carbohydrate Polymers 76 (2009) 68-73.
      https://doi.org/10.1016/j.carbpol.2008.09.021

[75] H. Xu, Y. Liu, Mechanisms of Cd 2+, Cu 2+ and Ni 2+ biosorption by aerobic
      granules, Separation and Purification Technology 58 (2008) 400-411.
      https://doi.org/10.1016/j.seppur.2007.05.018

[76] J. Ishii, A. Ono, Uncertainty estimation for emissivity measurements near room
      temperature with a Fourier transform spectrometer, Measurement science and
      technology 12 (2001) 2103. https://doi.org/10.1088/0957-0233/12/12/311

[77] T.M. Petrova, A. Solodov, A. Solodov, O. Lyulin, Y.G. Borkov, S. Tashkun, V.
      Perevalov, Measurements of CO2 line parameters in the 9250–9500 cm− 1 and
      10,700–10,860 cm− 1 regions, Journal of Quantitative Spectroscopy and Radiative
      Transfer 164 (2015) 109-116. https://doi.org/10.1016/j.jqsrt.2015.06.001

[78] A. Rohman, Y.C. Man, Fourier transform infrared (FTIR) spectroscopy for
      analysis of extra virgin olive oil adulterated with palm oil, Food research
      international 43 (2010) 886-892. https://doi.org/10.1016/j.foodres.2009.12.006

[79] X.-Y. Zhang, H.-P. Li, X.-L. Cui, Y. Lin, Graphene/TiO 2 nanocomposites:
      synthesis, characterization and application in hydrogen evolution from water
      photocatalytic splitting, Journal of Materials Chemistry 20 (2010) 2801-2806.
      https://doi.org/10.1039/b917240h

[80] V.R. Saasa, E. Mukwevho, B.W. Mwakikunga, Structural, optical and light
      sensing properties of carbon-ZnO films prepared by pulsed laser deposition,
      International Frequency Sensor Association Publishing (IFSA Publishing2016.

[81] L.W. Zhang, H.B. Fu, Y.F. Zhu, Efficient TiO2 photocatalysts from surface
      hybridization of TiO2 particles with graphite-like carbon, Advanced Functional
      Materials 18 (2008) 2180-2189. https://doi.org/10.1002/adfm.200701478

[82]   F. Spadavecchia, G. Cappelletti, S. Ardizzone, C.L. Bianchi, S. Cappelli, C. Oliva, P. Scardi, M. Leoni, P. Fermo, Solar photoactivity of nano-N-TiO 2 from tertiary amine: role of defects and paramagnetic species, Applied Catalysis B: Environmental 96 (2010) 314-322. https://doi.org/10.1016/j.apcatb.2010.02.027

[83]   N.M. Bahadur, T. Furusawa, M. Sato, F. Kurayama, N. Suzuki, Rapid synthesis, characterization and optical properties of TiO 2 coated ZnO nanocomposite particles by a novel microwave irradiation method, Materials Research Bulletin 45 (2010) 1383-1388. https://doi.org/10.1016/j.materresbull.2010.06.048

[84]   H. Du, G. Xu, W. Chin, L. Huang, W. Ji, Synthesis, characterization, and nonlinear optical properties of hybridized CdS− polystyrene nanocomposites, Chemistry of materials 14 (2002) 4473-4479. https://doi.org/10.1021/cm010622z

[85]   A. Nese, S. Sen, M.A. Tasdelen, N. Nugay, Y. Yagci, Clay-PMMA Nanocomposites by Photoinitiated Radical Polymerization Using Intercalated Phenacyl Pyridinium Salt Initiators, Macromolecular Chemistry and Physics 207 (2006) 820-826. https://doi.org/10.1002/macp.200500511

[86]   R. Vaiyapuri, B.W. Greenland, S.J. Rowan, H.M. Colquhoun, J.M. Elliott, W. Hayes, Thermoresponsive supramolecular polymer network comprising pyrene-functionalized gold nanoparticles and a chain-folding polydiimide, Macromolecules 45 (2012) 5567-5574. https://doi.org/10.1021/ma300796w

[87]   G.D. Bothun, Hydrophobic silver nanoparticles trapped in lipid bilayers: Size distribution, bilayer phase behavior, and optical properties, Journal of nanobiotechnology 6 (2008) 13. https://doi.org/10.1186/1477-3155-6-13

[88]   T. Cedervall, I. Lynch, S. Lindman, T. Berggård, E. Thulin, H. Nilsson, K.A. Dawson, S. Linse, Understanding the nanoparticle–protein corona using methods to quantify exchange rates and affinities of proteins for nanoparticles, Proceedings of the National Academy of Sciences 104 (2007) 2050-2055. https://doi.org/10.1073/pnas.0608582104

[89]   R.A. Sperling, W. Parak, Surface modification, functionalization and bioconjugation of colloidal inorganic nanoparticles, Philosophical Transactions of the Royal Society of London A: Mathematical, Physical and Engineering Sciences 368 (2010) 1333-1383. https://doi.org/10.1098/rsta.2009.0273

[90]   K. Takemoto, T. Matsuda, N. Sakai, D. Fu, M. Noda, S. Uchiyama, I. Kotera, Y. Arai, M. Horiuchi, K. Fukui, SuperNova, a monomeric photosensitizing

fluorescent protein for chromophore-assisted light inactivation, Scientific reports 3 (2013) 2629. https://doi.org/10.1038/srep02629

[91] J.K. Kim, S.Y. Yang, Y. Lee, Y. Kim, Functional nanomaterials based on block copolymer self-assembly, Progress in Polymer Science 35 (2010) 1325-1349. https://doi.org/10.1016/j.progpolymsci.2010.06.002

[92] V. Wagner, B. Hüsing, S. Gaisser, A.-K. Bock, Nanomedicine: Drivers for development and possible impacts, JRC-IPTS, EUR 23494 (2008).

[93] I. Olver, S. Shelukar, K.C. Thompson, Nanomedicines in the treatment of emesis during chemotherapy: focus on aprepitant, International journal of nanomedicine 2 (2007) 13. https://doi.org/10.2147/nano.2007.2.1.13

[94] S.V. Vinogradov, T.K. Bronich, A.V. Kabanov, Nanosized cationic hydrogels for drug delivery: preparation, properties and interactions with cells, Advanced drug delivery reviews 54 (2002) 135-147. https://doi.org/10.1016/S0169-409X(01)00245-9

[95] J.A. Champion, S. Mitragotri, Shape induced inhibition of phagocytosis of polymer particles, Pharmaceutical research 26 (2009) 244-249. https://doi.org/10.1007/s11095-008-9626-z

[96] Y. Hathout, Approaches to the study of the cell secretome, Expert review of proteomics 4 (2007) 239-248. https://doi.org/10.1586/14789450.4.2.239

[97] S. Inokuchi, T. Aoyama, K. Miura, C.H. Österreicher, Y. Kodama, K. Miyai, S. Akira, D.A. Brenner, E. Seki, Disruption of TAK1 in hepatocytes causes hepatic injury, inflammation, fibrosis, and carcinogenesis, Proceedings of the National Academy of Sciences 107 (2010) 844-849. https://doi.org/10.1073/pnas.0909781107

[98] D.G. Thomas, S. Gaheen, S.L. Harper, M. Fritts, F. Klaessig, E. Hahn-Dantona, D. Paik, S. Pan, G.A. Stafford, E.T. Freund, ISA-TAB-Nano: a specification for sharing nanomaterial research data in spreadsheet-based format, BMC biotechnology 13 (2013) 2. https://doi.org/10.1186/1472-6750-13-2

[99] Y. Wang, Z. Li, J. Wang, J. Li, Y. Lin, Graphene and graphene oxide: biofunctionalization and applications in biotechnology, Trends in biotechnology 29 (2011) 205-212. https://doi.org/10.1016/j.tibtech.2011.01.008

Carbonaceous Composite Materials

Materials Research Foundations **42** (2018) 309-334

Materials Research Forum LLC

doi: http://dx.doi.org/10.21741/9781945291975-11

Chapter 11

Biochar and its Composites

P. Senthil Kumar[*1], P.R. Yaashikaa[1] and Mu. Naushad[2]

[1]Department of Chemical Engineering, SSN College of Engineering, Chennai 603 110, India

[2]Department of Chemistry, College of Science, King Saud University, Riyadh, Saudi Arabia

senthilkumarp@ssn.edu.in*

Abstract

Biochar is actually a carbon-rich stable solid charcoal primarily used for soil modifications. Biochar can be produced using different methods like pyrolysis, gasification, etc. Biochar exists in many forms including biochar glomalin, terra preta, biochar hydrogel having potential to increase soil fertility at low pH, providing protection against infectious diseases and increasing the agricultural productivity. Biochar is an eco-friendly approach for replacing activated carbon and other carbon based materials. Physicochemical properties of biochar can be increased by chemical and physical activation methods. Biochar finds application as adsorbing materials in water and air pollution, production of biodiesel and soil conditioning.

Keywords

Biochar, Production, Characterization, Composites, Environmental Impacts, Applications

Contents

1.   Introduction

Biochar is a fine-grained, exceptionally permeable charcoal substance that is recognized from different charcoals in its proposed use as a soil alteration. The expression 'biochar' alludes to dark black carbon shaped by the pyrolysis of biomass i.e. by warming biomass in sans oxygen or low oxygen condition with the end goal that combustion does not occur. Conventional charcoal is one case of biochar delivered from wood. Biochar is charcoal that has been delivered under conditions that improve certain qualities considered helpful in agriculture, for example, high surface zone per unit of volume and low measures of remaining gums or resins. The specific heat treatment of natural biomass used to deliver biochar adds to its expansive surface region and its trademark capacity to hold on in soils with next to no organic rot. While crude natural materials supply supplements to plants and soil microorganisms, biochar fills in as an impetus or catalyst that improves plant take-up of supplements and water. Contrasted with other soil adaptations, the high surface range and porosity of biochar empower it to adsorb or hold supplements and water and furthermore give natural surroundings to advantageous microorganisms to thrive [1]. "Biochar" is a moderately new term, yet it is not another substance. Soils all through the world contain biochar stored through characteristic occasions, for example, woodland and meadow. Actually, those ranging high in normally occurring biochar are probably the most ripe or fertile soils on the planet. Because of lot of biochar fused into the surprisingly fruitful or healthy soils, certain districts still remains exceptionally fertile regardless of hundreds of years of draining from substantial tropical downpours. The utilization of biochar as soil added substance has been proposed as a way to and at the same time alleviate anthropogenic environmental change while enhancing rural soil richness. Biochar created from thermochemical transformation of biomass decreases ozone depleting substance outflows and is helpful for enhancing environmental frameworks in farming. Apart from lessening ozone depleting substance discharges, biochar enhances the physicochemical and microbial properties of soil and assimilates noxious and vindictive substances. Biochar delivered amid the thermochemical decay of biomass not just decreases the measure of carbon produced into the air, however it is likewise a domain neighbourly swap for enacted carbon and other carbon materials. Those components cause that biochar can be a conceivably decent material in the improvement of the way toward fertilizing the soil and lasting fertilizer quality. Being an eco-friendly method for soil alteration, it likewise offers standard systems for carbon catch and capacity [2].

2.   Biochar structure and composition

The intricate and heterogeneous chemical and physical organization of biochars gives a magnificent stage to contaminant expulsion. The chemical composition of biochars relies upon the kind of feedstock and pyrolysis conditions such as living arrangement time, temperature, rate of heating and reactor sort; in this manner, not all biochars are the same and it is hard to characterize the correct synthetic structure of biochar. Biochars are for the most part made out of carbon. The natural bit of biochar has high carbon content, and the inorganic segment essentially contains minerals for example, Calcium, Magnesium, Potassium and inorganic carbonates (carbonate particle), contingent upon its feedstock sort. The pyrolysis temperature utilized for biochar creation does not shape graphite to any noteworthy degree and fragrant rings in biochar are not orchestrated in flawlessly stacked and adjusted sheets, as they are in graphite. More sporadic courses of action of carbon are shaped amid biochar generation, and they contain oxygen and hydrogen. Sometimes, mineral arrangement relies upon the feedstock sort. By and large, biochars are not completely carbonized and display carbonized and non-carbonized stages. Expansion of biochar to soil brings about an increment in water penetration, soil water maintenance, particle trade limit, and supplement maintenance and a change in Nitrogen utilize proficiency [3, 4].

3.   Biomass conversion methodology

Biomass is a complex heterogeneous structure made out of cellulose, lignin, and other natural elements such as Carbon, Hydrogen, Nitrogen, Sulphur and Oxygen and inorganic parts that can surpass 40% of its dry weight. Inorganic substances such as silicon dioxide (SO_2), aluminium oxide, iron(III) oxide, titanium oxide, calcium oxide, sodium oxide, etc. may initially exist in biomass or be delivered amid thermal medications. The conversion techniques for biomass utilize can be divided into three classes: direct burning, thermochemical forms, and biochemical procedures [5]. Particular illustrations incorporate anaerobic assimilation, (bio) gasification and mixing with petroleum derivative sources. Vitality substance of crude and thermochemically handled biomass can be utilized for direct burning, cofiring with coal, or gasification. Target end-utilize alternatives stay to be (1) heat and control era, (2) transportation powers to supplant non-renewable energy sources, and (3) compound feedstocks. Biomass is a general term enveloping the natural issue in living beings and their deposits beginning from developing plants and farm manure, where compost can be thought of as a prepared type of plant materials. Biomass energy, or quickly bioenergy, is put away in plant and creature squander materials. Particular cases incorporate wood from characteristic backwoods, rural yield build-ups, mechanical squanders, for example, slop and paper squanders, and creature squander (excrement). Natural biomass is enormous and commonly has high dampness and basic earth metal substance, in this way entangling the immediate utilization of biomass as a fuel. Biomass can be dealt with thermochemically by means of pyrolysis to make materials with higher vitality densities for resulting bioenergy creation. A definite objective of biomass squander to-vitality change is to deliver non-renewable energy source substitutes for warming, power era, and motor operation [6].

4.   Biochar production

The current ubiquity in biochar change came about because of seriously researched conventional cultivating practices. Terra (Land) preta is one type of biochar that outcomes in high soil natural carbon, cation exchange limit, pH and soil fertility. These perceptions enlivened the biochar change to enhance soil quality and product yield. Today, asserted advantages of biochar alterations incorporate made strides overflow quality (lessened nutrient supplement filtering) and the carbon negative idea emerging from the unmanageability of pyrogenic carbon and the alleviation of ozone depleting substance discharges like greenhouse gases [7]. Biochar can be produced using conventional methods namely Pyrolysis, Gasification and Hydrothermal Carbonization.

Carbonaceous Composite Materials   Materials Research Forum LLC
Materials Research Foundations **42** (2018) 309-334   doi: http://dx.doi.org/10.21741/9781945291975-11

4.1   Pyrolysis

Pyrolysis, a thermal transformation process, is most encouraging advancements for sequestration of carbon and creation of bio-oil as feedstock for delivering second-range transportation powers (Figure 1). Pyrolysis is the procedure where biomass is heated in a domain with low oxygen level producing pyrolysis gas and singes which is biochar. Pyrolysis gas can be combusted to generate heat with low discharge and the biochar has a huge number of employments: soil change, creatures sustain supplements, channel material, carbon stockpiling, energy source, steel generation and so on. The qualities of the pyrolysis method are the adaptability to utilize diverse sorts of energizes, low outflow, low natural effect and the distinctive employments of the burn [8]. The char created in the pyrolysis procedure has numerous potential ranges of utilization, one of which is soil quality enhancer. Biochar heavily loaded with nutrients will greatly deliver supplements for the plants by increasing the nutritional composition of the soil finally resulting in increased productivity and yield rate. The carbon is put away, supposed carbon catch, in the soil and it works an indistinguishable route as carbon catch and capacity. Pyrolysis process can be classified as slow, batch, continuous, semi-batch and fast process.

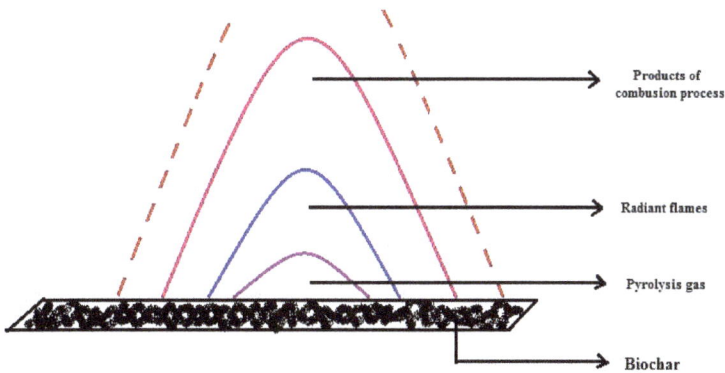

Figure 1 Process of Pyrolysis.

4.2.1 Materials required for pyrolysis process

It is imperative that the fuel in a pyrolysis procedure meets the clarification necessities for accomplishing the required nature of the finished results. Just natural material is permitted to be utilized for biochar processing. All materials that contain inorganic substances, for example, e.g. plastics, elastic and metal must be expelled from the fuel before pyrolyzing. The natural material must be untreated with e.g. paint and permeated. When utilizing agrarian waste as fuel is critical so ensure that this waste has been developed in a manageable way. Wood chips from forest service must originate from timberlands that are manageable delivered. Verification of the fuel is vital. The measure of substantial metals in the fuel is likewise managed since the biochar has a capacity to tie certain heavy metals the necessities contrast extraordinarily. The biomass likewise contains a few minerals and other minor natural atoms and polymers. The structure of these polymers and minerals fluctuates from various sorts of biomass and influences the pyrolysis procedure and its finished results. The minerals in the biomass, particularly alkali metals, can catalytically affect the pyrolysis response however expanding the yield of biochar. The creation of various sorts of biomass fluctuates depending on the development of biomass, environmental conditions, soil nature and farming operation. The carbon substance may shift in a similar kind of biomass as much as 10 %. Few major materials essential for pyrolysis process includes straw, type of wood material and its properties such as humidity, solid manure, etc. Pre-treatment of these materials is important such as drying, shredding and sorting [9].

4.2.2 Working principle

Pyrolysis is the immediate thermal decay of biomass without oxygen which delivers a blend of solids (the biochar appropriate), fluid (bio-oil), and gas (syngas) items (Figure 2). The particular yield from the pyrolysis is reliant on process condition, for example, temperature, and can be streamlined to deliver either vitality or biochar. Temperatures of range 400–500 °C (752–932 °F) deliver more biochar, while temperatures over 700 °C (1,292 °F) support the yield of fluid and gas fuel components. Pyrolysis happens all the more rapidly at the higher temperatures, normally requiring seconds rather than hours. High temperature pyrolysis is otherwise called gasification, and delivers principally syngas, which has been utilized as vehicle fuel in few times and places. The yield contains 60% bio-oil, 20% biochar, and 20% syngas. In pyrolysis process biomass is essential as a fuel. This biomass, that more often wood material, has ingested carbon from the climate amid its lifetime. At the point when the biomass is included the pyrolysis plant a start-up vitality source is utilized to start the procedure. Inside the reactor it is near oxygen free and temperatures reach up to 1000 °C. At these conditions approx. 50 % of the carbon present in the biomass is gasified and creates the pyrolysis gas. The rest of the measure of carbon is contained in the strong item, the biochar [10].

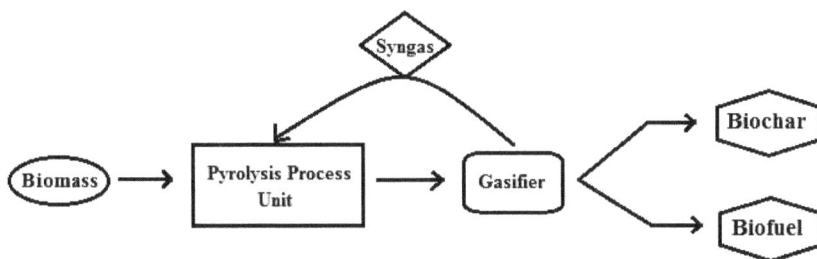

Figure 2: Working of Pyrolysis.

4.2.3 Factors influencing the process of pyrolysis

To get the correct quality of the finished products biochar and pyrolysis gas, the working factors can be balanced. The working parameters influencing the pyrolysis procedure are temperature, mass stream, molecule size, weight and dampness content. Under these optimized conditions, the pyrolysis procedure happens at quicker rate with expanded product yield and quality [11].

Temperature

The suitable temperature directly affects the carbon generation and the biochar attributes. Higher temperature gives less char in a wide range of pyrolysis responses. With a higher temperature one can envision that more unstable material is constrained out of the biomass and in this way the measures of biochar diminishes. Then again the level of carbon in the biochar shows increments. The temperature in the reactor influences the heat exchange rate and the gas stream rate.

Pressure

High pressure expands the gas stream rate in and on the particle surface, which give an optional carbonization. The impact is most present with pressure under 0.5 MPa. The pyrolysis forms which have been pressurized can deliver more burn resulting in biochar. The generation of bio-oil profits by vacuum in the process diminishing the measure of biochar. The response is exothermic, under high pressure and low mass streams.

Humidity

The humidity content in the material can effectively affect the creation of biochar and pyrolysis gas based upon nature in the reactor. Under pressure the procedure gives more biochar if the humidity content is low. In quick pyrolysis dryer material is required with a greatest of 10 % dampness substance to spare the measure of energy expected to dry the material before the pyrolysis begins. Moderate pyrolysis is more unbiased to humidity content yet again the proficiency of the reactor diminishes for high moisture content because of the requirement for drying the material. The humidity content influences the biochar's feature. By adjusting the humidity content activated carbon is synthesized with unique structural features.

4.3   Biomass gasification

Biomass gasification has a high potential for application in waste treating contrasted with other existing methods like land fill, burning, and so on, in light of the fact that it can acknowledge a wide assortment of sources of inputs and numerous valuable items can be created. Biomass gasification is a many-sided process including drying the feedstock took after by pyrolysis, fractional burning of intermediates, lastly gasification of the subsequent items (Figure 3 and 4). It is performed within the sight of a gasifying media which can be air, oxygen (O_2), steam (H_2O) or carbon dioxide (CO_2) present within a reactor called a gasifier. The gasifying medium additionally assumes an essential part of changing over strong char and substantial hydrocarbons (HC) to low-sub-atomic weight gasses like carbon monoxide (CO) and H2. The quality and properties of the item are reliant on the feedstock material, gasifying operator, feedstock measurements, temperature and weight inside the reactor, plan of reactor and the nearness of impetus and sorbent. There are numerous valuable items from the gasification of biomass like syngas, warm, control, bio-energizes, compost and bio-char. Syngas can be additionally handled by methods for the Fischer–Tropsch procedure into methanol, dimethyl ether and other compound feedstocks. Biomass feedstocks are characterized into four principle gatherings: woody biomass, herbaceous biomass, marine biomass and manures [12]. The gasifier is normally intended to create a desired end product; in any case, the feedstock material is an essential parameter to determine and improve where ever important. The gasification technique can be classified into following steps.

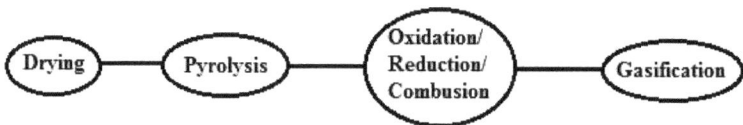

Figure 3: Steps in Gasification.

Working operation

Gasification is a procedure used to moderately combust biomass in a high temperature usually 700-1000 °C, oxygen inadequate condition to create a top standard fuel gases like hydrogen, carbon monoxide, etc. and biochar. This fuel gas might be utilized to immediately fuel a standard inside ignition motor. There are a wide range of gasifier outlines. The main gasifier that has demonstrated to deliver low tar content gas is the monetarily accessible High Temperature Downdraft Gasifier (HTDG). The HTDG is a constant operation gasifier that can achieve temperatures of 1500 °C. The biomass moves gradually downwards in the HTDG and is changed over into biochar as it experiences three primary response stages [13].

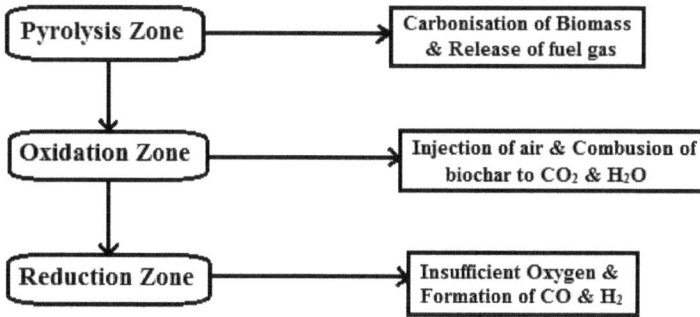

Figure 4 Gasification Process Stages.

Factors influencing the gasification process

The process of gasification depends on certain parameters which influences the whole process. The biochar yield and quality depends on those factors to a greater extent. The parameters affecting gasification process are quality and nature of feedstock, humidity, catalyst, steam to biomass ratio, sorbent to biomass ratio, etc. [14]. Few factors are discussed below.

Working conditions

The incomplete weight of the gasifying agent, temperature and thermal rate inside the gasifier are other indispensable parameters which can possibly impact the leave gas yield and general biomass change. Fractional weight of the gasifying operator has an immediate association with the reactivity of biomass burn while an expansion in temperature by and large expands the thermal rate of the feedstock particles by giving a more temperature differentiation. Besides, a slow thermal rate can really prompt lower gas yields and higher char yields. These two thermal rate operations to a great extent impact the plan of the gasifier and the expected outcome. The slower thermal rate prompts a higher char generation rate due to recombination of unpredictable hydrocarbons on the surface of the biochar particles. A higher temperature can prompt an expansion in the decomposition of the chars by changing them to the desired gasses.

Moisture capacity

Two sorts of humidity are contemplated in a biomass feedstock, in particular the inborn humidity, which is the water substance of the material without considering the effect of climate; and the extraneous dampness, which consolidates the impact of climate conditions. The qualities of the exit gasses and ideal operation of the gasifier rely upon the moisture substance to a huge degree. Woody and low-humid biomasses contain less than 15 wgt% humidity. This makes them more appropriate for heat change, since most gasifiers are intended to suit biomass feedstock with a moisture substance of 15–30 dry wgt%. The issue with high humid content is the power retribution related with drying the biomass before gasification. The low humid content is good since it has lower power retribution in the drying procedure preceding gasification.

Size of particle and steam to biomass ratio

It is noticed that leftover char yield is higher because of defective pyrolysis due of higher thermal exchange resistance offered by bigger particles. Upgrade in carbon change and measure of H2 was accounted for the molecule measurement was reduced. Furthermore, a reduction in molecule estimate enhances syngas effectiveness and reductions in biochar yields. In any case, it is considered that particle size should meet the requirement and not be less than needed, as molecule measure decrease requires exceptional power. The proportion of steam to biomass is a dominant parameter that influences the incoming power necessities, outlet gas quality and desired product yield. Low steam to biomass proportions result in higher measures of biochar and methane while expanding this proportion emphatically improves the changing responses by giving an oxidative situation, in this way raising the oxidized outcome gas yield. Power contained in the overabundance steam alongside enthalpy depletion in creating this steam, bring about diminishing procedure efficiencies. It likewise adversely impacts the temperature inside the gasifier which thus brings about low char breaking.

4.3   Hydrothermal carbonisation

Dry procedures including pyrolysis and gasification can accomplish high item yields with minimized power loss when the humid substance of the biomass is low. Then again, most biomass materials have high humid substance and in this way a different drying step is required to get high outcome yields and diminish the procedure energy. Aqueous procedures are expected to have the capacity to cure this inadequacy of dry procedures. The char delivered from an aqueous procedure is regularly called hydrochar likewise called biochar. In an aqueous procedure, biomass blended with water is put in a closed reactor and the temperature is raised after a specific time for stability. The pressure is likewise raised for water to keep up a fluid state over 100 °C. Based upon the temperature under immersed weight, biochar, bio-oil, and vaporous items, for example, CO, CO_2, CH_4, and H_2 are the fundamental outcomes of an aqueous procedure under 250 °C, at 250-400 °C, or more 400 °C, separately. In this manner, the aqueous procedure at every temperature extend is called hydrothermal carbonization (HTC), hydrothermal liquefaction (HTL), and hydrothermal gasification (HTG), individually. The biochar delivered from the HTC procedure has higher carbon content than the char created from dry procedures and response temperature, weight and water-biomass proportion are the main principle factors influencing the qualities of the desired products [15].

5.   Biochar characterization

The possible parts of biochar in carbon sequestration, depletion of ozone harming substance discharge, water maintenance, and sorption of contaminants are altogether generally subject to the surface properties of the biochar. Therefore, there have been different researches on surface physical and compound properties of biochar to explain potential components. The biochar physicochemical properties can cause changes in the soil supplement and carbon accessibility, and give physical preservation to microorganisms against predators and drying up; this may modify the microbial population and scientific categorization of the soil [16].

5.1   Physical characterization

Morphology

Morphology is a mass measure of the size, shape, and structure of the biochar. This trademark is normally got as a picture of the biochar surface and varies as a component of pyrolysis temperature and feedstock sort. Pictures of biochar surfaces of various feedstocks demonstrate various miniaturized scale and macropores. Increase in temperature shows huge variations in pore size when images using scanning electron microscopy. These pores are shaped from gas vesicles delivered amid pyrolysis. This leads to expanded porosity and surface region on the biochar surface. The wood-determined biochars for the most part have expansive level elements got from the ligno-cellulose plant tissue of the first feedstock that results in large sized pores. Nonwood-inferred biochars, as the almond shell biochar, commonly have smaller pores, not so much macrostructure.

Surface area

Various researches have demonstrated that biochar consists of high surface region and there is a positive relationship between particular surface area and pyrolysis temperature. In any case, for a few feedstocks, this relationship is not valid at higher temperatures, most likely because of loss of microporous structure from plastic disfigurement, sintering, or combination. This deviation is a component of the creation conditions that incorporate thermal rate, weight, and maintenance time. The capacity to give microbial living spaces, give soil amassing cores, preserving water and included supplements, and expel contaminants is reliant on surface area. Various factors are used in determining the surface area of biochar like temperature, time, sorption gas, etc.

Porosity

Porosity includes the surface area of the biochar molecule that is not filled by solids. Biochar porosity is obtained from two fundamental sources: macropores and micropores. The macropores emerge from the vascular structure of some biomass feedstock, and the abundance micropores emerge from gas vesicles that shape in the creation procedure because of gas discharge because of thermal rates and times and the pyrolysis temperature. Micropores are in charge of the high sorption limit of most biochars. They have additionally been appeared to give microhabitats to microorganisms, all the while giving security from predation and empowering biodiversity through the arrangement of various specialty conditions.

Cation exchange capacity

The cation exchange capacity (CEC) of biochars is generally high, to some extent because of their negative surface charge and affinity for soil cations, including most substantial metals. For this reason, several researches have been directed examining the adsorption of cations onto biochar which is the essential system by which biochar can be utilized for remediation of substantial metal pollution. CEC fluctuates between earthbound biomass from various feedstocks. Contrasted with numerous earthly biomass biochar, the CEC of biochar from green algal growth is moderately high with extractable metals like Ca, Mg, and K. The CEC of biochar-altered soil is emphatically subject to the age and surface useful properties and charge of the connected biochar.

5.2   Chemical characterization

Particular information with respect to the science and functional groups of biochar is essential in understanding in a better way how biochar reacts with the soil. The expansion of biochar to soil has been appeared to expand sorption of heavy metals. Since basic investigation is critical, it doesn't give coordinate data relating to the course of action of the components inside the specimen. To acquire data on the sub-atomic bonds inside biochar, a suite of spectroscopic analytical tools are normally utilized. These logical methodologies most usually incorporated are Fourier change infrared (FTIR), Nuclear Magnetic Resonance (NMR) and Raman spectroscopies. FTIR and Raman spectroscopies are both vibrational regularly used to investigate soil and biochar tests, and they give complimentary spectra peaks. NMR spectroscopy is a broadly utilized method that exploits the attractive properties of components to give data on the synthetic conditions of natural particles. Biochar possess high affinity for low solubility molecules with high hydrophobic and aromatic properties. The functional groups determine how biochars will communicate with soil, and influence its pH. Despite the fact that biochars commonly cause an expansion in soil pH, biochars that have various oxygen functional groups does not cause increase in pH.

6.   Biochar composites

Biochars fertilizers and composites can be created from an assortment of natural dissipate through appropriate preparing and plans. In the event that more natural squanders could be minimized to biochar fertilizers and composites indicating great business esteem, at that point both the issues of inaccessibility of compost contributions to rural creation and natural contamination can be mitigated. Biochar fertilizers and composites, when legitimately arranged and utilized, can help advance low information agricultural framework to wind up noticeably more supportable and gainful. Developed biochar manures and composites, even without microbial involvement, are as of now applicable. Studies show that formulation of fertilizers and composites with microbes tends to increase soil fertility and also crop productivity [17]. A composite is defined as basic material which comprises of at least two joined constituents that are consolidated at a perceptible level. One constituent is known as the building stage and the one in which it is implanted is known as the lattice. The building stage material for this situation is the biochar.

Biochar Terra-preta

Treating the soil with fertilizers is the natural deterioration and adjustment of natural issue determined from plants, creatures or people through the activity of differing microorganisms in presence of oxygen. The end result of this natural procedure is a humus-like, stable form, being free of pathogens which can be helpfully connected to arrive as an operator for soil improvement or as natural compost. Numerous soil fertilizing strategies and frameworks have been generated, shifting from little, home-made reactors utilized by singular family units, on-site reactors worked by agriculturists, to huge, high-innovation reactors utilized by proficient manure makers. Regardless of various process procedures, the principal organic, compound, physical parts of treating the soil continue as before. All legitimate treating the soil forms experience four phases: (1) mesophilic, (2) thermophilic, (3) cooling, at long last completion with (4) compost development. The length of each stage relies upon the underlying piece of the blend, its water substance, air circulation and amount and creation of microbial community. Its application to the land impacts a few organic, substance and physical soil properties in a successful and manageable way. Feedstock, compost development and fertilizer quality can impact the power and level of consequences for soil physical, compound and natural properties. Application may trigger here and now upgrades, for example, expanding microbial activity. Terra preta was in all probability shaped by blending of biochar with biogenic dissipate from human settlements (fertilizers and nourishment waste including bones and fiery debris) which were microbially changed over to a biochar-compost-like substrate. Hence, co-composting of biochar and new natural material is probably going to have various advantages contrasted with the minor blending of biochar or fertilizer with soil. Besides the significance of biochar consolidation, extra revisions like clay minerals can increase the value of the end composite item, e.g. by advancing an improved CEC or water holding limit because of their high adsorption or swelling limit. Besides, their joining into natural substrates advances the development of organo–mineral combination started by the organic action of soil fauna after consequent soil application [18].

Biochar hydrogel

Biochar polymer hydrogels are freely cross-connected, three-dimensional systems of adaptable polymer chains that convey separated, ionic utilitarian functional groups either created by the joining of polymer substrate with biochar or by physical blending. They are fundamentally the materials that can retain liquids of more than 15 times their own dried weight, either under load or without stack, for example, water, electrolyte arrangement, saline solutions, and natural liquids, for example, sweat and blood. They are polymer composites which are described by hydrophilicity containing carboxylic corrosive, carboxamide, hydroxyl, amine, and so forth, insoluble in water, and are cross-connected polyelectrolytes. As a result of their ionic nature and structure, they assimilate vast amounts of water and different watery arrangements without dissolving by solvation of water atoms by means of hydrogen bonds. Despite the fact that the established utilization of progressive alkalinity-creating frameworks is the prolongation of plant survival submerged anxiety, new information demonstrate that they likewise have an impact in soils which have water substance near field limit [19]. The benefits of the soil alteration with biochar hydrogels can be compressed as:

(i) Increment the plant accessible water in soils;

(ii) Generate quick growth and development of plants, even under ideal watering conditions;

(iii) Draw out the survival of plants submerged anxiety.

Therefore, the environmental scope of the plants has been enlarged significantly. In this way, with the guide of biochar hydrogels it is conceivable to change over contaminated land into a fruitful field.

Biochar mokusaku-eki

Mokusaku-eki is a Japanese expression which implies pyroligneous corrosive, which is as of now utilized generally by Japanese individuals as compost, as insect repellent and other numerous different applications. Biochar mokusaku-eki is in this way is called a biochar composite material. The impact of pyroligneous corrosive on seed germination is generally known and utilized as a part of different ways. Biochar has been considered as a key contribution for rising and managing generation and all the while minimizing contamination and reliance on composts. In Japan, biochar has been utilized since 1697, it was connected in agribusiness and cultivation, including for enhancing the force of old pine trees close places of worship. As a rule, bokashi mokusaku-eki and biochar have been found to effectively affect soil richness and yield generation. In any case, explore on the examination of impacts of biochar mokusaku-eki on soil properties and qualities of crop have made attention towards this composite by the researchers [20].

Biochar bokashi

Bokashi is a Japanese word, interpreted signifies "matured natural substance", consequently biochar bokashi is a composite material of biochar and matured natural substance, and is made by treating plant-related end-products with effective microorganisms (EM) and after that dried (for longer timeframe of realistic usability) before blending with biochar. Biochar bokashi can be effectively used to treat spoiling waste, which brings about a higher disintegration of natural substance and essentially no smell amid process. This delivers excellent composite, which adds to maintainable advancement over various segments of agriculture [21].

Biochar glomalin

Glomalin is a type of balanced out carbon present in soil. It is a glyco protein and acts like a paste, restricting carbon, nitrogen and other organic segments of soil to the mineral parts, mud and sand. It is yielded by mycorrhizal organisms (AMF). It is trusted that these parasites deliver glomalin keeping in mind the end goal to help the hyphae in the soil. Biochar glomalin guarantee that enough supplements will be put away in the soil to confirm that no additional application of fertilizers will be mandatory in the field. Biochar glomalin augments soil carbon recovery by utilizing biochar with suitable arbuscular mycorrhizal growths (AMF). The thought is that the adsorptive properties of biochar and also its interior surface territory will additionally support AMF action by giving appropriate environment for its growth and development [22].

7.   Environmental impacts of biochar

Feedstocks used for biochar creation like woody biomass, grasses, and excrements) has impacts on biochar qualities, including convergences of basic constituents, thickness, porosity, and hardness. Optimization of biochar for a particular application may require choice of a feedstock and pyrolysis generation system and conditions to deliver biochars with particular qualities. Studies focus on the connections between biochar generation conditions, biochar attributes, and potential end-applications of biochar. Biochars and Hydrochars can be adverse to edit yields. What is missing is an entire unthinking comprehension of how biochars cause yield diminishments or increments. Additionally information is to determine the impacts of connection between biochar qualities, environmental conditions and soil properties on supplement filtering, maintenance, and immobilization [23]. No data is at present accessible on how biochar attributes influence microbial-interceded supplement cycling and soil microbial groups. Longer-term field examine concentrating on supplement and water utilize effectiveness is expected to evaluate the heritage of biochar applications and survey their actual esteem. Biochar might be of an incentive for sequestering bioavailable metals and anti-infection agents in polluted soils, and also to capture and reusing supplements in water bodies. More research focusing on an expansive range of soils influenced by natural debasement is required. Particularly planned biochars could sorb less effectively stable metals, for example, Cd and Ni or sequester portable natural stages. Contaminant sorption techniques and energy should be recognized, stacking limits of biochars should be evaluated, and a definitive destiny of contaminants in biochar-corrected soils should be archived before substantial scale biochar field applications in polluted soils start. Cost and potency correlations for biochars with respect to other contaminant moderation innovations would likewise be useful [24].

8.   Applications

The high surface zone and cation exchange capacity of biochar resolved to a huge degree by source materials, temperature during pyrolysis and any after generation preparing. A high surface region is vital for the removal of both natural and inorganic contaminants on biochar the essential component by which contamination versatility is decreased in polluted soils. Investigations about on bigger scales and longer time of biochar represented that the expanded sorption of contaminants in biochar corrected soils could be significant regarding diminishing soil contaminants in crops along with lakes, wetlands, and polluted soils and residue. Additional advantages of biochar contrasted with activated charcoal as a therapeutic technique are its ease, negligible site unsettling influence as a latent, in situ treatment, and potential cost reserve funds if clearing of polluted residue can be accomplished without digging or creating hollow and transfer [25, 26].

8.1   Biochar as sorbents

Carbonaceous materials have been utilized for quite a while as sorbents for natural and inorganic contaminants in soil and water. Presently, the activated carbon, which is charcoal that has been dealt with oxygen for increasing microporosity and surface area is the most generally utilized carbonaceous sorbent. Biochar is very like activated carbon as for production through pyrolysis, with medium to high surface areas. Activation or treatment is not possible with biochar. Moreover, the biochar contains a non-carbonized part that may connect with soil contaminants. In particular, the degree of oxygen containing carboxyl, hydroxyl, and phenolic surface groups in biochar could successfully tie with soil contaminants. These multi-utilitarian qualities of biochar demonstrate the potential as an exceptionally successful natural sorbent for natural and inorganic contaminants in soil and water [27].

Removal of water contaminants

Heavy metals are essential poisonous water contaminations with antagonistic impacts on the digestion of people, creatures and plants. Scopes of biochars have been connected to the evacuation of heavy metals. Biochars delivered from rice husk and excrement can be utilized to expel lead, copper, zinc and cadmium from contaminated fluid environment.

Defecation char had higher metal expulsion efficiencies than rice husk-char for all the heavy metals. At the point when excrement char was utilized, heavy metals were not just adsorbed on the ionized hydroxyl-oxygen functional groups while the adsorption on ionized phenolic oxygen groups is the main heavy metal expulsion system of rice husk-char. The investigation demonstrated that oxygen containing functional groups assume imperative parts in the sorption of heavy metals on biochar. Sorption of natural contaminants from water onto biochar happens because of its high surface zone and microporosity. Biochars delivered at temperature >400°C are more powerful for natural contaminant sorption as a result of their high surface territory and micropore improvement. The biochar has also been used as support for synthesis of photocatalysts used for photodegradation of persistent organic pollutants [28-30].

Removal of air pollutants

Contaminants in air have unfavourable intense or perpetual effects on human wellbeing relying upon their nature, fixation, and time of exposure. CO_2 is the most critical an unnatural weather change gas, despite the fact that its immediate wellbeing effect is not huge. Adsorption is a standout amongst the most broadly utilized techniques for storing and capturing of CO_2. The adsorbents utilized broadly to catch CO_2 incorporate carbon materials such as activated carbon, carbon nanotube, zeolites, etc. Broad investigations have been led to grow new carbon adsorbents from biochar. The biochar delivered at high temperatures demonstrated a vast CO_2 adsorption limit which was assigned to the physisorption of CO_2. The removal effciency is directly proportional to the surface area of biochar. At the point when the biochar particular surface region was adequately huge, the amount of nitrogenous functional groups assumed a more critical part in the adsorption of CO_2 on the biochar surface. Pyrolyzed rice straw utilizing microwaves were used to deliver biochar and can be utilized it to adsorb CO_2. The CO_2 adsorption limit expanded with expanding pyrolysis temperature. Contrasted with the biochar created utilizing an ordinary pyrolysis process at the same temperature, the biochar delivered utilizing microwave pyrolysis demonstrated a higher CO_2 adsorption limit when the pyrolysis temperature was 400°C exhibiting the benefits of microwave pyrolysis as far as cost and time [31].

Mechanism of pollutants removal using biochar

Studies related in understanding the mechanism of removal of pollutants using biochar demonstrated four major mechanisms as follows:

- Electrostatic attraction/repulsion between charges of surface of biochar and pollutants

- Sorption of pollutants by biochar
- Precipitation/Co-precipitation of metals
- Surface complexation of metals

Adsorption

The techniques for metal removal by biochar incorporates adsorption, oxidation, reduction and in particular, immobilization of heavy metals. The physical structure of biochar may likewise impact the conduct of metals. Large scale, smaller scale and nanoporous structures all through biochar's surface could make conditions that guide the lessening of metals into less portable ones. The tiny graphene moieties in biochar may fill in as both adsorption and redox response sides because of their high bonding for contaminants and their capacity to exchange electrons. The reducible and oxidizable segments of heavy metals rises especially in treatment with biochar however the immobilization impacts reduced at higher rate of heavy metals. In conditions with negative redox potential gave by biochar, it would be adsorbed less promptly to soils, and would be more versatile. Likewise, under reducing rate, Fe and Mn oxides are solubilized so their fixations would both be required to rise in pore water under negative redox potential. It is believed that oxidation of biochar happens most quickly on its external surfaces, accompanied by inside pores. This contrast between the pores explains the different removal capacity of biochar to heavy metals.

Electrostatic attraction

Electrostatic attraction/repulsion between natural contaminants and biochar is another convenient adsorption technique. Biochar surfaces are mostly negatively charged, which could encourage the electrostatic attraction of positively charged cationic natural substances. Both electron rich and poor functional groups are available in high temperature determined biochars; subsequently, they are hypothetically equipped for associating with both electron donors and electron acceptors. An electrostatic repulsion between negatively charged anionic natural groups and biochars could advance hydrogen holding and enhance adsorption of heavy metals [32].

8.2   Biochar in agriculture

Biochar, a soil alteration, has potential as a profitable device for the agricultural sector with its interesting capacity to enable form to soil, save water, create sustainable power source and sequester carbon. Biochar can be made from a wide assortment of feedstocks, including wood and plant matter, in addition to excrement. Biochar immediately changes the physical and compound properties of soil and thus the microorganism movement.

Thus, the change in microorganism movement can impact the physical and compound properties of the soil. The enormous and entangled pores of biochar are held after its application. A mass of small scale pores is framed, giving spaces to organism development and spread and securing microorganisms from an adverse outer condition, in this way decreasing the survival rate among organisms. There are supplements required by different organisms for their development and advancement in the smaller scale pores of biochar [33].

Improving soil fertility and productivity

As a soil modifier, biochar enhances the earth's natural soil resource by expanding efficiency and product yields, lessening soil acidity, minimizing the requirement for some chemical fertilizer and manure inputs and giving other soil benefits. Biochar can enhance soil strength, prompting higher harvest yields. When applied in the soil in a finely ground shape, biochar's huge surface zone and pore structure is cordial to the microbes and other growths that plants need to retain supplements from the soil. In this way, biochar gives a safe environment to useful microbial movement that is significant for crop generation to thrive. Biochar has appeared to lessen the soil discharges of nitrous oxide and enhance the take-up of methane. Biochar has additionally been appeared to decrease soil pH. Soils in extensive parts of the world experience the ill effects of low pH. The negative charges on the surface area of biochar are produced during nutrients binding and are used for maintaining soil pH. Biochar can possibly expand the world's farming profitability by enhancing contaminated soils [34].

Bioenergy production

The technique produces bioenergy, for example, syngas and bio-oils. The bonds that hold the molecules in biomass together should be broken. The energy that is discharged is a blend of particles together called syngas. Syngas is made essentially out of carbon monoxide and hydrogen, and now carbon dioxide and other minor particles. It is flammable and has a large portion of the energy thickness of gaseous petrol. Syngas can be utilized specifically as an energy source or for generation of manufactured flammable gas, ammonia, etc. The other bioenergy outcome is bio-oil. Bio-oil can be utilized as a substitute for fuel oil or thermal oil. It additionally can possibly be utilized as a part of a bio-refinery where important chemicals are removed [35].

8.3   Carbon sequestration

Late green developments have brought natural eco-friendly activities and minimizing carbon emanations to the cutting edge of political, financial, and social movement. This gives a need to modest methods for catching or expelling carbon from the air. Biochar, derived from biomass which is in wealth, fills this void while being a type of waste transfer and reusing. By an adjustment of the current natural carbon cycle, the biochar is delivered from biomass and half comes back to the soil as charcoal and the other half comes back to commercialization for a natural fuel. All yields retain CO_2 amid development and discharge. The objective of agrarian carbon expulsion is to utilize the product and its connection to the carbon cycle to for all time sequester carbon inside the soil. This is finished by choosing cultivating techniques that arrival biomass to the soil and upgrades the conditions in which the carbon inside the plants will be decreased to its basic nature and put away in a steady state. The impacts of soil sequestration can be switched. Soil will remain as source of greenhouse gas emission if soil disruption or rejection of tillage practices occurs. After few years of sequestration, soil winds up plainly soaked and stops to retain carbon. This infers there is a worldwide point of confinement to the measure of carbon that soil can hold. Many variables influence the expenses of carbon sequestration including soil quality, exchange costs and different externalities, for example, spillage and unanticipated natural harm. Since decrease of climatic CO_2 is a long period concern, farmers can be hesitant to receive more costly agricultural systems when there is not a reasonable product, soil, or monetary advantage [36].

Conclusion

Biochar has been used in an assortment of applications, for example, adsorbents, as catalysts, and soil modifications, based upon their attributes. Biochars are being utilized as a part of an expanding number of fields. The foundation of a nonstop supply framework will be expected to advance the utilization of biochar to higher esteem included areas. Specifically, an expansion in the temperature enhanced the adsorption properties, for example, surface zone, porosity, etc. in woody biochars. With expanding CO_2 levels, expanding consideration is being given to environmental change at a worldwide level. Ozone harming substance like greenhouse gas release from farmland biological communities assume an imperative part and should be comprehended to illustrate their commitment to worldwide environmental change. To decrease CO_2 outflows, the usage of biochar from farmland biological systems is winding up more essential. Biochar enhances soil compaction, porosity, soil thickness, water substance, and mass thickness. Biochar advances mineralization, obsession and change of natural nitrogen in soils to give a nitrogen source to soil microorganisms and improve the soil nitrogen content. Biochar can likewise enhance the pH and cation exchange capacity of soil. The soil natural substance rises with biochar application. Inexhaustible and complex pores in biochar are held in soils and give space for microorganism development and proliferation, shielding organisms from a negative outside condition and lessening survival rate among organisms. Biochar can be added to contaminated soil to retain toxic and malicious substances in the soil. Biochar can minimize the harmful impact of natural contaminants on plants and decrease the gathering of natural contaminations in plants. Biochar can likewise enhance the action of soil organisms, initiating expanded decomposition of toxins.

References

[1]   J. Lee, K.H. Kim, E.E. Kwon. Biochar as a catalyst. Renewable and Sustainable Energy Reviews. 77 (2017) 70-79. https://doi.org/10.1016/j.rser.2017.04.002

[2]   P.P. Otte, J. Vik. Biochar Systems: Developing a socio-technical system framework for biochar production in Norway. Technology in Society. 51 (2017) 34-45. https://doi.org/10.1016/j.techsoc.2017.07.004

[3]   Rawal, S.D. Joseph, J.M. Hook, et al. Mineral-Biochar Composites: Structure and Porosity. Environmental Science and Technology. 50 (14) (2016) 7706-7714. https://doi.org/10.1021/acs.est.6b00685

[4]   S.V. Vassilev, D. Baxter, L.K. Anderson, C.G. Vassileva. An overview of chemical composition of biomass. Fuel. 89 (2010) 913-933. https://doi.org/10.1016/j.fuel.2009.10.022

[5]   J.S. Cha, S.H. Park, S.C. Jung, C. Ryu, J.K. Jeon, M.C. Shin, Y.K. Park. Production and Utilization of Biochar: A Review. Journal of Industrial and Engineering Chemistry. 40 (2016) 1-15. https://doi.org/10.1016/j.jiec.2016.06.002

[6]   P. Godlewska, H.P. Schmidt, Y.S. Ok, P. Oleszczuk. Biochar for composting improvement and contaminants reductions: A review. Bioresource Technology. (2017) https://doi.org/10.1016/j.biortech.2017.07.095. https://doi.org/10.1016/j.biortech.2017.07.095

[7]   K.L. Yu, B.F. Lau, P.L. Show, H.C. Ong, T.C. Ling, W.H. Chen, N.E. Poh, J.S. Chang. Recent developments on algal biochar production and characterization. BioresourceTechnology.(2017). https://doi.org/10.1016/j.biortech.2017.08.009. https://doi.org/10.1016/j.biortech.2017.08.009

[8]   Y.S. Ok, S.M. Uchimiya, S.X. Chang, N. Bolan. Biochar Production, Characterization and Applications. New York: CRC Press, Taylor and Francis Group.

[9]   L.V. Zwieten, S. Kimber, S. Morris, K. Chan, A. Downie, J. Rust, S. Joseph, A. Cowie. Effects of biochar from slow pyrolysis of papermill waste on agronomic performance and soil fertility. Plant and Soil. 327 (2010) 235-246. https://doi.org/10.1007/s11104-009-0050-x

[10] D. Ozcimen, A.E. Mericboyu. Characterization of biochar and bio-oil samples obtained from carbonization of various biomass materials. Renewable Energy. 35 (6) (2010) 1319-1324. https://doi.org/10.1016/j.renene.2009.11.042

[11] T. Xie, K.R. Reddy, C. Wang, E. Yargicoglu, K. Spokas. Characteristics and Applications of Biochar for Environmental Remediation: A Review. Critical Reviews in Environmental Science and Technology. (2014). http://dx.doi.org/10.1080/10643389.2014.924180. https://doi.org/10.1080/10643389.2014.924180

[12] V.S. Sikarwar, M. Zhao, P. Clough, J. Yao, X. Zhong, M.Z. Memon, N. Shah, E.J. Anthony, P.S. Fennell. An overview of advances in biomass gasification. Energy and Environmental Sciences. 9 (2016) 2939-2977. https://doi.org/10.1039/C6EE00935B

[13] R.K. Xu, S.C. Xiao, J.H. Yuan, A.Z. Zhao. Adsorption of methyl violet from aqueous solutions by the biochars derived from crop residues. Bioresource Technology. 102 (2011) 10293–10298. https://doi.org/10.1016/j.biortech.2011.08.089

[14] E.N. Yargicoglu, B.Y. Sadasivam, K.R. Reddy, K. Spokas. Physical and Chemical Characterization of waste wood derived biochars. Waste Management. 36 (2015) 256-268. https://doi.org/10.1016/j.wasman.2014.10.029

[15] Z. Liu, F.S. Zhang. Removal of lead from water using biochars prepared from hydrothermal liquefaction of biomass. Journal of Hazardous Materials. 167 (2009) 933–939. https://doi.org/10.1016/j.jhazmat.2009.01.085

[16] K. Jindo, H. Mizumoto, Y. Sawada, M.A.S. Monedero, T. Sonoki. Physical and Chemical Characterization of Biochar derived from different agricultural residues. Biogeosciences. 11 (2014) 6613-6621. https://doi.org/10.5194/bg-11-6613-2014

[17] O. Kostov and J.M. Lynch. Composted sawdust as a carrier for Bradyrhizobium, Rhizobium and Azospirillum in crop production. World Journal of Microbiology and Biotechnology. 14 (1998) 389 – 397. https://doi.org/10.1023/A:1008869329169

[18]   B. Glaser and J.J. Birk. State of the scientific knowledge on properties and genesis of Anthropogenic Dark Earths in Central Amazonia (terra preta de Índio). Geochimica et Cosmochimica Acta. 82 (2011) 39-51. https://doi.org/10.1016/j.gca.2010.11.029

[19]   L.O. Ekebafe, D.E. Ogbeifun and F.E. Okieimen. Effect of native cassava starch-poly (sodium acrylateco-acrylamide) hydrogel on the growth performance of maize (Zea may) seedlings. American Journal of Polymer Science. 1 (2011) 1 – 6.

[20]   C.J. Barrow. Biochar: potential for countering land degradation and for improving agriculture. Applied Geography. 34 (2012) 21 – 28. https://doi.org/10.1016/j.apgeog.2011.09.008

[21]   M.O. Ekebafe, L.O. Ekebafe, S.O. Ugbesia. Biochar composts and composites. Science Progress. 98 (2) (2015) 169-176. https://doi.org/10.3184/003685015X14301544319061

[22]   T. Ishii and K. Kadoya. Effects of charcoal as a soil conditioner on citrus growth and vesicular-arbuscular mycorrhizal development. Journal of the Japanese Society for Horticultural Science. 63 (1994) 529-535. https://doi.org/10.2503/jjshs.63.529

[23]   G. Xu, Y. Lv, H. Shao, L. Wei. Recent advances in biochar applications in agricultural soils: Benefits and environmental implications. Clean – Soil, Air, Water. 40 (2012) 1093-1098. https://doi.org/10.1002/clen.201100738

[24]   A.V. McBeath, R.J. Smernik, E.S. Krull, J. Lehmann. The influence of feedstock and production temperature on biochar carbon chemistry: A solid-state 13C NMR study. Biomass Bioenergy. 60 (2013) 121–129. https://doi.org/10.1016/j.biombioe.2013.11.002

[25]   A.Pimchaui, A. Dutta, P. Basu. Torrefaction of agriculture residue to enhance combustible properties. Energy and Fuels. 24 (2010) 4638-4645. https://doi.org/10.1021/ef901168f

[26]   J. Lehmann, M.C. Rillig, J. Thies, C.A. Masiello, W.C. Hockaday, D. Crowley. Biochar effects on soil biota – A review. Soil Biology and Biochemistry. 43 (2011) 1812–1836. https://doi.org/10.1016/j.soilbio.2011.04.022

[27]   M. Ahmad, A.U. Rajapaksha, J.E. Lim, M. Zhang, N. Bolan, D. Mohan, M. Vithanage, S.S. Lee, Y.S. Ok. Biochar as a sorbent for contaminant management in soils and water: A review. Chemosphere. 99 (2014) 19-33. https://doi.org/10.1016/j.chemosphere.2013.10.071

[28]   A. Kumar, A. Kumar, G. Sharma, A.a.H. Al-Muhtaseb, M. Naushad, A.A. Ghfar, C. Guo, F.J. Stadler, Biochar-templated g-C_3N_4/$Bi_2O_2CO_3$/$CoFe_2O_4$ nano-assembly for visible and solar assisted photo-degradation of paraquat, nitrophenol reduction and CO_2 conversion, Chemical Engineering Journal, 339 (2018) 393-410. https://doi.org/10.1016/j.cej.2018.01.105

[29]   A. Kumar, Shalini, G. Sharma, M. Naushad, A. Kumar, S. Kalia, C. Guo, G.T. Mola, Facile hetero-assembly of superparamagnetic Fe3O4/BiVO4 stacked on biochar for solar photo-degradation of methyl paraben and pesticide removal from soil, Journal of Photochemistry and Photobiology A: Chemistry, 337 (2017) 118-131. https://doi.org/10.1016/j.jphotochem.2017.01.010

[30]   A. Kumar, A. Kumar, G. Sharma, M. Naushad, F.J. Stadler, A.A. Ghfar, P. Dhiman, R.V. Saini, Sustainable nano-hybrids of magnetic biochar supported g-C_3N_4/$FeVO_4$ for solar powered degradation of noxious pollutants- Synergism of adsorption, photocatalysis & photo-ozonation, Journal of Cleaner Production, 165 (2017) 431-451. https://doi.org/10.1016/j.jclepro.2017.07.117

[31]   H. Lyu, Y. Gong, R. Gurav, J. Tang. Chapter 9 – Potential Application of Biochar for Bioremediation of Contaminated Soils. Biochar Application (Essential Soil Microbial Ecology). (2016) 221-246.

[32]   H. Li, X. Dong, E.B. Silva, L.M. Oliviera, Y. Chen, L.Q. Ma. Mechanism of metal sorption by biochars: Biochar characteristics and modifications. Chemosphere. 178 (2017) 466-478. https://doi.org/10.1016/j.chemosphere.2017.03.072

[33]   Z. Tan, C.S.K. Lin, X. Ji, T.J. Rainey. Returning Biochar to Fields: A review. Applied Social Ecology. 116 (2017) 1-11. https://doi.org/10.1016/j.apsoil.2017.03.017

[34]   L.D. Burrell, F. Zehetner, N. Rampazzo, B. Wimmer, G. Soja. Long-term effects of biochar on soil physical properties. Geoderma. 282 (2016) 96-102. https://doi.org/10.1016/j.geoderma.2016.07.019

[35]   K. Homagain, C. Shahi, N. Luckai, M. Sharma. Biochar-based bioenergy and its environmental impact in Northwestern Ontario Canada: A Review. Journal of Forestry Research. 25 (4) (2014) 737-748. https://doi.org/10.1007/s11676-014-0522-6

[36]   J. Lehmann, J. Gaunt, M. Rondon. Biochar sequestration in terrestrial ecosystems: a review. Mitigation and Adaption Stratergies for Global Change. 11 (2006) 403–427. https://doi.org/10.1007/s11027-005-9006-5

Keyword Index

www.ingramcontent.com/pod-product-compliance
Lightning Source LLC
Chambersburg PA
CBHW071322210326
41597CB00015B/1315